An Introduction to Variational Inequalities and Their Applications

This is a volume in
PURE AND APPLIED MATHEMATICS

A Series of Monographs and Textbooks

Editors: SAMUEL EILENBERG AND HYMAN BASS

A list of recent titles in this series appears at the end of this volume.

An Introduction to Variational Inequalities and Their Applications

David Kinderlehrer
SCHOOL OF MATHEMATICS
UNIVERSITY OF MINNESOTA
MINNEAPOLIS, MINNESOTA

Guido Stampacchia†
SCUOLA NORMALE SUPERIORE
PISA

1980

ACADEMIC PRESS
A Subsidiary of Harcourt Brace Jovanovich, Publishers
New York London Toronto Sydney San Francisco

ACADEMIC PRESS, INC.
111 Fifth Avenue, New York, New York 10003

United Kingdom Edition published by
ACADEMIC PRESS, INC. (LONDON) LTD.
24/28 Oval Road, London NW1 7DX

Library of Congress Cataloging in Publication Data

Kinderlehrer, David.
An introduction to variational inequalities and their applications.

(Pure and applied mathematics, a series of monographs and textbooks ;)
Includes index.
1. Inequalities (Mathematics) 2. Calculus of variations. I. Stampacchia, Guido, joint author.
II. Title. III. Series.
QA3.P8 [QA295] 510'.8s [512.9'7] 79–52793
ISBN 0–12–407350–6

PRINTED IN THE UNITED STATES OF AMERICA

80 81 82 9 8 7 6 5 4 3 2 1

Questo volume è dedicato alla memoria
del nostro diletto amico e collega

Nestor Riviere

Contents

Preface

The rapid development of the theory of variational inequalities and the prolific growth of its applications made evident to us the need of an introduction to the field. This is our response; we hope that it will be found useful and enlightening. We drew an outline of our enterprise in July of 1976, confident that it would be completed by August or, perhaps, September. Our conception of the necessary labor was optimistic; nonetheless, the finished book follows the original plan.

Many of the chapters have been adapted from courses we gave at the Scuola Normale Superiore, the University of Minnesota, the University of Paris, the Collège de France, the Mittag-Leffler Institut, and Northwestern University. A few suggestions about the use of this book in courses are discussed in the introduction.

The life and work of the late Guido Stampacchia are discussed by Jacques Louis Lions and Enrico Magenes in the *Bollettino della Unione Matematica Italiana* **15** (1978), 715–756.

I wish to express my profound indebtedness to our many friends and collaborators who so generously assisted us during these years. Most important, without the nurturing and encouragement of Mrs. Sara Stampacchia, it would not have been possible for us to conceive this project nor for me to bring it to its completion.

In addition, I would like to thank especially Haim Brezis for his many useful suggestions and his constant interest in our progress. Silvia Mazzone,

Michael Crandall, and David Schaeffer offered useful contributions and assistance in various stages of the writing, all of which are sincerely appreciated. Finally, I would like to thank Matania Ben Artzi and Bevan Thompson for their careful reading of portions of the material.

Ronchi and Minneapolis
1976–1979

Glossary of Notations

Common Notations

$\mathbb{R}^N$ Euclidean N-dimensional space, the product of N copies of the real line $\mathbb{R}$

$\mathbb{C}^N$ complex N-dimensional space, the product of N copies of the complex numbers $\mathbb{C}$

$x = (x_1, \ldots, x_N)$, $y = (y_1, \ldots, y_N)$, etc. coordinates in $\mathbb{R}^N$

$xy = x_1 y_1 + \cdots + x_N y_N = x \cdot y = (x, y)$ scalar product in $\mathbb{R}^N$

$|x| = (\sum_1^N x_i^2)^{1/2}$ length of $x \in \mathbb{R}^N$

$B_r(x)$ the open ball of radius r and center $x \in \mathbb{R}^N$

Ω an open, generally bounded and connected, subset of $\mathbb{R}^N$

$\partial\Omega$ the boundary of Ω

$\overline{\Omega} = \Omega \cup \partial\Omega$ the closure of Ω

$\operatorname{int} U = \mathring{U}$ the interior of U

$\operatorname{supp} u$ the support of the function u, which is the smallest closed set outside of which $u = 0$

$\partial u/\partial x_j$ partial derivative of u with respect to x_j, also u_{x_j}, $\partial_j u$, or u_j

$u_x = \operatorname{grad} u = (u_{x_1}, \ldots, u_{x_n})$

$\alpha = (\alpha_1, \ldots, \alpha_n)$, $\alpha_i \geq 0$ integers, a multi-index of length $|\alpha| = \sum_1^N \alpha_i$

$$D^\alpha u = \frac{\partial^{\alpha_1}}{\partial x_1^{\alpha_1}} \cdots \frac{\partial^{\alpha_N}}{\partial x_N^{\alpha_N}} u$$

$$\left(\text{except in Chapter VI where } D^\alpha u = (-i)^{|\alpha|} \frac{\partial^{\alpha_1}}{\partial x_1^{\alpha_1}} \cdots \frac{\partial^{\alpha_N}}{\partial x_N^{\alpha_N}} u\right)$$

$dx = dx_1 \cdots dx_N$ Lebesgue measure on $\mathbb{R}^N$

$\Delta = \sum_1^N \partial^2/\partial x_j^2$ Laplace operator or Laplacian

$\langle \cdot, \cdot \rangle$ a pairing between a (real) Banach space V and its dual V'; $\langle \cdot, \cdot \rangle : V' \times V \to \mathbb{R}$

$D_j = \dfrac{1}{i}\dfrac{\partial}{\partial y_j}$, $D_t = \dfrac{1}{i}\dfrac{d}{dt}$ (in Chapter VI only)

$a_i b_i = \sum_{i=1}^N a_i b_i$
$a_{ij}\xi_i\eta_j = \sum_{i,j=1}^N a_{ij}\xi_i\eta_j$ summation convention: repeated indices are summed over their ranges, which are usually $1 \le i, j \le N$.

Function Spaces

$C(\Omega)(C(\bar{\Omega}))$ the functions continuous in $\Omega(\bar{\Omega})$

$C^m(\Omega)(C^{m,\lambda}(\Omega))$ the functions m times continuously differentiable in Ω (whose mth derivatives satisfy a Hölder condition with exponent λ) cf. Chapter II, Section 4

$C^m(\bar{\Omega})(C^{m,\lambda}(\bar{\Omega}))$ the functions of $C^m(\Omega)$, each of which is m times continuously differentiable in some neighborhood of $\bar{\Omega}$ (and whose mth derivatives satisfy a Hölder condition with exponent λ)

$L^s(\Omega)$ the Lebesgue measurable functions u of Ω for which

$$\|u\|_{L^s(\Omega)} = \left(\int_\Omega |u|^s\,dx\right)^{1/s} < \infty \quad (1 \le s < \infty)$$

$L^\infty(\Omega)$ the Lebesgue measurable functions u of Ω which are essentially bounded,

$$\|u\|_{L^\infty(\Omega)} = \inf\{M : |u| \le M \quad \text{a.e. in} \quad \Omega\}$$

$H^{m,s}(\Omega)$, $H_0^{m,s}(\Omega)$, $H^m(\Omega)$, $H_0^m(\Omega)$, $H^{-m}(\Omega)$ Sobolev spaces, see Chapter II, Section 4

$H^1_{\text{loc}}(\Omega)$ functions in $H^1(\Omega')$ for each bounded $\Omega' \subset \Omega$ whose closure $\bar{\Omega}' \subset \Omega$

$\mathscr{D}'(\Omega)$ distributions defined on Ω

$L^2(0, T : H^s(0, R))$ see Chapter VIII, Section 1

Introduction

In this work we intend to present the elements of variational inequalities and free boundary problems with several enticing examples of their application. It is a textbook, and, moreover, we have attempted to draw a broad audience by writing the first few sections of each chapter in a manner accessible to any informed reader including economists, engineers, and our other scientific colleagues.

Offered here is a view of the variational inequality problems and methods that have evolved in recent years. Their comprehension will bring the reader to the level of research in much of the subject. Our perspective will have its strongest appeal to the reader acquainted with partial differential equations. Indeed, one of our aims is to motivate the study of this subject too. A description of prerequisites and a few suggestions about the use of this book as a textbook are given at the end of the introduction.

We now offer some description, however rudimentary, of the nature of a variational inequality and the free boundary problem it suggests. We also outline our contents. Without attempting to formulate any precise definitions, let us discuss several examples.

Example 1. Let f be a smooth real valued function on the closed interval $I = [a, b]$. We seek the points $x_0 \in I$ for which

$$f(x_0) = \min_{x \in I} f(x).$$

Three cases can occur:

(i) if $a < x_0 < b$, then $f'(x_0) = 0$;
(ii) if $x_0 = a$, then $f'(x_0) \geq 0$; and
(iii) if $x_0 = b$, then $f'(x_0) \leq 0$.

These statements may be summarized by writing

$$f'(x_0)(x - x_0) \geq 0 \qquad \text{for all} \quad x \in I.$$

Such an inequality will be referred to as a *variational inequality*.

Example 2. Let f be a smooth real valued function defined on the closed convex set $\mathbb{K}$ of Euclidean N dimensional space $\mathbb{R}^N$. Again we shall characterize the points $x_0 \in \mathbb{K}$ such that

$$f(x_0) = \min_{x \in \mathbb{K}} f(x).$$

Assume x_0 is a point where the minimum is achieved and let $x \in \mathbb{K}$. Since $\mathbb{K}$ is convex, the segment $(1 - t)x_0 + tx = x_0 + t(x - x_0), 0 \leq t \leq 1$, lies in $\mathbb{K}$. The function

$$\Phi(t) = f(x_0 + t(x - x_0)), \qquad 0 \leq t \leq 1,$$

attains its minimum at $t = 0$; so, as in Example 1,

$$\Phi'(0) = \operatorname{grad} f(x_0)(x - x_0) \geq 0 \qquad \text{for any} \quad x \in \mathbb{K}.$$

Consequently, the point x_0 satisfies the variational inequality

$$x_0 \in \mathbb{K}: \quad \operatorname{grad} f(x_0)(x - x_0) \geq 0 \qquad \text{for all} \quad x \in \mathbb{K}.$$

If $\mathbb{K}$ is bounded, the existence of at least one x_0 is immediate.

Example 3. Let $\Omega \subset \mathbb{R}^N$ be a bounded domain with boundary $\partial\Omega$ and let ψ be a given function on $\bar{\Omega} = \Omega \cup \partial\Omega$ satisfying

$$\max_{\Omega} \psi \geq 0 \qquad \text{and} \qquad \psi \leq 0 \quad \text{on} \quad \partial\Omega.$$

Define

$$\mathbb{K} = \{v \in C^1(\bar{\Omega}) : v \geq \psi \text{ in } \Omega \text{ and } v = 0 \text{ on } \partial\Omega\},$$

a convex set of functions which we assume is not empty. We seek a function $u \in \mathbb{K}$ for which

$$\int_{\Omega} |\operatorname{grad} u|^2 \, dx = \min_{v \in \mathbb{K}} \int_{\Omega} |\operatorname{grad} v|^2 \, dx.$$

Assuming such a u to exist, we argue analogously to our previous example relying again on the convexity of $\mathbb{K}$. For any $v \in \mathbb{K}$, the sequence $u + t(v - u) \in \mathbb{K}$, $0 \le t \le 1$, whence the function

$$\Phi(t) = \int_\Omega |\operatorname{grad}(u + t(v - u))|^2 \, dx, \qquad 0 \le t \le 1,$$

attains its minimum at $t = 0$. This implies that $\Phi'(0) \ge 0$, which leads to the variational inequality

$$u \in \mathbb{K}: \quad \int_\Omega \operatorname{grad} u \operatorname{grad}(v - u) \, dx \ge 0 \qquad \text{for all} \quad v \in \mathbb{K}.$$

Intervening here is the point set

$$I = \{x \in \Omega : u(x) = \psi(x)\},$$

which is called the set of coincidence. Its presence distinguishes u from the solution of a boundary value problem.

One interprets u as the height function of the equilibrium position of a thin membrane constrained to lie above the body $\{(x, x_{N+1}) : x_{N+1} \le \psi(x), x \in \Omega\}$ and with fixed height zero on $\partial\Omega$.

Example 4. With Ω a bounded open set as before, let $\varphi^1, \varphi^2, \lambda_1, \lambda_2, f_1$, and f_2 be suitably smooth functions in $\bar{\Omega}$ with $\varphi^1 \le \varphi^2$. Define the convex set of pairs of functions

$$\mathbb{K} = \{v = (v^1, v^2) : v^1 \le v^2 \text{ in } \Omega, v^i = \varphi^i \text{ on } \partial\Omega, v^i \in C^1(\bar{\Omega}), i = 1, 2\}.$$

We search for the solution $u \in \mathbb{K}$ of

$$\sum_i \int_\Omega \{\operatorname{grad} u^i \operatorname{grad}(v^i - u^i) + \lambda_i u^i(v^i - u^i)\} \, dx$$
$$\ge \sum_i \int_\Omega f_i(v^i - u^i) \, dx \qquad \text{for all} \quad v \in \mathbb{K}.$$

In this case u is a point of $\mathbb{K}$ where the functional

$$F(v) = \frac{1}{2} \sum_i \int_\Omega \{|\operatorname{grad} v^i|^2 + \lambda_i (v^i)^2\} \, dx - \sum_i \int_\Omega f_i v^i \, dx$$

attains a minimum.

The solution pair $u = (u^1, u^2)$ represents the equilibrium position of two membranes subject to applied forces constrained by not being able to pass through each other. In this case the coincidence set is

$$I = \{x \in \Omega : u^1(x) = u^2(x)\}.$$

Example 5. With Ω and $\mathbb{K}$ as in Example 3, we ask for the element $u \in \mathbb{K}$ of least area, namely,

$$\int_\Omega \sqrt{1 + |\text{grad } u|^2}\, dx = \min_{v \in \mathbb{K}} \int_\Omega \sqrt{1 + |\text{grad } v|^2}\, dx.$$

The associated variational inequality is

$$u \in \mathbb{K}: \quad \int_\Omega \frac{\text{grad } u \text{ grad}(v - u)}{\sqrt{1 + |\text{grad } u|^2}}\, dx \geq 0 \qquad \text{for all} \quad v \in \mathbb{K}.$$

Among the examples considered, the first two may be solved by means of calculus for they depend on a finite number of variables. The other examples are similar to problems of the calculus of variations since the competing functions lie in an infinite dimensional space.

As usual in the study of calculus of variations, a problem formulated in a naive class of admissible functions need not admit a solution in that class. It is now well known that this difficulty may be overcome by broadening the class of competing functions to a suitable subset of a Sobolev space or of a more general space of distributions. In this context, the existence of solutions of Examples 3 and 4 may be demonstrated easily, in essence from the properties of the projection operator of a closed convex set in a Hilbert space. At first glance, Example 5 is more complicated since it suggests consideration of a Banach space which is not reflexive. An alternative approach will be to establish suitable a priori estimates.

The general problem of the existence of solutions is treated in the first three chapters. The first chapter is devoted to variational inequalities in $\mathbb{R}^N$ and the related topic of fixed points of mappings. Chapter II concerns variational inequalities in Hilbert space, covering problems like Examples 3 and 4. The material in its appendices includes basic properties of Sobolev spaces and a self-contained proof of the regularity of solutions of second order equations with bounded measurable coefficients. More general existence theorems are considered in Chapter III.

Having weakened the notion of solution, we are now led to problems of regularity, which is the exploration of the smoothness of the solution. It is worthwhile emphasizing that this question for variational inequalities differs from its analog in the theory of boundary value problems. In the latter, the smoother one assumes the data, the smoother, in general, the solution is found to be. In our situation, however, the constraints defining the convex set may impede the regularity of the solution, permitting it a certain amount of smoothness which cannot be surpassed regardless of the smoothness of the data.

A typical example of this situation is the "obstacle problem," Example 3.

In this case, the smoothness of ψ notwithstanding, we are able to conclude only that the first derivatives of u are continuous and the second derivatives of u are bounded but not continuous. This limitation is evident even in one dimension. The investigation of these properties is the subject of Chapter IV.

So the presence of the obstacle is reflected in the limited regularity of the solution and, as we noted in Example 3, the presence of the coincidence set I. Now I is determined by the variational condition of Example 3, so it is not arbitrary. In fact, we shall observe that u is the solution of the Laplace equation with given Cauchy data on ∂I, the boundary of I. This suggests that ∂I itself should possess some additional properties. It is called a "free boundary." The aspects of this topic especially relevant in two independent variables are discussed in Chapter V. For example, it is shown there that if Ω is convex and ψ is analytic and strictly concave, then ∂I is an analytic Jordan curve.

More general free boundary problems for arbitrary dimension are analyzed in Chapter VI. Here we introduce hodograph, or partial hodograph, transformations which permit straightening the free boundary but necessitate the study of highly nonlinear equations. For certain questions such a theory remains inadequate. Indeed, to resolve Example 4 we are led to the investigation of an elliptic system whose unknowns are defined on different sides of the free boundary. We describe this situation as well as introduce the theory of elliptic equations and systems with coercive boundary conditions. Among the many subjects which may be studied in this framework are plasma confinement and elliptic equations which share Cauchy data on a given hypersurface.

The remainder of the book, Chapters VII and VIII, is devoted to applications of variational inequalities in problems of engineering and physics. These include the lubrication of a journal bearing, the filtration of a liquid through a porous medium, and the motion of a fluid past a given profile. Filtration is considered also in its three dimensional form, thus providing the basis for the analysis of a free boundary in higher dimensions. In Chapter VIII we discuss the Stefan problem.

Frequent reference is made to results in partial differential equations and other fields of analysis. Items which we have found appropriate to include in the text are listed in the table of contents. They comprise Section 4 and the appendices of Chapter II, the appendix of Chapter IV, Section 3 of Chapter VI, and others. Certain facts, on the other hand, we believe to be readily accessible even to the reader who is not expert in differential equations. These we merely cite as known or state without proof. They include such items as the classical maximum principle, the Hopf boundary point lemma, and the Calderon–Zygmund inequalities. Among the references we have found convenient for this material are Courant and Hilbert [1], "Methods of Mathematical Physics," Vol. II, and Bers *et al.* [1], "Partial Differential

Equations." More extensive treatments are found in the books of Morrey [1] and Lions and Magenes [1] cited in the bibliography. In particular, such treatises or the literature are of avail when we require the conclusions of the elliptic regularity theory. We regret that we are unable to give a complete exposition of this material.

The reader is not presumed to have any prior knowledge of variational inequalities. This is an opportune place to mention a few of the papers and books which have appeared in recent years which the reader may wish to consult for collateral or historical information: Baiocchi [2, 3, 4], Baiocchi and Capelo [1], Brezis [1, 3], Kinderlehrer [5], Lions [1], Lions and Duvaut [1], Magenes [1], and Stampacchia [3, 4, 5]. In the bibliography there are additional references.

It has been our experience that a one-quarter course may be based on Chapters I, II, the first part of Chapter III, and the first part of Chapter IV. We have often found it useful to include the first appendix of Chapter II. A course which also serves as an introduction to elliptic equations with bounded measurable coefficients might also include Appendices B, C, and D of Chapter II. Additional material may then be drawn from Chapters V and VI if interest is toward free boundary problems, or from Chapter VII if applications are the principal emphasis.

CHAPTER **I**

Variational Inequalities in $\mathbb{R}^N$

1. Fixed Points

Many problems of nonlinear analysis may be solved by means of fixed point theorems. Let F be a mapping of a set A into itself:

$$F: A \to A.$$

A point $x \in A$ is called a fixed point of F if

$$F(x) = x.$$

In other words, fixed points of F are the solutions of the equation $F(x) = x$. The first theorem about fixed points, called the contraction mapping theorem, is the abstract formulation of Picard's method of successive approximations.

Let S be a metric space with metric, or distance function d.

Definition 1.1. A mapping $F: S \to S$ is a *contraction mapping* if

$$d(F(x), F(y)) \leq \alpha d(x, y), \qquad x, y \in S, \tag{1.1}$$

for some α, $0 \leq \alpha < 1$. When we allow $\alpha = 1$, the mapping F is called *nonexpansive*.

We state the contraction mapping theorem.

Theorem 1.2. *Let S be a complete metric space and let $F: S \to S$ be a contraction mapping. Then there exists a unique fixed point of F.*

Note that the theorem is not generally true if F is nonexpansive. For example a translation of a linear space into itself does not admit fixed points.

Another fundamental theorem is that of Brouwer. There are many proofs of this theorem connected with many branches of mathematics and yet its statement is very simple. We denote by $\mathbb{R}^N$ the (real) Euclidean space of dimension $N \geq 1$.

Theorem 1.3 (Brouwer). *Let F be a continuous mapping of a closed ball $\Sigma \subset \mathbb{R}^N$ into itself. Then F admits at least one fixed point.*

The proofs of these theorems may be found in many books (cf. Massey [1]). We point out that in Brouwer's theorem, the ball Σ may be replaced by a compact convex subset of $\mathbb{R}^N$. One object of the next section is to present this extension.

2. The Characterization of the Projection onto a Convex Set

In this section we consider the projection onto a convex set in a real Hilbert space H, since it is in this context that it will be useful to us in the sequel. The demonstrations are identical in the case H is finite dimensional. The reader may choose to read $\mathbb{R}^N$ for H.

Lemma 2.1. *Let $\mathbb{K}$ be a closed convex subset of a Hilbert space H. Then for each $x \in H$ there is a unique $y \in \mathbb{K}$ such that*

$$\|x - y\| = \inf_{\eta \in \mathbb{K}} \|x - \eta\|. \tag{2.1}$$

Remark 2.2. *The point y satisfying* (2.1) *is called the projection of x on $\mathbb{K}$ and we write*

$$y = \mathrm{Pr}_{\mathbb{K}}\, x.$$

Notice that $\mathrm{Pr}_{\mathbb{K}}\, x = x$, for all $x \in \mathbb{K}$.

Proof of the lemma. Let $\eta_k \in \mathbb{K}$ be a minimizing sequence, namely,

$$\lim_{k \to \infty} \|\eta_k - x\| = d = \inf_{\eta \in \mathbb{K}} \|\eta - x\|. \tag{2.2}$$

From the parallelogram law,

$$\|x + y\|^2 + \|x - y\|^2 = 2\|x\|^2 + 2\|y\|^2, \qquad x, y \in H,$$

a consequence of the existence of an inner product on H, we compute that

$$\|\eta_k - \eta_h\|^2 = 2\|x - \eta_k\|^2 + 2\|x - \eta_h\|^2 - 4\|x - \tfrac{1}{2}(\eta_k + \eta_h)\|^2. \quad (2.3)$$

Now $\mathbb{K}$ is convex, so $\frac{1}{2}(\eta_k + \eta_h) \in \mathbb{K}$ and

$$d^2 \leq \|x - \tfrac{1}{2}(\eta_k + \eta_h)\|^2.$$

Therefore

$$\|\eta_k - \eta_h\|^2 \leq 2\|x - \eta_k\|^2 + 2\|x - \eta_h\|^2 - 4d^2,$$

and we conclude from (2.2) that

$$\lim_{h,k \to \infty} \|\eta_k - \eta_h\| = 0.$$

Hence, since H is complete, there is an element $y \in \mathbb{K}$ such that

$$\lim_{k \to \infty} \eta_k = y.$$

Moreover,

$$\|x - y\| = \lim_{k \to \infty} \|x - \eta_k\| = d.$$

To see that y is unique, merely observe that any two elements $y, y' \in \mathbb{K}$ which satisfy (2.1) may be inserted in (2.3) in place of η_k, η_h. This yields

$$\begin{aligned} \|y - y'\|^2 &= 2\|x - y\|^2 + 2\|x - y'\|^2 - 4\|x - \tfrac{1}{2}(y + y')\|^2 \\ &\leq 4d^2 - 4d^2 = 0, \end{aligned}$$

or $y = y'$. Q.E.D.

We proceed to characterize the projection.

Theorem 2.3. *Let $\mathbb{K}$ be a closed convex set of a Hilbert space H. Then $y = \mathrm{Pr}_{\mathbb{K}}\, x$, the projection of x on $\mathbb{K}$, if and only if*

$$y \in \mathbb{K}: \quad (y, \eta - y) \geq (x, \eta - y) \qquad \textit{for all} \quad \eta \in \mathbb{K}. \quad (2.4)$$

Proof. Let $x \in H$ and $y = \mathrm{Pr}_{\mathbb{K}} x \in \mathbb{K}$. Since $\mathbb{K}$ is convex

$$(1 - t)y + t\eta = y + t(\eta - y) \in \mathbb{K} \qquad \text{for any} \quad \eta \in \mathbb{K}, \qquad 0 \leq t \leq 1,$$

and hence, by (2.1), the function

$$\Phi(t) = \|x - y - t(\eta - y)\|^2 = \|x - y\|^2 - 2t(x - y, \eta - y) + t^2\|\eta - y\|^2$$

attains its minimum at $t = 0$. So $\Phi'(0) \geq 0$, namely,

$$(x - y, \eta - y) \leq 0 \qquad \text{for} \quad \eta \in \mathbb{K},$$

or

$$(y, \eta - y) \geq (x, \eta - y) \qquad \text{for} \quad \eta \in \mathbb{K}.$$

On the other hand, if

$$y \in \mathbb{K}: \quad (y, \eta - y) \geq (x, \eta - y) \qquad \text{for} \quad \eta \in \mathbb{K},$$

then

$$0 \leq (y - x, (\eta - x) + (x - y)) \leq -\|x - y\|^2 + (y - x, \eta - x).$$

Therefore

$$\|y - x\|^2 \leq (y - x, \eta - x) \leq \|y - x\| \|\eta - x\|,$$

so, finally,

$$\|y - x\| \leq \|\eta - x\| \qquad \text{for} \quad \eta \in \mathbb{K}. \quad \text{Q.E.D.}$$

Corollary 2.4. *Let $\mathbb{K}$ be a closed convex set of a Hilbert space H. Then the operator $\Pr_{\mathbb{K}}$ is nonexpansive, that is,*

$$\|\Pr_{\mathbb{K}} x - \Pr_{\mathbb{K}} x'\| \leq \|x - x'\| \qquad \textit{for} \quad x, x' \in H. \tag{2.5}$$

Proof. Given $x, x' \in H$, let $y = \Pr_{\mathbb{K}} x$ and $y' = \Pr_{\mathbb{K}} x'$. Then

$$y \in \mathbb{K}: \quad (y, \eta - y) \geq (x, \eta - y), \qquad \eta \in \mathbb{K},$$

$$y' \in \mathbb{K}: \quad (y', \eta - y') \geq (x', \eta - y'), \qquad \eta \in \mathbb{K}.$$

We choose $\eta = y'$ in the first inequality and $\eta = y$ in the second. Adding we obtain

$$\|y - y'\|^2 = (y - y', y - y') \leq (x - x', y - y') \leq \|x - x'\| \|y - y'\|,$$

or

$$\|y - y'\| \leq \|x - x'\|. \quad \text{Q.E.D.}$$

Notice that the proof of the uniqueness of the projection follows again.

We conclude this section with the proof of Brouwer's theorem for a compact convex set.

Theorem 2.5 (Brouwer). *Let $\mathbb{K} \subset \mathbb{R}^N$ be compact and convex and let $F: \mathbb{K} \to \mathbb{K}$ be continuous. Then F admits a fixed point.*

Proof. Let Σ be a closed ball in $\mathbb{R}^N$ such that $\mathbb{K} \subset \Sigma$. From Corollary 2.4, $\Pr_{\mathbb{K}}$ is continuous; hence the mapping

$$F \circ \Pr_{\mathbb{K}} : \Sigma \to \mathbb{K} \subset \Sigma$$

is a continuous mapping of Σ into itself. It admits a fixed point x by Theorem 1.3, namely,

$$F \circ \Pr_{\mathbb{K}} x = x \in \mathbb{K}.$$

In particular, $\Pr_{\mathbb{K}} x = x$ so $F(x) = x$. Q.E.D.

3. A First Theorem about Variational Inequalities

In the study of variational inequalities we are frequently concerned with a mapping F from a linear space X, or a convex subset $\mathbb{K} \subset X$, into its dual X'. This will be particularly evident in Chapters II and III. Recall that the dual $(\mathbb{R}^N)'$ of $\mathbb{R}^N$ is the space of all linear forms

$$a: \mathbb{R}^N \to \mathbb{R}, \qquad x \to \langle a, x \rangle,$$

defined on $\mathbb{R}^N$. Indeed the bilinear mapping

$$\mathbb{R}^{N'} \times \mathbb{R}^N \to \mathbb{R}, \qquad a, x \to \langle a, x \rangle,$$

is referred to as a pairing. On the other hand we may always identify $(\mathbb{R}^N)'$ with $\mathbb{R}^N$, for example, we may identify $a \in (\mathbb{R}^N)'$ with the element $\pi a \in \mathbb{R}^N$ such that $\langle a, x \rangle = (\pi a, x)$. Neither the identification nor the realization of the pairing it determines is unique, but we always assume that

$$\langle a, x \rangle = (\pi a, x), \qquad a \in (\mathbb{R}^N)', \quad x \in \mathbb{R}^N,$$

where $\pi: (\mathbb{R}^N)' \to \mathbb{R}^N$ is the identification and $(\cdot, \cdot)$ is the scalar product on $\mathbb{R}^N$. Finally, a function

$$F: \mathbb{R}^N \to (\mathbb{R}^N)'$$

is continuous if each of the functions $F_1(x), \ldots, F_N(x)$ determined by the relation

$$\langle F(x), y \rangle = (\pi F(x), y) = \sum_j F_j(x) y_j$$

is continuous. The reader may verify that this is equivalent to the "natural" definition of continuity.

Theorem 3.1. *Let $\mathbb{K} \subset \mathbb{R}^N$ be compact and convex and let*

$$F: \mathbb{K} \to (\mathbb{R}^N)'$$

be continuous. Then there is an $x \in \mathbb{K}$ such that

$$\langle F(x), y - x \rangle \geq 0 \quad \text{for all} \quad y \in \mathbb{K}. \tag{3.1}$$

Proof. Proving the theorem is equivalent to showing there exists

$$x \in \mathbb{K}: \quad (x, y - x) \geq (x - \pi F(x), y - x) \qquad \text{for all} \quad y \in \mathbb{K}.$$

Now the mapping

$$\text{Pr}_{\mathbb{K}}\,(I - \pi F): \mathbb{K} \to \mathbb{K}$$

where $Ix = x$ is continuous; hence, by Theorem 2.5, it admits a fixed point $x \in \mathbb{K}$, namely,

$$x = \text{Pr}_{\mathbb{K}}(I - \pi F)x.$$

Consequently by Theorem 2.3, which is the characterization of the projection,

$$(x, y - x) \geq (x - \pi F(x), y - x) \qquad \text{for} \quad y \in \mathbb{K}. \quad \text{Q.E.D.}$$

Corollary 3.2. *Let x be a solution to* (3.1) *and suppose that $x \in \mathring{\mathbb{K}}$, the interior of $\mathbb{K}$. Then $F(x) = 0$.*

Proof. If $x \in \mathring{\mathbb{K}}$, then the points $y - x$ describe a neighborhood of the origin, i.e., for any $\xi \in \mathbb{R}^N$, there is an $\varepsilon \geq 0$ and $y \in \mathbb{K}$ such that $\xi = \varepsilon(y - x)$. Consequently

$$\langle F(x), \xi \rangle = \varepsilon \langle F(x), y - x \rangle \geq 0 \qquad \text{for all} \quad \xi \in \mathbb{R}^N,$$

from which it follows that $F(x) = 0$. Q.E.D.

Definition 3.3. Let $\mathbb{K}$ be a convex set of $\mathbb{R}^N$ and $x \in \partial\mathbb{K}$. A hyperplane

$$\langle a, y - x \rangle = 0, \qquad a \in (\mathbb{R}^N)' - \{0\},$$

is said a *hyperplane of support*, or a *supporting hyperplane*, of $\mathbb{K}$ if

$$\langle a, y - x \rangle \geq 0 \qquad \text{for all} \quad y \in \mathbb{K}.$$

Corollary 3.4. *Let x be a solution of* (3.1) *and suppose that $x \in \partial\mathbb{K}$. Then $F(x)$ determines a hyperplane of support for $\mathbb{K}$, provided $F(x) \neq 0$.*

Namely, the affine function $f(y) = \langle F(x), y - x \rangle$ is nonnegative for $y \in \mathbb{K}$.

4. Variational Inequalities

We consider

Problem 4.1. *Given $\mathbb{K}$ closed and convex in $\mathbb{R}^N$ and*

$$F: \mathbb{K} \to (\mathbb{R}^N)'$$

continuous, find

$$x \in \mathbb{K}: \quad \langle F(x), y - x\rangle \geq 0 \qquad \text{for} \quad y \in \mathbb{K}.$$

If $\mathbb{K}$ is bounded, we have proved the existence of a solution to Problem 4.1 in the previous section. On the other hand, notice that the problem does not always admit a solution. For example, if $\mathbb{K} = \mathbb{R}$,

$$f(x)(y - x) \geq 0 \qquad \text{for all} \quad y \in \mathbb{R}$$

has no solution for $f(x) = e^x$. The following theorem gives a necessary and sufficient condition for the existence of solutions. Given a convex set $\mathbb{K}$, we set $\mathbb{K}_R = \mathbb{K} \cap \Sigma_R$ where Σ_R is the closed ball of radius R and center $0 \in \mathbb{R}^N$. Returning to our $F: \mathbb{K} \to (\mathbb{R}^N)'$ we notice that there exists at least one

$$x_R \in \mathbb{K}_R: \quad \langle F(x_R), y - x_R\rangle \geq 0 \qquad \text{for} \quad y \in \mathbb{K}_R \tag{4.1}$$

whenever $\mathbb{K}_R \neq \varnothing$ by the previous theorem.

Theorem 4.2. *Let $\mathbb{K} \subset \mathbb{R}^N$ be closed and convex and*

$$F: \mathbb{K} \to (\mathbb{R}^N)'$$

be continuous. A necessary and sufficient condition that there exist a solution to Problem 4.1 is that there exist an $R > 0$ such that a solution $x_R \in \mathbb{K}_R$ of (4.1) satisfies

$$|x_R| < R. \tag{4.2}$$

Proof. It is clear that if there exists a solution to Problem 4.1, then x is a solution to (4.1) whenever $|x| < R$.

Suppose now that $x_R \in \mathbb{K}_R$ satisfies (4.2). Then x_R is also a solution to Problem 4.1. Indeed, since $|x_R| < R$, given $y \in \mathbb{K}$, $w = x_R + \varepsilon(y - x_R) \in \mathbb{K}_R$ for $\varepsilon \geq 0$ sufficiently small. Consequently

$$x_R \in \mathbb{K}_R \subset \mathbb{K}: \quad 0 \leq \langle F(x_R), w - x_R\rangle = \varepsilon\langle F(x_R), y - x_R\rangle \qquad \text{for} \quad y \in \mathbb{K},$$

which means that x_R is a solution to Problem 4.1. Q.E.D.

From this theorem we may deduce many sufficient conditions for existence. We state one which is both useful and introduces the notion of coerciveness.

Corollary 4.3. *Let* $F: \mathbb{K} \to (\mathbb{R}^N)'$ *satisfy*

$$\frac{\langle F(x) - F(x_0), x - x_0 \rangle}{|x - x_0|} \to +\infty \quad \text{as} \quad |x| \to +\infty, \quad x \in \mathbb{K}, \tag{4.3}$$

for some $x_0 \in \mathbb{K}$. *Then there exists a solution to Problem* 4.1.

Proof. Choose $H > |F(x_0)|$ and $R > |x_0|$ such that

$$\langle F(x) - F(x_0), x - x_0 \rangle \geq H|x - x_0| \quad \text{for} \quad |x| \geq R, \quad x \in \mathbb{K}.$$

Then

$$\begin{aligned} \langle F(x), x - x_0 \rangle &\geq H|x - x_0| + \langle F(x_0), x - x_0 \rangle \\ &\geq H|x - x_0| - |F(x_0)|\,|x - x_0| \\ &\geq (H - |F(x_0)|)(|x| - |x_0|) > 0 \quad \text{for} \quad |x| = R. \end{aligned} \tag{4.4}$$

Now let $x_R \in \mathbb{K}_R$ be the solution of (4.1). Then

$$\langle F(x_R), x_R - x_0 \rangle = -\langle F(x_R), x_0 - x_R \rangle \leq 0$$

so, in view of (4.4), $|x_R| \neq R$. In other words, $|x_R| < R$. Q.E.D.

Generally, the solution to the variational inequality is not unique. There is however a very natural condition which insures uniqueness. Suppose that $x, x' \in \mathbb{K}$ are two distinct solutions to Problem 4.1. Then

$$x \in \mathbb{K}: \quad \langle F(x), y - x \rangle \geq 0, \qquad y \in \mathbb{K},$$

$$x' \in \mathbb{K}: \quad \langle F(x'), y - x' \rangle \geq 0, \qquad y \in \mathbb{K},$$

so, setting $y = x'$ in the first inequality, $y = x$ in the second, and adding the two we obtain

$$\langle F(x) - F(x'), x - x' \rangle \leq 0.$$

Hence a natural condition for uniqueness is that

$$\langle F(x) - F(x'), x - x' \rangle > 0 \quad \text{whenever} \quad x, x' \in \mathbb{K}, \quad x \neq x'. \tag{4.5}$$

It turns out that the conditions of Corollary 4.3 and those implying uniqueness are useful in the study of existence in the infinite dimensional problem. We give the following definitions.

Definition 4.4. Condition (4.3) of Corollary 4.3 is a *coerciveness* condition.

Definition 4.5. By analogy with (4.5), we call a mapping $F: \mathbb{K} \to (\mathbb{R}^N)'$ *monotone* if

$$\langle F(x) - F(x'), x - x' \rangle \geq 0 \qquad \text{for all} \quad x, x' \in \mathbb{K}.$$

It is called *strictly monotone* if equality holds only when $x = x'$, that is, when condition (4.5) is valid.

As a related application of strictly monotone mappings we state

Proposition 4.6. *Let $F: \mathbb{K}_1 \to (\mathbb{R}^N)'$ be a continuous strictly monotone mapping of the closed convex $\mathbb{K}_1 \subset \mathbb{R}^N$. Let $\mathbb{K}_2 \subset \mathbb{K}_1$ be closed and convex. Suppose there exist solutions of the problems*

$$x_j \in \mathbb{K}_j: \quad \langle F(x_j), y - x_j \rangle \geq 0 \qquad \textit{for} \quad y \in \mathbb{K}_j, \quad j = 1, 2$$

(i) *If $F(x_2) = 0$, then $x_1 = x_2$.*

(ii) *If $F(x_2) \neq 0$ and $x_1 \neq x_2$, then the hyperplane $\langle F(x_2), y - x_2 \rangle = 0$ separates x_1 from $\mathbb{K}_2$.*

The proof is left as an exercise.

5. Some Problems Which Lead to Variational Inequalities

In this section we lightly touch on some elementary problems that are associated to variational inequalities. In particular, we discuss the connection between convex functions and monotone operators.

Let $f \in C^1(\mathbb{K})$, $\mathbb{K} \subset \mathbb{R}^N$, be a closed convex set, and set

$$F(x) = \operatorname{grad} f(x).$$

At this point we do not distinguish between $\mathbb{R}^N$ and $(\mathbb{R}^N)'$.

Proposition 5.1. *Suppose there exists an $x \in \mathbb{K}$ such that*

$$f(x) = \min_{y \in \mathbb{K}} f(y).$$

Then x is a solution of the variational inequality

$$x \in \mathbb{K}: (F(x), y - x) \geq 0 \qquad \textit{for} \quad y \in \mathbb{K}.$$

(See Example 2 of the Introduction.)

Proof. If $y \in \mathbb{K}$, then $z = x + t(y - x) \in \mathbb{K}$ for $0 \leq t \leq 1$; therefore the function $\varphi(t) = f(x + t(y - x))$, $0 \leq t \leq 1$, attains its minimum when $t = 0$. Consequently,

$$0 \leq \varphi'(0) = (\text{grad}\, f(x), y - x) = (F(x), y - x). \quad \text{Q.E.D.}$$

The converse holds if f is convex:

Proposition 5.2. *Suppose f is convex and x satisfies*

$$x \in \mathbb{K}: \quad (F(x), y - x) \geq 0 \qquad \textit{for all} \quad y \in \mathbb{K}.$$

Then

$$f(x) = \min_{y \in \mathbb{K}} f(y).$$

Proof. Indeed, since f is convex,

$$f(y) \geq f(x) + (F(x), y - x) \qquad \text{for any} \quad y \in \mathbb{K}.$$

But $(F(x), y - x) \geq 0$, so

$$f(y) \geq f(x). \quad \text{Q.E.D.}$$

Proposition 5.3. *Let $f: E \to \mathbb{R}^1$, $E \subset \mathbb{R}^N$, be a continuously differentiable convex (strictly convex) function. Then $F(x) = \text{grad}\, f(x)$ is monotone (strictly monotone).*

Proof. Given $x, x' \in E$,

$$f(x) \geq f(x') + (F(x'), x - x')$$

and

$$f(x') \geq f(x) + (F(x), x' - x).$$

Adding these we obtain

$$(F(x') - F(x), x' - x) \geq 0, \qquad x, x' \in E.$$

Hence F is monotone. The proof that F is strictly monotone provided f is strictly convex is identical. Q.E.D.

However, not all monotone operators arise as gradients of convex functions. For example, consider the vector field which is not exact

$$F(x) = (x_1, x_2 + \varphi(x_1)), \qquad x = (x_1, x_2) \in \mathbb{R}^2,$$

where φ is a smooth function of the single variable $x_1 \in \mathbb{R}^1$ such that

$$|\varphi(x_1) - \varphi(x_1')| \leq |x_1 - x_1'| \qquad \text{for} \quad x_1, x_1' \in \mathbb{R}^1.$$

We calculate that

$$\begin{aligned}(F(x) - F(x'), x - x') &= ((x_1 - x_1', x_2 - x_2' + \varphi(x_1) - \varphi(x_1')),\\ &\qquad (x_1 - x_1', x_2 - x_2'))\\ &= |x - x'|^2 + (x_2 - x_2')(\varphi(x_1) - \varphi(x_1'))\\ &\geq |x - x'|^2 - |x_2 - x_2'|\,|\varphi(x_1) - \varphi(x_1')|\\ &\geq |x - x'|^2 - \tfrac{1}{2}|x_2 - x_2'|^2 - \tfrac{1}{2}|\varphi(x_1) - \varphi(x_1')|^2\\ &\geq \tfrac{1}{2}|x - x'|^2.\end{aligned}$$

Conditions for a monotone operator to be given as the gradient of a convex function have been studied at length by Rockafellar [1].

To complete this chapter, we mention a problem of mathematical programming which can be reduced to a variational inequality.

Complementarity Problem 5.4. *Let*

$$\mathbb{R}_+^N = \{x = (x_1, \ldots, x_N) \in \mathbb{R}^N : x_i \geq 0\}$$

be a closed convex of $\mathbb{R}^N$, *and let* $F: \mathbb{R}_+^N \to \mathbb{R}^N$. *Find* $x_0 \in \mathbb{R}_+^N$ *such that* $F(x_0) \in \mathbb{R}_+^N$ *and* $(F(x_0), x_0) = 0$.

Theorem 5.5. *The point* $x_0 \in \mathbb{R}_+^N$ *is a solution to complementarity Problem* 5.4 *if and only if*

$$x_0 \in \mathbb{R}_+^N: \quad (F(x_0), y - x_0) \geq 0 \qquad \textit{for} \quad y \in \mathbb{R}_+^N.$$

Proof. First note that if x_0 is a solution to Complementarity Problem 5.4, $(F(x_0), y) \geq 0$ for any $y \in \mathbb{R}_+^N$, so

$$(F(x_0), y - x_0) = (F(x_0), y) - (F(x_0), x_0) = (F(x_0), y) \geq 0.$$

On the other hand suppose that $x_0 \in \mathbb{R}_+^N$ is a solution to the variational inequality. Then

$$y = x_0 + e_i, \qquad e_i = (0, \ldots, 0, 1, 0, \ldots, 0) \qquad (1 \text{ in } i\text{th place})$$

is an element of $\mathbb{R}_+^N$, so

$$0 \leq (F(x_0), x_0 + e_i - x_0) = (F(x_0), e_i) = F_i(x_0)$$

or $F(x_0) \in \mathbb{R}_+^N$. Hence, since $y = 0 \in \mathbb{R}_+^N$,

$$(F(x_0), x_0) \leq 0.$$

But x_0, $F(x_0) \in \mathbb{R}^N_+$ implies that $(F(x_0), x_0) \geq 0$, so

$$(F(x_0), x_0) = 0. \qquad \text{Q.E.D.}$$

Comments and Bibliographical Notes

The characterization of the projection onto a convex set by a variational inequality is contained in Section 3. It is also used in Lions and Stampacchia [1]. Theorem 3.1 was proved in Hartman and Stampacchia [1]. Another proof, assuming F to be monotone, is due to Browder [1]. The proof given here employs a simplification due to H. Brezis. Theorem 4.2 may be found in Hartman and Stampacchia [1] or Stampacchia [4]. The condition of Corollary 4.3 has been used frequently beginning with Browder [1, 2] and Minty [1, 2].

The relations between convex functions and variational inequalities are considered in Rockafellar [1] and Moreau [1]. The concept of subdifferential (subgradient) was developed by the latter author.

The fact that the complementarity problem may be reduced to a variational inequality was noticed by Karamardian. For the connections between mathematical programming and variational inequalities in $\mathbb{R}^N$ see Mancino and Stampacchia [1].

Exercises

1. Prove Proposition 4.6.

2. A mapping F from $\mathbb{R}^N$ into $(\mathbb{R}^N)'$ is called *cyclically monotone* if one has

$$\langle F(x_0), x_1 - x_0 \rangle + \langle F(x_1), x_2 - x_1 \rangle + \cdots + \langle F(x_n), x_0 - x_n \rangle \leq 0$$

for any set of points $\{x_0, x_1, \ldots, x_n\}$ (n arbitrary).

Show that $F(x) = \operatorname{grad} f(x)$ is cyclically monotone if $f(x)$ is a C^1 convex function.

3. (i) Deduce Brouwer's fixed point from Theorem 3.1.

(ii) Let F be a continuous mapping of a closed ball $\Sigma \subset \mathbb{R}^N$ into itself.

Assume that the vector $F(x)$ never has the same direction x for $x \in \partial\Sigma$. Then there exists a point x_0 of Σ where $F(x_0) = x_0$.

(Proof. (i) Applying Theorem 3.1 for the mapping $I - F$ we get

$$x_0 \in \mathbb{K}: \quad (x_0 - F(x_0), y - x_0) \geq 0 \qquad \text{for all} \quad y \in \mathbb{K}.$$

Because of the assumption of Theorem 2.5. we may choose $y = F(x_0)$ and therefore $F(x_0) = x_0$.

(ii) From Theorem 3.1. there exists an $x_0 \in \Sigma$ such that

$$(F(x_0), y - x_0) \geq 0 \qquad \text{for all} \quad y \in \Sigma.$$

But $x_0 \notin \partial\Sigma$; otherwise $F(x_0)$ would have the same direction of x_0. Then, by Corollary 3.2., it follows that $F(x_0) = 0$.)

4. State and solve the complementarity problem when F is a continuous mapping from $\mathbb{R}^N$ into $(\mathbb{R}^N)'$.

5. Prove the following lemma of Knaster–Kuratowski–Mazurkiewicz (Brézis [4]): "Let X be an arbitrary set in $\mathbb{R}^N$. To each $x \in X$ let a closed set $F(x)$ in $\mathbb{R}^N$ be given satisfying

(i) For at least one point x_0 of X, the set $F(x_0)$ is compact.

(ii) The convex hull of every finite subset $\{x_1, x_2, \ldots, x_n\}$ of X is contained in the corresponding union $\bigcup_{i=1}^n F(x_i)$. Then $\bigcap_{x \in X} F(x) \neq \varnothing$."

(Proof. Since the sets $F(x) \cap F(x_0)$ are compacts, as closed subsets of a compact, in order to prove the lemma it is enough to prove that the family $F(x)_{x \in X}$ has the finite intersection property.

Assume, by contradiction, that there exists a finite number of points $x_1, x_2, \ldots, x_k$ such that

$$\bigcap_{i=1}^{k} F(x_i) = \varnothing.$$

Let U_i denote the complement of $F(x_i)$; then

$$\bigcup_{i=1}^{k} U_i = \mathbb{R}^N.$$

Denote by $\psi_i(x)$ a partition of the unity in $\mathbb{R}^N$ associated to the covering U_i, i.e., $0 \leq \psi_i(x) \leq 1$, $\psi_i \in C^\infty$, supp $\psi_i \subset U_i$, and $\sum_{i=1}^k \psi_i(x) = 1$ for all $x \in \mathbb{R}^N$.

Consider the mapping that to any x associates

$$\varphi(x) = \sum_{i=1}^{k} \psi_i(x) x_i.$$

Since $\Phi(x)$ is contained in the convex hull of $\{x_1, \ldots, x_k\} = \mathbb{K}$, the mapping $\Phi(x)$ maps $\mathbb{K}$ in itself. Therefore from Brouwer's Theorem 1.2 there exists at least a fixed point $\bar{x} \in \mathbb{K}$, i.e.,

$$\bar{x} = \sum_{i=1}^{k} \psi_i(\bar{x}) x_i.$$

Arranging the sets U_i we may assume, for some s, $s \leq k$, that

$$\psi_i(\bar{x}) \begin{cases} \neq 0 & \text{for} \quad i \leq s \\ = 0 & \text{for} \quad i > s. \end{cases}$$

Then $\bar{x}$ belongs to the convex hull of the set $\{x_1, \ldots, x_s\}$ and thus, $\bar{x} \in \bigcup_{i \leq s} F(x_i)$, i.e., $\bar{x} \in F(x_i)$ for some $i \leq s$. This implies $\bar{x} \notin U_i$ and hence $\psi_i(\bar{x}) = 0$ for some $i \leq s$. This leads to a contradiction.)

6. Let $\mathbb{K}$ be a nonempty convex set in $\mathbb{R}^N$ that is not a point. Show that there exists an integer k, $0 < k \leq N$, such that $\mathbb{K}$ is embedded in $\mathbb{R}^k$ and there $\mathbb{K}$ has an interior point.

7. Consider a convex function f from $\mathbb{R}^N$ in $\mathbb{R} \cup \{+\infty\}$ that is proper, i.e., not merely the constant function $+\infty$. An element $x^* \in (\mathbb{R}^N)'$ is said to be a subgradient of f at x if

$$f(y) \geq f(x) + \langle x^*, y - x \rangle \qquad \text{for all} \quad y \in \mathbb{R}^N.$$

Geometrically, this condition means that the graph of the affine function

$$g \to f(x) + \langle x^*, y - x \rangle$$

is a supporting hyperplane to the convex set, called the epigraph of f,

$$\{(x, \lambda) \in \mathbb{R}^{N+1} : \lambda \geq f(x)\},$$

at the point $(x, f(x))$. The set of all subgradients x^* of f at x is denoted by $\partial f(x)$. The multivalued mapping

$$\partial f \colon x \to \partial f(x) \subset \mathbb{R}^N$$

is called the subdifferential of f.

A multivalued mapping A from $\mathbb{R}^N$ into $(\mathbb{R}^N)'$ is called monotone if

$$(y_1 - y_2, x_1 - x_2) \geq 0 \qquad \text{for} \quad x_1, x_2 \in \mathbb{R}^N, \quad y_1 \in Ax_1, \quad y_2 \in Ax_2,$$

where Ax denotes the set (possibly empty) corresponding to x.

If f is a proper convex function on $\mathbb{R}^N$, then the minimum of f occurs at the point x if and only if $0 \in \partial f(x)$.

8. Define the multivalued map from a closed convex set $\mathbb{K} \subset \mathbb{R}^N$ into $(\mathbb{R}^N)'$: set, for $x \in \mathbb{K}$, $\chi(x) = 0$ and for $x \in \partial\mathbb{K}$ let $\chi(x)$ be the set of elements of $(\mathbb{R}^N)'$ such that

$$\langle \chi(x), y - x \rangle \geq 0, \qquad \langle \chi(x), y - x \rangle = 0$$

are the supporting planes to $\mathbb{K}$ in x.

Then the variational inequality (3.1) can be written as

$$x \in \mathbb{K}: \quad F(x) \in \chi(x).$$

9. Let $f_1(x), \ldots, f_n(x)$ be smooth convex functions on $\mathbb{R}^N$ such that grad $f_i(x) \neq 0$, $i = 1, \ldots, n$, and suppose that

$$\mathbb{K} = \{x \in \mathbb{R}^N : f_i(x) \leq 0, 1 \leq i \leq n\}, \ \overset{\circ}{\mathbb{K}} \neq \varnothing. \tag{P.1}$$

Show that the planes

$$\left(\sum_{j=1}^{n} \lambda_j \operatorname{grad} f_j(x_0), y - x_0 \right) = 0,$$

with

$$\lambda_j \geq 0, \quad f_j(x_0) \leq 0, \qquad \text{and} \qquad \lambda_j f_j(x_0) = 0, \qquad 1 \leq j \leq n,$$

are the planes of support of $\mathbb{K}$ at x_0.

Now let $F: \mathbb{K} \to \mathbb{R}^N$ be continuous and suppose that

$$x_0 \in \mathbb{K}: \quad (F(x_0), y - x_0) \geq 0 \qquad \text{for all} \quad y \in \mathbb{K}.$$

Then, if $F(x_0) \neq 0$, we have seen that

$$(F(x_0), y - x_0) = 0$$

is a plane of support of $\mathbb{K}$, so by the above

$$F(x_0) + \sum_{j=1}^{n} \lambda_j \operatorname{grad} f_j(x_0) = 0$$

for some choice of $(\lambda_1, \ldots, \lambda_n) \in \mathbb{R}^n_+$. If $F(x_0) = 0$, choose $\lambda_j = 0$. The numbers $\lambda_1, \ldots, \lambda_n$ are called Lagrange multipliers and this result is the *Kuhn–Tucker condition.* We summarize:

Let $\mathbb{K} \subset \mathbb{R}^N$ *be defined by* (P.1), *let*

$$F: \mathbb{K} \to \mathbb{R}^N$$

be continuous, and suppose that

$$x_0 \in \mathbb{K}: \quad (F(x_0), y - x_0) \geq 0 \qquad \textit{for} \quad y \in \mathbb{K}.$$

Then

$$F(x_0) + \sum \lambda_j \operatorname{grad} f_j(x_0) = 0$$

where

$$\lambda_j \geq 0, \qquad f_j(x_0) \leq 0 \qquad \textit{and} \qquad \lambda_j f_j(x_0) = 0, \qquad 1 \leq j \leq n.$$

CHAPTER **II**

Variational Inequalities in Hilbert Space

1. Bilinear Forms

Many interesting questions in the theory of variational inequalities may be formulated in terms of bilinear forms on Hilbert spaces. This theory is a generalization of the variational theory of boundary value problems for linear elliptic equations.

Let H be a real Hilbert space and let H' denote its dual. We set $(\cdot, \cdot)$ the inner product on H, and $\|\cdot\|$ its norm, and

$$H' \times H \to \mathbb{R}.$$

$$f, x \to \langle f, x \rangle$$

the pairing between H and H'.

Let $a(u, v)$ be a (real) bilinear form on H, i.e., $a\colon H \times H \to \mathbb{R}$ is continuous and linear in each of the variables u, v. A bilinear form $a(u, v)$ is symmetric if

$$a(u, v) = a(v, u) \qquad \text{for} \quad u, v \in H.$$

A linear and continuous mapping

$$A\colon H \to H'$$

determines a bilinear form via the pairing

$$a(u, v) = \langle Au, v \rangle. \tag{1.1}$$

The conditions of linearity are satisfied and $|a(u, v)| \le c\|u\| \cdot \|v\|$ for a $c > 0$, which implies that a is continuous. And vice versa, given a bilinear form $a(u, v)$, the linear mapping

$$v \to a(u, v) \qquad \text{for} \quad v \in H$$

determines a continuous linear transformation $A: H \to H'$ which satisfies (1.1).

Definition 1.1. The bilinear form $a(u, v)$ is *coercive on* H if there exists $\alpha > 0$ such that

$$a(v, v) \ge \alpha \|v\|^2 \qquad \text{for} \quad v \in H. \tag{1.2}$$

The bilinear form $a(u, v)$ is coercive if and only if the mapping A defined by (1.1) is coercive in the sense of Chapter I, Definition 4.4. Evidently, a coercive symmetric bilinear form $a(u, v)$ defines a norm $(a(v, v))^{1/2}$ on H equivalent to $\|v\|$. We consider now:

Problem 1.2. *Let $\mathbb{K} \subset H$ be closed and convex and $f \in H'$. To find*:

$$u \in \mathbb{K}: \quad a(u, v - u) \ge \langle f, v - u \rangle \qquad \textit{for all} \quad v \in \mathbb{K}. \tag{1.3}$$

2. Existence of a Solution

The purpose of this section is to solve Problem 1.2 and to prove

Theorem 2.1. *Let $a(u, v)$ be a coercive bilinear form on H, $\mathbb{K} \subset H$ closed and convex and $f \in H'$. Then there exists a unique solution to Problem 1.2. In addition, the mapping $f \to u$ is Lipschitz, that is, if u_1, u_2 are solutions to Problem 1.2 corresponding to $f_1, f_2 \in H'$, then*

$$\|u_1 - u_2\| \le (1/\alpha)\|f_1 - f_2\|_{H'}. \tag{2.1}$$

Observe that the mapping $f \to u$ is linear if $\mathbb{K}$ is a subspace of H.

Proof. We begin with the demonstration of (2.1). Suppose there exist $u_1, u_2 \in H$ solutions of the variational inequalities

$$u_i \in \mathbb{K}: \quad a(u_i, v - u_i) \ge \langle f_i, v - u_i \rangle \qquad \text{for} \quad v \in \mathbb{K}, \quad i = 1, 2.$$

Setting $v = u_2$ in the variational inequality for u_1 and $v = u_1$ in that for u_2 we obtain, upon adding,

$$a(u_1 - u_2, u_1 - u_2) \leq \langle f_1 - f_2, u_1 - u_2 \rangle.$$

Hence by the coerciveness of a,

$$\alpha \|u_1 - u_2\|^2 \leq \langle f_1 - f_2, u_1 - u_2 \rangle \leq \|f_1 - f_2\|_{H'} \cdot \|u_1 - u_2\|,$$

and therefore (2.1) holds.

It remains to show the existence of u, which we present in several steps. First suppose that $a(u, v)$ is symmetric, and define the functional

$$I(u) = a(u, u) - 2\langle f, u \rangle, \qquad u \in H.$$

Let $d = \inf_{\mathbb{K}} I(u)$. Since

$$\begin{aligned} I(u) &\geq \alpha \|u\|^2 - 2\|f\|_{H'} \cdot \|u\|_H \\ &\geq \alpha \|u\|^2 - (1/\alpha)\|f\|_{H'}^2 - \alpha \|u\|^2 \\ &\geq -(1/\alpha)\|f\|_{H'}^2, \end{aligned}$$

we see that

$$d \geq (1/\alpha)\|f\|_{H'}^2 > -\infty.$$

Let u_n be a minimizing sequence of I in $\mathbb{K}$ such that

$$\{u_n \in \mathbb{K} : d \leq I(u_n) \leq d + (1/n)\}.$$

Applying the parallelogram law, and keeping in mind that $\mathbb{K}$ is convex, we see that

$$\begin{aligned} \alpha \|u_n - u_m\|^2 &\leq a(u_n - u_m, u_n - u_m) \\ &= 2a(u_n, u_n) + 2a(u_m, u_m) - 4a(\tfrac{1}{2}(u_n + u_m), \tfrac{1}{2}(u_n + u_m)) \\ &= 2I(u_n) + 2I(u_m) - 4I(\tfrac{1}{2}(u_n + u_m)) \\ &\leq 2[(1/n) + (1/m)]. \end{aligned}$$

We have used

$$4\langle f, u_n \rangle + 4\langle f, u_m \rangle - 8\langle f, \tfrac{1}{2}(u_n + u_m) \rangle = 0.$$

Hence the sequence $\{u_n\}$ is Cauchy and the closed set $\mathbb{K}$ contains an element u such that $u_n \to u$ in H and $I(u_n) \to I(u)$. So $I(u) = d$.

Now for any $v \in \mathbb{K}$, $u + \varepsilon(v - u) \in \mathbb{K}$, $0 \leq \varepsilon \leq 1$, and $I(u + \varepsilon(v - u)) \geq I(u)$. Then $(d/d\varepsilon)I(u + \varepsilon(v - v))|_{\varepsilon=0} \geq 0$. In other words,

$$2\varepsilon a(u, v - u) + \varepsilon^2 a(v - u, v - u) - 2\varepsilon \langle f, v - u \rangle \geq 0$$

or

$$a(u, v - u) \geq \langle f, v - u \rangle - \tfrac{1}{2}\varepsilon a(v - u, v - u) \qquad \text{for any} \quad \varepsilon, \quad 0 \leq \varepsilon \leq 1.$$

Setting $\varepsilon = 0$ we see that u is a solution to Problem 1.2.

We treat now the general case as a perturbation of the symmetric one. Introduce the coercive bilinear form

$$a_t = a_0(u, v) + tb(u, v), \qquad 0 \le t \le 1,$$

where

$$a_0(u, v) = \tfrac{1}{2}(a(u, v) + a(v, u))$$

and

$$b(u, v) = \tfrac{1}{2}(a(u, v) - a(v, u))$$

are the symmetric and antisymmetric parts of a. Observe that $a_1(u, v) = a(u, v)$ and that $a_t(u, v)$ is coercive with the same constant α.

Lemma 2.2. *If Problem 1.2 is solvable for $a_\tau(u, v)$ and all $f \in H'$, then it is solvable for $a_t(u, v)$ and all $f \in H'$, where $\tau \le t \le \tau + t_0$, $t_0 < \alpha/M$, and*

$$M = \sup \frac{|b(u, v)|}{\|u\| \cdot \|v\|} < +\infty.$$

Proof. Define the mapping

$$T: H \to \mathbb{K}$$

by $u = Tw$ if

$$u \in \mathbb{K}: \quad a_\tau(u, v - u) \ge \langle F_t, v - u \rangle \qquad \text{for all} \quad v \in \mathbb{K},$$

where

$$\langle F_t, v \rangle = \langle f, v \rangle - (t - \tau)b(w, v) \qquad \text{and} \qquad \tau \le t \le \tau + t_0.$$

By hypothesis, T is well defined. Given $u_1 = Tw_1$ and $u_2 = Tw_2$ we may apply (2.1) so that

$$\|u_1 - u_2\| \le (1/\alpha)(t - \tau)M\|w_1 - w_2\| \le (1/\alpha)t_0 M\|w_1 - w_2\|$$

with $t_0 M/\alpha < 1$. Hence T is a contraction mapping and admits a unique fixed point by Theorem 1.2 of Chapter I. For this $u = w$,

$$u \in \mathbb{K}: \quad a_t(u, v - u) \ge \langle f, v - u \rangle \qquad \text{for} \quad v \in \mathbb{K}$$

and every t, $\tau \le t \le \tau + t_0$.

To complete the proof of the theorem, it suffices to observe that Problem 1.2 may be solved for $a_0(u, v)$, which is symmetric. Applying Lemma 2.2 a finite number of times, we see that Problem 1.2 admits a solution for $t = 1$.
Q.E.D.

As in the proof of Theorem 3.1 of Chapter I, the variational inequality was resolved by considering a minimization problem and applying a fixed point theorem. Note that in the case $\mathbb{K} = H$, Theorem 2.1 reduces to the Lax–Milgram lemma, which asserts that any linear function $\langle f, v\rangle$ on H may be represented by $a(u, v)$ with a suitable $u \in H$ when $a(u, v)$ is a bilinear coercive form on H.

3. Truncation

Let $E \subset \mathbb{R}^N$ be (Lebesgue) measurable and choose $\varphi \in L^2(E)$. We set

$$\mathbb{K} = \{v \in L^2(E) : v \geq \varphi \text{ a.e. in } E\} \subset L^2(E),$$

evidently a closed convex set. Let us fix

$$a(u, v) = (u, v) = \int_E u(x)v(x)\, dx,$$

the scalar product on $L^2(E)$, which is a coercive bilinear form. For a given $f \in L^2(E)$, there exists a unique

$$u \in \mathbb{K}: \quad \int_E u(v - u)\, dx \geq \int_E f(v - u)\, dx \qquad \text{for all} \quad v \in \mathbb{K} \tag{3.1}$$

by Theorem 2.1. We claim that

$$u = \max(\varphi, f) = \begin{cases} \varphi(x) & \text{if} \quad f(x) \leq \varphi(x) \\ f(x) & \text{if} \quad f(x) \geq \varphi(x). \end{cases} \tag{3.2}$$

In fact, defining u by (3.2), we compute

$$\begin{aligned} \int_E u(v - u)\, dx &= \int_{\{f < \varphi\}} \varphi(v - \varphi)\, dx + \int_{\{\varphi \leq f\}} f(v - f)\, dx \\ &\geq \int_{\{f \leq \varphi\}} f(v - \varphi)\, dx + \int_{\{\varphi \leq f\}} f(v - f)\, dx \end{aligned}$$

since $v - \varphi \geq 0$ for any $v \in \mathbb{K}$. Consequently,

$$\int_E u(v - u)\, dx \geq \int_E f(v - u)\, dx \qquad \text{for all} \quad v \in \mathbb{K},$$

so u is the solution to the variational inequality.

The interested reader will realize at this point that it is possible to raise questions of regularity of the solution to Problem 1.2. So, one result we mention in this connection is

Theorem 3.1. *Let $\Omega \subset \mathbb{R}^N$ be open and let $\varphi, f \in H^{1,p}(\Omega)$, $2N/(N-2) \leq p \leq \infty$. Let*

$$\mathbb{K} = \{v \in L^2(\Omega) : v \geq \varphi \text{ a.e. in } \Omega\}.$$

Then the solution u of (3.1) *satisfies $u \in H^{1,p}(\Omega)$.*

The proof of the theorem is reserved for Appendix A. The definition of the space $H^{1,p}(\Omega)$ and its properties are summarized in the next section.

4. Sobolev Spaces and Boundary Value Problems

The use of Sobolev spaces is essential to our method. Here we do not intend to develop in detail the properties of these spaces but only to recall their definitions. Some aspects of them relevant for our study are described in Appendix A of this chapter. More information is available in many books and papers. In particular, we refer frequently to the book of Morrey [1], where such spaces are investigated very deeply.

As applications we give weak formulations of some classical boundary value problems and observe that Theorem 2.1 may be invoked to obtain their solutions.

We recall the definitions of some function spaces.

Definition 4.1. Let $\Omega \subset \mathbb{R}^N$ be a bounded open set with closure $\overline{\Omega}$ and boundary $\partial\Omega$. By $C^k(\overline{\Omega})$ we denote the space of k times differentiable (real valued) functions on $\overline{\Omega}$. By $C^{k,\lambda}(\overline{\Omega})$, $0 < \lambda < 1$, we indicate the functions k times continuously differentiable in $\overline{\Omega}$ whose derivatives of order k are Hölder continuous with exponent λ, $0 < \lambda < 1$. Recall that $u \in C^{0,\lambda}(\overline{\Omega})$, or u is *Hölder continuous* with exponent λ in $\overline{\Omega}$, if

$$[u]_\lambda = \sup_{x, x' \in \overline{\Omega}} \frac{|u(x) - u(x')|}{|x - x'|^\lambda} < +\infty.$$

If we allow $\lambda = 1$, then u is called a *Lipschitz function.*

The N-tuple of nonnegative integers $\alpha = (\alpha_1, \ldots, \alpha_N)$ is called a multi-index of length $|\alpha| = \alpha_1 + \cdots + \alpha_N \geq 0$. We set $D^\alpha = (\partial/\partial x_1)^{\alpha_1} \cdots (\partial/\partial x_N)^{\alpha_N}$, a differential operator of order $|\alpha|$. Here $(\partial^0/\partial x_i^0)u = u$, $1 \leq i \leq N$.

Definition 4.2. In the linear space $C^m(\overline{\Omega})$ we introduce the norm

$$\|u\|_{H^{m,s}(\Omega)} = \sum_{0 \le |\alpha| \le m} \|D^\alpha u\|_{L^s(\Omega)}, \qquad 1 \le s < \infty, \tag{4.1}$$

and we denote by $H^{m,s}(\Omega)$ the completion of $C^m(\overline{\Omega})$ in this norm. Usually, we write $H^m(\Omega) = H^{m,2}(\Omega)$.

If we assume that $\partial\Omega$ is Lipschitz, by which we mean that every point of $\partial\Omega$ has a neighborhood U such that $\partial\Omega \cap U$ is the graph of a Lipschitz function, then one may check (cf. Morrey [1], who calls such domains "strongly Lipschitz") that the elements of $H^{m,s}(\Omega)$ are the functions $u \in L^s(\Omega)$ for which there exist functions $g_\alpha \in L^s(\Omega)$, $|\alpha| \le m$, such that

$$\int_\Omega u(x) D^\alpha \zeta(x)\, dx = (-1)^{|\alpha|} \int_\Omega g_\alpha(x)\zeta(x)\, dx, \qquad \zeta \in C_0^\infty(\Omega), \quad 0 \le |\alpha| \le m.$$

Here $C_0^\infty(\Omega)$ denotes the infinitely differentiable functions having compact support in Ω.

Definition 4.3. $H_0^{m,s}(\Omega)$ is the closure of $C_0^\infty(\Omega)$ in the norm (4.1).

Definition 4.4. $H^{m,\infty}(\Omega)$ is the class of functions of $C^{m-1}(\overline{\Omega})$ whose derivatives of order $m-1$ satisfy a Lipschitz condition in $\overline{\Omega}$.

In the definition of $H^{1,s}(\Omega)$ we could have replaced the space $C^1(\overline{\Omega})$ by $C^{0,1}(\overline{\Omega}) = H^{1,\infty}(\Omega)$, namely, Lipschitz functions in $\overline{\Omega}$. In the case for which $\partial\Omega$ itself is Lipschitz, this is easily seen; for given $u \in H^{1,\infty}(\Omega)$, u admits an extension to $\tilde{u} \in H_0^{1,\infty}(\mathbb{R}^N)$ (= Lipschitz functions in $\mathbb{R}^N$ with compact support). Since $\tilde{u}$ may be approximated in $H_0^{1,s}(\mathbb{R}^N)$, $1 \le s < \infty$, by smooth functions, for example, by mollification, it follows that u is the limit in $H^{1,s}(\Omega)$ of smooth functions in $\overline{\Omega}$. Hence $C^1(\overline{\Omega}) \subset H^{1,\infty}(\Omega) \subset H^{1,s}(\Omega)$, $1 \le s < \infty$, and Lipschitz functions are dense in $H^{1,s}(\Omega)$, $1 \le s < \infty$. The only feature of this reasoning particular to Lipschitz domains Ω is the existence of the extension $\tilde{u}$. But it is a known variant of Tietze's extension theorem that $u \in H^{1,\infty}(\Omega)$ admits an extension to $\tilde{u} \in H_0^{1,\infty}(\mathbb{R}^N)$ for *any* bounded open domain $\Omega \subset \mathbb{R}^N$ (cf. the book (Stein [1]), for example). Hence $H^{1,\infty}(\Omega)$ is dense in $H^{1,s}(\Omega)$, $1 \le s < \infty$, for any bounded open domain Ω.

Suppose once again that $\partial\Omega$ is Lipschitz. Then given a domain Ω_0, $\Omega \subset \overline{\Omega} \subset \Omega_0$, any function $u \in H^{1,s}(\Omega)$, $1 \le s < \infty$, admits an extension $\tilde{u} \in H^{1,s}(\Omega_0)$. Moreover, $u \ge 0$ a.e. in Ω implies that $\tilde{u} \ge 0$ a.e. in Ω_0.

According to Poincaré's inequality, there is a $\beta > 0$ such that

$$\int_\Omega \zeta^2\, dx \le \beta \int_\Omega \zeta_x^2\, dx \qquad \text{for} \quad \zeta \in H_0^1(\Omega),$$

where

$$\zeta_x = (\zeta_{x_1}, \ldots, \zeta_{x_N}) = \left(\frac{\partial}{\partial x_1}\zeta, \ldots, \frac{\partial}{\partial x_N}\zeta\right).$$

Hence $\|\zeta_x\|_{L^2(\Omega)}$ is a norm equivalent to (4.1) for $H_0^1(\Omega)$. We shall always understand that

$$\|\zeta\|_{H_0^1(\Omega)} = \|\zeta_x\|_{L^2(\Omega)}.$$

For $1 < s < \infty$, $H^{m,s}(\Omega)$ and $H_0^{m,s}(\Omega)$ are reflexive Banach spaces and $H^m(\Omega)$ and $H_0^m(\Omega)$ are Hilbert spaces.

Definition 4.5. We denote the dual of $H_0^{m,s}(\Omega)$ by $H^{-m,s'}(\Omega)$, $(1/s) + (1/s') = 1$, or simply $H^{-m}(\Omega)$ when $s = 2$.

Elements of $H^{-1,s'}(\Omega)$ may be characterized as the derivatives of functions $f_i \in L^{s'}(\Omega)$ in the distributional sense, namely, for $f \in H^{-1,s'}(\Omega)$ there exist $f_0, f_1, \ldots, f_N \in L^{s'}(\Omega)$ such that

$$\langle f, \zeta\rangle = \int_\Omega \left\{ f_0\zeta - \sum_1^N f_i\zeta_{x_i} \right\} dx, \qquad \zeta \in H_0^{1,s}(\Omega).$$

As further examples of the application of Theorem 2.1, we show the existence of weak solutions to classical boundary value problems for the Laplace equation.

Dirichlet's Problem. *Given $f \in H^{-1}(\Omega)$ and $g \in H^1(\Omega)$, find $u \in H^1(\Omega)$ such that*

$$-\Delta u = f \quad in \quad \Omega; \qquad u = g \quad on \quad \partial\Omega. \tag{4.2}$$

When $f = f_0$, g and u are smooth, the meaning of (4.2) is clear. To obtain its meaning in the present weak context, assume this smoothness and multiply both sides of the equation by $\zeta \in C_0^\infty(\Omega)$. After an integration by parts, there results

$$(u, \zeta)_{H_0^1(\Omega)} = \int_\Omega u_{x_i}\zeta_{x_i}\, dx = \int_\Omega f\zeta\, dx.$$

Here and in the sequel we use the summation convention: iterated indices are summed. Similarly, we have only a generalized notion of what "$u = g$ on $\partial\Omega$" means since we have only a generalized notion of the meaning of the statement "$\zeta = 0$ on $\partial\Omega$," namely, $\zeta \in H_0^1(\Omega)$. So $u \in H^1(\Omega)$ is called a *weak solution* of (4.2) if

$$\int_\Omega u_{x_i}\zeta_{x_i}\,dx = \langle f, \zeta\rangle \qquad \text{for} \quad \zeta \in H_0^1(\Omega), \quad u - g \in H_0^1(\Omega). \tag{4.3}$$

Finally, we write this problem as a variational inequality. Introduce the affine space

$$M = M_g = \{v \in H^1(\Omega) : v - g \in H_0^1(\Omega)\}.$$

Now (4.2) becomes the weak Dirichlet problem.

Problem 4.6. *To find*

$$u \in M: \quad \int_\Omega u_{x_i}(v - u)_{x_i}\,dx \geq \langle f, v - u\rangle \qquad \textit{for} \quad v \in M.$$

Of course, the bilinear form

$$u, v \to \int_\Omega u_{x_i} v_{x_i}\,dx,$$

which is the scalar product on $H_0^1(\Omega)$, is not coercive on the entire space $H^1(\Omega)$. We argue in this way. Inasmuch as

$$\int_\Omega u_{x_i}(v - u)_{x_i}\,dx = \int_\Omega (u - g)_{x_i}(v - u)_{x_i}\,dx + \int_\Omega g_{x_i}(v - u)_{x_i}\,dx,$$

u is a solution to the problem if and only if $w = u - g$ is a solution to

$$\int_\Omega w_{x_i}(v - u)_{x_i}\,dx \geq \langle f, v - u\rangle - \int_\Omega g_{x_i}(v - u)_{x_i}\,dx. \tag{4.4}$$

Recalling that $g_{x_i} \in L^2(\Omega)$, $1 \leq i \leq N$, Problem 4.6 becomes equivalent to finding

$$w \in H_0^1(\Omega): \quad \int_\Omega w_{x_i}(\zeta - w)_{x_i}\,dx \geq \langle F, \zeta - w\rangle \qquad \text{for} \quad \zeta \in H_0^1(\Omega) \tag{4.5}$$

with $F \in H^{-1}(\Omega)$ defined by

$$\langle F, \zeta\rangle = \langle f, \zeta\rangle - \int_\Omega g_{x_i}\zeta_{x_i}\,dx.$$

By Theorem 2.1, (4.5) admits a solution w. It is clear that $u = w + g$ is a solution to Problem 4.6.

In general, the reader is invited to obtain this condition:

If $a(u, v)$ is a bilinear form on $H_0^1(\Omega)$, coercive on $H_0^1(\Omega)$, and $\mathbb{K}$ is closed convex with $\mathbb{K} - \mathbb{K} = \{u - v : u, v \in \mathbb{K}\} \subset H_0^1(\Omega)$, then there exists a solution

$$u \in \mathbb{K}: \quad a(u, v - u) \geq \langle f, v - u \rangle \qquad \text{for } v \in \mathbb{K}$$

for any $f \in H^{-1}(\Omega)$.

Returning to Problem 4.6, since the convex set appearing there is the entire space $H_0^1(\Omega)$,

$$\int_\Omega w_{x_i} \zeta_{x_i}\, dx = \langle F, \zeta \rangle \qquad \text{for} \quad \zeta \in H_0^1(\Omega),$$

it follows that $w + g = u$ is a solution to (4.3).

A more general Dirichlet problem is to solve

$$-\Delta u + \lambda u = f \qquad \text{in} \quad \Omega,$$
$$u = g \qquad \text{on} \quad \partial\Omega.$$

Once again multiplying the equation by $\zeta \in H_0^1(\Omega)$ and integrating by parts leads to the problem

Problem 4.7 *Find $u \in M = \{v \in H^1(\Omega) : v - g \in H_0^1(\Omega)\}$ such that*

$$a(u, \zeta) = \langle f, \zeta \rangle \qquad \text{for} \quad \zeta \in H_0^1(\Omega) \tag{4.6}$$

where

$$a(v, \zeta) = \int_\Omega v_{x_j} \zeta_{x_j}\, dx + \lambda \int_\Omega v\zeta\, dx \qquad \text{for} \quad v, \zeta \in H^1(\Omega).$$

We claim that $a(v, \zeta)$ is coercive on $H_0^1(\Omega)$ provided that $\lambda > -1/\beta$, β the constant of Poincaré's inequality. In fact, for $t \in (0, 1)$,

$$\begin{aligned} a(v, v) &= t \int_\Omega v_x^2\, dx + (1 - t) \int_\Omega v_x^2\, dx + \lambda \int_\Omega v^2\, dx \\ &\geq t \int_\Omega v_x^2\, dx + [(1/\beta) - (t/\beta) + \lambda] \int_\Omega v^2\, dx. \end{aligned}$$

Choose t, $0 < t < 1$, so that $0 < t < \beta[(1/\beta) + \lambda] = 1 + \beta\lambda$. Then

$$a(v, v) \geq t \|v\|^2_{H_0^1(\Omega)}.$$

From this it follows, as in the previous example, that there is a unique weak solution to the Dirichlet problem.

We consider now the

Neumann Problem 4.8. *Again, when $\partial\Omega$, f, g are smooth, the classical problem is to find a function u such that*

$$-\Delta u + \lambda u = f \quad \textit{in} \quad \Omega,$$
$$\partial u/\partial n = g \quad \textit{on} \quad \partial\Omega,$$

with $\partial/\partial n$ the outward directed normal to $\partial\Omega$.

Multiplying the equation by $\zeta \in C^1(\overline{\Omega})$ and integrating by parts we obtain

$$a(u, \zeta) = \int_\Omega f\zeta \, dx + \int_{\partial\Omega} g\zeta \, d\sigma \tag{4.7}$$

where

$$a(u, \zeta) = \int_\Omega u_{x_i}\zeta_{x_i} \, dx + \lambda \int_\Omega u\zeta \, dx.$$

By choosing $\lambda > 0$ and $\alpha = \min(1, \lambda)$, we have

$$a(v, v) \geq \alpha \|v\|^2_{H^1(\Omega)}.$$

However, if $\lambda = 0$, the form $a(u, v)$ is not coercive, and in this case the Neumann problem does not always have a solution; and when it does, it is not unique.

Finally we consider the

Mixed Problem 4.9. *Suppose $\partial\Omega = \partial_1\Omega \cup \overline{\partial_2\Omega}$ with $\partial_1\Omega \cap \partial_2\Omega = \varnothing$. We wish to solve the problem*

$$-\Delta u + \lambda u = f \quad \textit{in} \quad \Omega,$$
$$u = 0 \quad \textit{on} \quad \partial_1\Omega, \tag{4.8}$$
$$\frac{\partial u}{\partial n} = 0 \quad \textit{on} \quad \partial_2\Omega.$$

For $\lambda > 0$, the operator $-\Delta + \lambda$ leads to a coercive bilinear form on $H^1(\Omega)$ as we have seen above. If $\lambda = 0$, the form is still coercive provided that $\partial_1\Omega$ is sufficiently large, namely, if $\partial_1\Omega$ is large enough to guarantee the existence of a $\beta > 0$ such that

$$\int_\Omega \zeta^2 \, dx \leq \beta \int_\Omega \zeta_x^2 \, dx \qquad \text{for} \quad \zeta \in C^\infty(\overline{\Omega})$$

satisfying $\zeta = 0$ on $\partial_1\Omega$. A sufficient condition for this to hold when Ω is smooth and connected is that $\partial_1\Omega$ have nonvoid interior in $\partial\Omega$.

A natural question at this point is the sense in which the weak solution to these problems correspond to more classical ones. About the Dirichlet problems (4.6) and (4.7) and the Neumann problem (4.8) it is possible to show that they are as smooth as is compatible with the data of the problem (Bers *et al.* [1]). The solution to the mixed problem (4.9) is generally Hölder continuous, but information about higher differentiability depends on certain compatibility conditions satisfied by the data.

At this point we give a precise example of a regularity theorem for weak solutions of the Dirichlet problem which will be useful to us in our study of obstacle problems, which we begin in Section 6.

Theorem 4.10. *Let Ω be a domain of $\mathbb{R}^N$ whose boundary is suitably smooth. Suppose that*

$$u \in H_0^1(\Omega): \quad -\Delta u = f \qquad \textit{in} \quad \Omega.$$

If $f \in H^{m,s}(\Omega)$, then $u \in H^{m+2,s}(\Omega)$, $1 < s < \infty$.

Of course, the expression "$-\Delta u = f$ in Ω" stands for

$$\int_\Omega u_{x_i}\zeta_{x_i}\,dx = \int_\Omega f\zeta\,dx \qquad \text{for} \quad \zeta \in C_0^\infty(\Omega).$$

For the proof of this theorem we refer to Morrey.

We have considered in this section coercive bilinear forms. For the sake of simplicity we confined ourselves to bilinear forms connected to the special second order operator $-\Delta$.

In the next sections we will consider more general second order differential operators. We have in mind divergence form differential operators with bounded measurable coefficients.

Let $a_{ij}(x) \in L^\infty(\Omega)$, $i, j = 1, 2, \ldots, N$, satisfy

$$(1/\Lambda)\xi^2 \le a_{ij}(x)\xi_i\xi_j \le \Lambda\xi^2 \qquad \text{for} \quad \xi \in \mathbb{R}^N \quad \text{a.e.} \quad x \in \Omega \tag{4.9}$$

for some $\Lambda \ge 1$ and define

$$a(u, v) = \int_\Omega a_{ij}(x)u_{x_j}(x)v_{x_i}(x)\,dx, \qquad u, v \in H^1(\Omega). \tag{4.10}$$

Then the form (4.10) is coercive in $H_0^1(\Omega)$ since

$$a(v, v) \ge (1/\Lambda)\|v_x\|_{L^2(\Omega)}^2 = (1/\Lambda)\|v\|_{H_0^1(\Omega)}^2 \qquad \text{for} \quad v \in H_0^1(\Omega).$$

5. The Weak Maximum Principle

The weak maximum principle is formulated in terms of inequality in the sense of $H^1(\Omega)$. We begin with a discussion of this notion. Suppose that Ω is a bounded connected domain of $\mathbb{R}^N$ with boundary $\partial\Omega$.

Definition 5.1. Let $u \in H^1(\Omega)$ and $E \subset \overline{\Omega}$. The function u *is nonnegative on E in the sense of* $H^1(\Omega)$, or briefly, $u \geq 0$ *on E in* $H^1(\Omega)$, if there exists a sequence $u_n \in H^{1,\infty}(\Omega)$ such that

$$u_n(x) \geq 0 \quad \text{for} \quad x \in E; \qquad u_n \to u \quad \text{in} \quad H^1(\Omega). \tag{5.1}$$

If $-u \geq 0$ on E in $H^1(\Omega)$, then u is *nonpositive* on E in $H^1(\Omega)$ or $u \leq 0$ *on E in* $H^1(\Omega)$. If $u \geq 0$ on E in $H^1(\Omega)$ and $u \leq 0$ on E in $H^1(\Omega)$ we say that $u = 0$ *on E in* $H^1(\Omega)$. Similarly, we say that $u \leq v$ *on E in* $H^1(\Omega)$ if $v - u \geq 0$ on E in $H^1(\Omega)$ for two elements $u, v \in H^1(\Omega)$. In particular, v may be a constant, which leads to the definition

$$\sup_E u = \inf\{M \in R : u \leq M \text{ on } E \text{ in } H^1(\Omega)\}.$$

It is useful to observe that the subset of functions $u \in H^1(\Omega)$ satisfying $u \geq 0$ on E in $H^1(\Omega)$ for a given $E \subset \Omega$ is a closed convex cone. A particular consequence of this is that in Definition 5.1 it suffices to choose a sequence $u_m \to u$ weakly in $H^1(\Omega)$ by the Banach–Saks theorem.

We compare the notion of $\geq$ in $H^1(\Omega)$ with $\geq$ a.e. in Ω.

Proposition 5.2. *Let $\Omega \subset \mathbb{R}^N$ be bounded, $E \subset \overline{\Omega}$, and $u \in H^1(\Omega)$.*

(i) *If $u \geq 0$ on E in $H^1(\Omega)$, then $u \geq 0$ on E a.e.*
(ii) *If $u \geq 0$ on Ω a.e., then $u \geq 0$ on Ω in $H^1(\Omega)$.*
(iii) *If $u \in H_0^1(\Omega)$ and $u \geq 0$ on Ω a.e., then there exists a sequence $u_n \in H_0^{1,\infty}(\Omega)$ such that $u_n \geq 0$ in Ω and $u_n \to u$ in $H_0^1(\Omega)$.*
(iv) *If E is open in Ω and $u \geq 0$ on E a.e., then $u \geq 0$ on K in the sense of $H^1(\Omega)$ for any compact $K \subset E$.*

Proof. (i) This is a consequence of the convergence almost everywhere to u of a subsequence of any sequence which tends to u in $L^2(\Omega)$.

(ii) Let $v_n \in H^{1,\infty}(\Omega)$ satisfy $v_n \to u$ in $H^1(\Omega)$ and in Ω pointwise a.e. Then $\max(v_n, 0) \geq 0$ and $u = \max(u, 0)$ in Ω a.e., so

$$\begin{aligned}\|\max(v_n, 0) - u\|_{L^2(\Omega)} &= \|\max(v_n, 0) - \max(u, 0)\|_{L^2(\Omega)} \\ &\leq \|v_n - u\|_{L^2(\Omega)} \to 0 \qquad \text{as} \quad n \to \infty.\end{aligned}$$

Since

$$\int_\Omega \max(v_n, 0)_x^2\, dx \le \int_\Omega v_{nx}^2\, dx \le C,$$

the sequence $\max(v_n, 0)$ contains a subsequence which converges weakly in $H^1(\Omega)$ to an element which must be u by the foregoing. In view of the remark preceding the proposition, $u \ge 0$ on Ω in $H^1(\Omega)$.

(iii) The proof is the same as (ii) starting with $v_n \in H_0^{1,\infty}(\Omega)$.

(iv) Given $K \subset E$ compact, let $\zeta \in C_0^\infty(\mathbb{R}^N)$ satisfy

$$\zeta = 1 \quad \text{on} \quad \{x : \operatorname{dist}(x, K) \le \tfrac{1}{2} \operatorname{dist}(\mathbb{R}^N - E, K)\},$$

$$\zeta = 0 \quad \text{on} \quad \mathbb{R}^N - E,$$

and $0 \le \zeta \le 1$. Let $\varphi_\varepsilon(x)$ be a family of mollifiers with support $\varphi_\varepsilon \subset B_\varepsilon(0)$ and set

$$w_n(x) = u * \varphi_{\varepsilon_n}(x) = \int_{B_{\varepsilon_n}(0)} u(y)\varphi_{\varepsilon_n}(x - y)\, dy,$$

where $\varepsilon_n \to 0$ monotonically and $\varepsilon_0 < \frac{1}{2} \operatorname{dist}(K, \mathbb{R}^N - E)$. So $w_n \le 0$ on K. Let $v_n \in H^{1,\infty}(\Omega)$ satisfy $v_n \to u$ in $H^1(\Omega)$. Then

$$u_n = \zeta w_n + (1 - \zeta) v_n$$

is the desired sequence. Q.E.D.

The reader, however, should not be misled into thinking that our new definition of inequality reduces to inequality in the almost everywhere sense. In fact, as we shall see in Section 6, inequality in the sense of $H^1(\Omega)$ serves to determine the capacity of a set. A set of measure zero, for example a closed interval in $\mathbb{R}^2$, may have positive capacity. The next assertion illustrates the role of inequality in $H^1(\Omega)$ in the weak maximum principle.

Proposition 5.3. *Let $u \in H^1(\Omega)$ and suppose that*

$$\sup_{\partial\Omega} u = M = \inf\{m \in \mathbb{R} : u \le m \text{ on } \partial\Omega \text{ in } H^1(\Omega)\} < +\infty.$$

Then for any $k \ge M$, $\max(u - k, 0) \in H_0^1(\Omega)$ and $\max(u - k, 0) \ge 0$ in Ω in $H^1(\Omega)$.

Proof. In order to prove that $\max(u - k, 0) \in H_0^1(\Omega)$ it suffices to prove the existence of a sequence $v_n \in H_0^{1,\infty}(\Omega)$ such that

$$v_n \to \max(u - k, 0) \quad \text{weakly in} \quad H^1(\Omega).$$

According to our hypothesis, for any $\varepsilon > 0$, we may find a sequence $u_n^\varepsilon \in C^1(\overline{\Omega})$ such that

$$k - u_n^\varepsilon + \varepsilon > 0 \qquad \text{on} \quad \partial\Omega$$

and

$$k - u_n^\varepsilon + \varepsilon \to k - u + \varepsilon \qquad \text{in} \quad H^1(\Omega)$$

as n tends to infinity. From this, we may choose $\varepsilon = 1/n$, $u_n^\varepsilon = u_n$, so

$$k + (1/n) - u_n > 0 \qquad \text{on} \quad \partial\Omega$$

and

$$u_n \to u \qquad \text{in} \quad H^1(\Omega).$$

Therefore the Lipschitz function $v_n = \max(u_n - (1/n) - k, 0)$ has compact support in Ω. Moreover (cf. Appendix A),

$$v_n \to \max(u - k, 0) \qquad \text{weakly in} \quad H^1(\Omega).$$

Hence, $\max(u - k, 0) \in H_0^1(\Omega)$.

Note finally that $\max(u - k, 0) \geq 0$ a.e. in Ω, so $\max(u - k, 0) \geq 0$ on Ω in $H^1(\Omega)$ by Proposition 5.2(ii). Q.E.D.

Our maximum principle is about divergence form equations with bounded measurable coefficients. Let $a_{ij}(x) \in L^\infty(\Omega)$ satisfy

$$(1/\Lambda)\xi^2 \leq a_{ij}(x)\xi_i\xi_j \leq \Lambda\xi^2 \qquad \text{for} \quad \xi \in \mathbb{R}^N \quad \text{a.e.} \quad x \in \Omega \tag{5.2}$$

for some $\Lambda \geq 1$ and define

$$a(u, v) = \int_\Omega a_{ij}(x)u_{x_j}(x)v_{x_i}(x)\,dx, \qquad u, v \in H^1(\Omega).$$

This determines the operator

$$L\colon H_0^1(\Omega) \to H^{-1}(\Omega)$$

given by

$$\langle Lu, v\rangle = a(u, v), \qquad u, v \in H_0^1(\Omega).$$

We frequently write

$$Lu = -(\partial/\partial x_i)(a_{ij}u_{x_j}) \in H^{-1}(\Omega).$$

It is not necessary to assume that (a_{ij}) is symmetric. From Theorem 2.1 we deduce that the Dirichlet problem for L has a unique solution, just as in our discussion of Problem 4.6. Precisely, we state

Theorem 5.4. *Let*

$$Lu = -(\partial/\partial x_i)(a_{ij}u_{x_j})$$

where a_{ij} *satisfy* (5.2). *Then for each* $\varphi \in H^1(\Omega)$ *and* $f \in H^{-1}(\Omega)$, *there is a unique solution to*

$$u \in H^1(\Omega): \quad Lu = f \quad \text{in} \quad \Omega,$$
$$u = \varphi \quad \text{on} \quad \partial\Omega.$$

Finally we turn to our maximum principle.

Theorem 5.5. *Let* $u \in H^1(\Omega)$ *be a solution to*

$$Lu = 0 \quad \text{in} \quad \Omega,$$
$$u = \varphi \quad \text{on} \quad \partial\Omega,$$

where $Lu = -(\partial/\partial x_i)(a_{ij}u_{x_j})$ *with* a_{ij} *satisfying* (5.2) *and* $\varphi \in H^1(\Omega)$. *Then*

$$\|u\|_{L^\infty(\Omega)} \leq \sup_{\partial\Omega} |\varphi|.$$

The expression $\sup_{\partial\Omega} |\varphi|$ is intended in the sense of $H^1(\Omega)$. We may assume it is finite for otherwise there is nothing to prove.

Proof. We shall prove that

$$u \leq \sup_{\partial\Omega} \varphi \quad \text{in} \quad H^1(\Omega).$$

Because $-u$ is a solution to the equation with boundary values $-\varphi$, it follows then that

$$-u \leq \sup_{\partial\Omega} (-\varphi) \quad \text{in} \quad H^1(\Omega).$$

The conclusion of the theorem then follows from Proposition 5.2(i).

According to Proposition 5.3, $\zeta = \max(u - M, 0) \in H_0^1(\Omega)$, so

$$0 = a(u, \zeta) = \int_\Omega a_{ij}u_{x_j}\zeta_{x_i}\,dx. \tag{5.3}$$

We divide Ω into the two sets, which are determined within a set of measure zero, $\{x : \zeta(x) > 0\}$ and $\{x : \zeta(x) = 0\}$. Observe (cf. Appendix A) that

$$\zeta_{x_i} = 0 \quad \text{a.e. in} \quad \{x : \zeta(x) = 0\},$$
$$\zeta_{x_i} = u_{x_i} \quad \text{a.e. in} \quad \{x : \zeta(x) > 0\}.$$

Consequently, $u_{x_j}(x)\zeta_{x_i}(x) = \zeta_{x_j}(x)\zeta_{x_i}(x)$ a.e. in Ω. Substituting in (5.3) we find

$$0 = a(u, \zeta) = \int_\Omega a_{ij}\zeta_{x_j}\zeta_{x_i}\,dx = a(\zeta, \zeta)$$

$$\geq (1/\Lambda)\int_\Omega \zeta_{x_i}^2\,dx = (1/\Lambda)\|\zeta\|^2_{H_0^1(\Omega)};$$

therefore $\zeta = 0$ in $H_0^1(\Omega)$, so $u - M \leq 0$ in $H^1(\Omega)$. Q.E.D.

One extension of this maximum principle, which we prove in Appendix B, is that if

$$Lu = f \quad \text{in} \quad \Omega,$$
$$u = 0 \quad \text{on} \quad \partial\Omega,$$

with $f = f_0 + \sum_1^N (f_i)_{x_i}, f_i \in L^s(\Omega)$ for $0 \leq i \leq N$, then

$$\|u\|_{L^\infty(\Omega)} \leq C_s \sum_0^N \|f_i\|_{L^s(\Omega)}$$

provided $s > N$.

Definition 5.6. A function $u \in H^1(\Omega)$ is a *supersolution to* L, or an L *supersolution*, if

$$a(u, \zeta) \geq 0 \qquad \text{for all} \quad \zeta \in H_0^1(\Omega) \quad \text{with } \zeta \geq 0 \text{ in } \Omega. \tag{5.4}$$

Analogously, u is an L *subsolution* provided that

$$a(u, \zeta) \leq 0 \qquad \text{for all} \quad \zeta \in H_0^1(\Omega) \quad \text{with } \zeta \geq 0.$$

For supersolutions, there is a minimum principle. Its proof, which follows the lines of Theorem 5.5, is omitted.

Theorem 5.7. *Let $u \in H^1(\Omega)$ be an L supersolution. Then*

$$u(x) \geq \inf_{\partial\Omega} u \qquad \text{a.e. in} \quad \Omega.$$

An interesting property of supersolutions is that if u, v are supersolutions to L, then $\min(u, v)$ is also a supersolution. We prove this in Section 6.

6. The Obstacle Problem: First Properties

In this section we consider the obstacle problem, primarily for the Laplace operator. We begin with a more general case. As before, let $\Omega \subset \mathbb{R}^N$ be bounded and connected with smooth boundary $\partial\Omega$ and let $a_{ij} \in L^\infty(\Omega)$ satisfy

$$(1/\Lambda)\xi^2 \le a_{ij}(x)\xi_i\xi_j \le \Lambda\xi^2 \qquad \text{for} \quad \xi \in \mathbb{R}^N \quad \text{a.e.} \quad x \in \Omega. \tag{6.1}$$

Set

$$a(u, v) = \int_\Omega a_{ij}(x)u_{x_j}(x)v_{x_i}(x)\,dx, \qquad u, v \in H^1(\Omega), \tag{6.2}$$

and define the mapping

$$L\colon H_0^1(\Omega) \to H^{-1}(\Omega)$$

by

$$\langle Lu, v\rangle = a(u, v), \qquad u, v \in H_0^1(\Omega).$$

Recall that if $a_{ij}(x) \in C^1(\overline{\Omega})$, then

$$Lu(x) = -(\partial/\partial x_i)(a_{ij}(x)u_{x_j}(x)), \qquad u \in C^2(\Omega),$$

is an elliptic equation in the classical sense.

Consider now a function $\psi \in H^1(\Omega)$ which satisfies $\psi \le 0$ on $\partial\Omega$ [in $H^1(\Omega)$] and define

$$\mathbb{K}_\psi = \mathbb{K} = \{v \in H_0^1(\Omega) : v \ge \psi \text{ on } \Omega \text{ in } H^1(\Omega)\}.$$

Note that by Proposition 5.2, $v \in \mathbb{K}$ if and only if $v \in H_0^1(\Omega)$ and $v \ge \psi$ a.e. in Ω.

Problem 6.1. *Given $f \in H^{-1}(\Omega)$, find*

$$u \in \mathbb{K}\colon \quad a(v, v - u) \ge \langle f, v - u\rangle \qquad \textit{for all} \quad v \in \mathbb{K}.$$

Theorem 6.2. *There exists a unique solution to Problem 6.1.*

The proof is immediate from Theorem 2.1 since (6.1) implies

$$a(v, v) \ge (1/\Lambda)\|v\|^2_{H_0^1(\Omega)}, \qquad v \in H_0^1(\Omega);$$

that is, $a(u, v)$ is coercive. Note that when $a_{ij} = a_{ji}$, the solution to Problem 6.1 is the solution to the minimization problem

$$\min_{v \in \mathbb{K}} \{a(v, v) - 2\langle f, v\rangle\}. \tag{6.3}$$

Now we give a useful characterization of our solution.

Definition 6.3. We say that $g \in H^1(\Omega)$ is a *supersolution of* $L - f$ if

$$\langle Lg - f, \zeta \rangle \equiv a(g, \zeta) - \langle f, \zeta \rangle \geq 0 \qquad \text{for} \quad 0 \leq \zeta \in H_0^1(\Omega).$$

In particular, the solution u of Problem 6.1 is an $L - f$ supersolution since $u + \zeta \in \mathbb{K}$ whenever $\zeta \geq 0$, $\zeta \in \mathrm{H}_0^1(\Omega)$. In the sequel we shall not distinguish between the statements "$u \geq 0$ on Ω in $\mathrm{H}^1(\Omega)$" and "$u \geq 0$ a.e. in Ω" (see Proposition 5.2).

Theorem 6.4. *Let u be the solution of Problem* 6.1 *and suppose that g is a supersolution of $L - f$ satisfying $g \geq \psi$ in Ω and $g \geq 0$ on $\partial\Omega$ (in $H^1(\Omega)$). Then*

$$u \leq g \qquad \textit{in} \quad \Omega.$$

Proof. We set $\zeta = \min(u, g) \in \mathbb{K}$ since $g \geq \psi$ in Ω and $g \geq 0$ on $\partial\Omega$. Then

$$a(u, \zeta - u) \geq \langle f, \zeta - u \rangle.$$

Now $\zeta - u = \min(u, g) - u \leq 0$ in Ω, so

$$a(g, \zeta - u) \leq \langle f, \zeta - u \rangle;$$

whence

$$a(g - u, \zeta - u) \leq 0.$$

Computing explicitly, using the definition of ζ,

$$\begin{aligned} 0 \geq a(g - u, \zeta - u) &= \int_\Omega a_{ij}(g - u)_{x_j}(\zeta - u)_{x_i}\, dx \\ &= \int_{\{x : g < u\}} a_{ij}(g - u)_{x_j}(\zeta - u)_{x_i}\, dx \\ &= \int_\Omega a_{ij}(\zeta - u)_{x_j}(\zeta - u)_{x_i}\, dx \\ &= a(\zeta - u, \zeta - u) \geq (1/\Lambda)\|\zeta - u\|^2_{H_0^1(\Omega)}. \end{aligned}$$

Hence $\zeta = u$, which means that $u \leq g$ in Ω. Q.E.D.

From Theorem 6.4 follows

Corollary 6.5. *Let u be the solution of Problem 6.1 for $f = 0$ and suppose that*

$$\psi \leq M \quad \text{in} \quad \Omega \quad \text{where} \quad M \geq 0.$$

Then

$$u \leq M \quad \text{in} \quad \Omega.$$

To prove this, it suffices to choose $g = M$ in Theorem 6.4.

For simplicity we have resolved Problem 6.1 with null Dirichlet data on $\partial\Omega$. We observe, however, that the variational inequality may also be solved in the convex

$$\mathbb{K} = \{v \in H^1(\Omega) : v \geq \psi \text{ in } \Omega, v - \varphi \in H_0^1(\Omega)\},$$

where $\varphi \in H^1(\Omega)$ is given and $\varphi \geq \psi$ on $\partial\Omega$. This proceeds in a manner identical to the solution of Dirichlet problem 4.7. We verify the case at hand. As the new convex set we choose

$$\mathbb{K}_0 = \{\eta \in H_0^1(\Omega) : \eta \geq \psi - \varphi \text{ in } \Omega\}.$$

Suppose that $u \in \mathbb{K}$ is a solution to

$$a(u, v - u) \geq \langle f, v - u \rangle \quad \text{for} \quad v \in \mathbb{K} \tag{6.4}$$

and write $u = \zeta + \varphi$, $\zeta \in \mathbb{K}_0$, $v = \eta + \varphi$, $\eta \in \mathbb{K}_0$. Then $a(u, v - u) = a(\zeta + \varphi, \eta - \zeta)$ so

$$\zeta \in \mathbb{K}_0 \colon \quad a(\zeta, \eta - \zeta) \geq \langle f, \eta - \zeta \rangle - a(\varphi, \eta - \zeta), \qquad \eta \in \mathbb{K}_0. \tag{6.5}$$

Since $\langle F, \xi \rangle = \langle f, \xi \rangle - a(\varphi, \xi)$, $\xi \in H_0^1(\Omega)$, defines an element $F \in H^{-1}(\Omega)$, (6.5) admits a unique solution ζ by Theorem 6.2. It is easy to check that $\zeta + \varphi = u$ is a solution to (6.4).

Finally, note that Theorem 6.4 and Corollary 6.5 extend easily to this case.

Theorem 6.6. *Let u and v be two $L - f$ supersolutions. Then $w = \min(u, v)$ is an $L - f$ supersolution.*

Proof. Let us define the convex

$$\mathbb{K} = \{\eta \in H^1(\Omega) : \eta - w \in H_0^1(\Omega), \eta \geq w \text{ a.e. in } \Omega\}$$

and let ζ be the solution of the variational inequality

$$\zeta \in \mathbb{K} \colon \quad \int_\Omega a_{ij} \zeta_{x_j} (\eta - \zeta)_{x_i} \, dx \geq \langle f, \eta - \zeta \rangle \quad \text{for} \quad \eta \in \mathbb{K}. \tag{6.6}$$

Since u and v are $L - f$ supersolutions, we have by Theorem 6.4 that

$$\zeta \leq u \quad \text{and} \quad \zeta \leq v \quad \text{in} \quad \Omega.$$

Thus

$$\zeta \leq \min(u, v) \quad \text{in} \quad \Omega.$$

But since $\zeta \in \mathbb{K}$, $\zeta = \min(u, v) = w$. To complete the proof of the theorem, recall that any solution of (6.6) is an $L - f$ supersolution. Q.E.D.

Definition 6.7. Let $u \in H^1(\Omega)$. Let us agree to say that $u(x) > 0$ at $x \in \Omega$ in the sense of $H^1(\Omega)$ provided there exist a neighborhood $B_\rho(x)$ and $\varphi \in H_0^{1,\infty}(B_\rho(x))$, $\varphi \geq 0$ and $\varphi(x) > 0$, such that $u - \varphi \geq 0$ on $B_\rho(x)$ in the sense of $H^1(\Omega)$. The set $\{x \in \Omega : u(x) > 0\}$ is open.

Let u be the solution of Problem 6.1 with "obstacle" ψ. We divide Ω into the sets $\{x \in \Omega : u(x) > \psi(x)\}$, which is open, and its complement $I = I[u]$, which is closed in Ω. Formally, I is the set of points x where $u(x) = \psi(x)$.

Definition 6.8. The set I is called the coincidence set of the solution u.

Consider a point $x_0 \in \Omega - I$. Then there is a ball $B_\rho(x_0)$ and $\varphi(x) \in C_0^\infty(B_\rho(x_0))$, $\varphi(x) > 0$ on $B_{\rho/2}(x_0)$, such that

$$u \geq \psi + \varphi \quad \text{in} \quad H^1(B_\rho(x_0)).$$

Hence for any $\zeta \in C_0^\infty(B_{\rho/2}(x_0))$, we may find an $\varepsilon > 0$ such that

$$u + \varepsilon\zeta \geq \psi + \tfrac{1}{2}\varphi \quad \text{in} \quad H^1(B_{\rho/2}(x_0)).$$

Consequently, $v = u + \varepsilon\zeta \in \mathbb{K}$. Substituting this v in the variational inequality and dividing by ε, we find that

$$a(u, \zeta) \geq \langle f, \zeta \rangle \quad \text{for all} \quad \zeta \in C_0^\infty(B_{\rho/2}(x_0)).$$

Since this is particularly true of $-\zeta$, we obtain

$$a(u, \zeta) = \langle f, \zeta \rangle \quad \text{in} \quad B_{\rho/2}(x_0).$$

In other words,

$$Lu = f \quad \text{in} \quad \Omega - I.$$

At this point we have obtained the properties

$$Lu = f \quad \text{in} \quad \Omega - I$$

and, formally,

$$u = \psi \quad \text{on} \quad I. \tag{6.7}$$

Of course (6.7) does not characterize the solution u. On the other hand, $Lu - f$ is nonnegative, that is,

$$a(u, \zeta) - \langle f, \zeta \rangle \geq 0 \qquad \text{whenever} \quad \zeta \geq 0, \quad \zeta \in H_0^1(\Omega).$$

Consequently, by the Riesz–Schwarz theorem, Schwarz [1], $a(u, \zeta) - \langle f, \zeta \rangle$ is given by a nonnegative Radon measure μ on Ω, namely,

$$a(u, \zeta) - \langle f, \zeta \rangle = \int_\Omega \zeta \, d\mu \qquad \text{for} \quad \zeta \in H_0^1(\Omega).$$

In view of (6.7) the support of μ is contained in I. To summarize:

Theorem 6.9. *Let u be the solution to Problem* 6.1. *Then there exists a nonnegative Radon measure μ such that*

$$Lu = f + \mu \qquad \text{in} \quad \Omega$$

with

$$\operatorname{supp} \mu \subset I = \{x \in \Omega : u(x) = \psi(x)\}.$$

In particular,

$$Lu = f \qquad \text{in} \quad \Omega - I.$$

We want to establish a relation between the measure μ and the capacity. Given a compact subset $E \subset \Omega$, we define the closed convex of $H_0^1(\Omega)$

$$\mathbb{K}_E = \{v \in H_0^1(\Omega) : v \geq 1 \text{ on } E \text{ in } H^1(\Omega)\}.$$

Definition 6.10. The *capacity of* E with respect to Ω, $\operatorname{cap}_\Omega E$, or simply, cap E, is defined by

$$\operatorname{cap} E = \inf_{v \in \mathbb{K}_E} \int_\Omega v_x^2 \, dx. \tag{6.8}$$

A set $F \subset \Omega$ is of *capacity zero* if cap $E = 0$ for every compact $E \subset F$.

The minimum problem (6.8) admits a solution $\alpha(x)$. Indeed, $\alpha(x)$ is the unique solution of the variational inequality

$$\alpha \in \mathbb{K}_E: \quad \int_\Omega \alpha_{x_i}(v - \alpha)_{x_i} \, dx \geq 0 \qquad \text{for} \quad v \in \mathbb{K}_E \tag{6.9}$$

and is called the conductor potential or capacitary potential of E (with respect to Ω). If $\zeta \in H_0^1(\Omega)$ with $\zeta \geq 0$, then $\alpha + \zeta \in \mathbb{K}_E$, so

$$\int_\Omega \alpha_{x_i} \zeta_{x_i} \, dx \geq 0.$$

Hence α is a $-\Delta$ supersolution. By Theorem 5.7, $\alpha \geq 0$ in Ω. On the other hand, $\min(\alpha, 1) \in \mathbb{K}_E$ and

$$\int_\Omega \min(\alpha, 1)_x^2 \, dx = \int_{\{x : \alpha(x) < 1\}} \alpha_x^2 \, dx \leq \int_\Omega \alpha_x^2 \, dx = \operatorname{cap} E$$

(where $\{x \ \alpha(x) < 1\}$ is understood in the pointwise sense appropriate to the application of Theorem A.1). Since α is the unique element of smallest Dirichlet integral in $\mathbb{K}_E$, $\alpha = \min(\alpha, 1)$. Thus

$$0 \leq \alpha \leq 1 \qquad \text{in} \quad \Omega.$$

Suppose, in fact, that $v_n \in \mathbb{K}_E \cap H^{1,\infty}(\Omega)$ satisfy $v_n \to \alpha$ in $H^1(\Omega)$. Define

$$\theta(t) = \begin{cases} 0, & t \leq 0 \\ t, & 0 \leq t \leq 1 \\ 1, & t > 1, \end{cases}$$

and set $w_n = \theta(v_n) \in \mathbb{K}_E \cap H^{1,\infty}(\Omega)$ with the property that $w_n = 1$ on E. Now

$$\int_\Omega w_{nx}^2 \, dx \leq \int_{\{x : 0 < v_n(x) < 1\}} v_{nx}^2 \, dx \leq \int_\Omega v_{nx}^2 \, dx \to \operatorname{cap} E.$$

It follows from the parallelogram law that $\{w_n\}$ also enjoys the property that $w_n \to \alpha$ in $H_0^1(\Omega)$. By Definition 5.1, $\alpha = 1$ on E in the sense of $H^1(\Omega)$.

Finally, if $x \in \Omega - E$, there is a neighborhood $U \subset \Omega - E$ of x and $\alpha + \zeta \in \mathbb{K}_E$ for any $\zeta \in C_0^\infty(U)$. Consequently

$$\int_\Omega \alpha_{x_i} \zeta_{x_i} \, dx = 0 \qquad \text{for} \quad \zeta \in C_0^\infty(\Omega - E).$$

We have shown that $\alpha \in \mathrm{H}_0^1(\Omega)$ satisfies

$$\begin{aligned} -\Delta\alpha &\geq 0 && \text{in} \quad \Omega, \\ -\Delta\alpha &= 0 && \text{in} \quad \Omega - E, \\ \alpha &= 1 && \text{on} \quad E \quad \text{in the sense of} \quad H^1(\Omega), \\ 0 \leq \alpha &\leq 1 && \text{in} \quad \Omega. \end{aligned}$$

Again applying the theorem of Riesz–Schwarz, there is a measure $m = m_E$ such that $-\Delta\alpha = m$ in Ω, namely,

$$\int_\Omega \alpha_{x_i}\zeta_{x_i}\,dx = \int_\Omega \zeta\,dm \qquad \text{for} \quad \zeta \in H_0^1(\Omega).$$

Denote by $g(x, y) = g_\Omega(x, y)$ the Green's function of $-\Delta$ for Ω. It is a part of the Riesz representation theorem that

$$\alpha(x) = \int_\Omega g(x, y)\,dm(y), \qquad x \in \Omega,$$

and hence, formally,

$$\text{cap } E = \int_\Omega \alpha_x^2\,dx = \int_\Omega \alpha\,dm(x) = \int_\Omega\int_\Omega g(x, y)\,dm(y)\,dm(x).$$

This formula is easily justified by a direct computation, with due regard to the singularity of the Green function.

Theorem 6.11. *Let $f_0, f_1, \ldots, f_N \in L^2(\Omega)$ and $\psi \in H^1(\Omega)$. Let u be the solution of Problem 6.1 in $\mathbb{K}_\psi$ for $f = f_0 + \sum (\partial/\partial x_i) f_i \in H^{-1}(\Omega)$, and let μ denote the measure determined by u. Then there exists a constant $C > 0$ such that*

$$\mu(E) \le C(\text{cap } E)^{1/2} \qquad \textit{for} \quad E \subset \Omega \qquad \textit{compact}.$$

In particular, if $\text{cap } E = 0$, *then* $\mu(E) = 0$.

Proof. We assume, of course, that $\mathbb{K}_\psi \ne \emptyset$, that is, that $\psi \le 0$ on $\partial\Omega$. Let $0 \le \zeta \in C_0^1(\Omega)$ satisfy $\zeta \ge 1$ on E. Then

$$\begin{aligned}
\mu(E) &\le \int_\Omega \zeta\,d\mu \\
&= \int_\Omega a_{ij}u_{x_j}\zeta_{x_i}\,dx - \int_\Omega \left\{ f_0\zeta - \sum_1^N f_i\zeta_{x_i} \right\} dx \\
&\le n^2\Lambda \|u_x\|_{L^2(\Omega)}\|\zeta_x\|_{L^2(\Omega)} + \sqrt{\beta}\,\|f_0\|_{L^2(\Omega)}\|\zeta_x\|_{L^2(\Omega)} \\
&\quad + \left(\sum_i^N \|f_i\|_{L^2(\Omega)}\right)\|\zeta_x\|_{L^2(\Omega)} \\
&= C\|\zeta_x\|_{L^2(\Omega)},
\end{aligned}$$

where Poincaré's inequality was used to estimate $\int_\Omega f_0 \zeta \, dx$. In particular, since $C_0^1(\Omega) \cap \mathbb{K}_E$ is dense in $\mathbb{K}_E$,

$$\mu(E) \le C \inf_{\zeta \in \mathbb{K}_E} \|\zeta_x\|_{L^2(\Omega)} = C(\text{cap } E)^{1/2}. \quad \text{Q.E.D.}$$

Corollary 6.12. *Let u be the solution to Problem* 6.1 *and let I denote its set of coincidence. If* cap $I = 0$, *then*

$$Lu = f \quad \textit{in} \quad \Omega.$$

One might observe that the hypotheses of Theorem 6.11 may be weakened since the best constant C satisfies

$$C \le C_0\{\|u_x\|_{L^2(\Omega)} + \|f\|_{H^{-1}(\Omega)}\},$$

where C_0 depends on the bilinear form a and Ω, and therefore C depends on ψ only through $\|u_x\|_{L^2(\Omega)}$. In particular, the conclusion holds for obstacles $\psi \in L^\infty(\Omega)$ for which $\mathbb{K}_\psi \ne \varnothing$.

Once again we may raise the question of the smoothness of the solution u. In Chapter IV we provide an extensive discussion of this topic. For the present, let us note that the solution cannot be expected to be in $C^2(\Omega)$. It suffices to consider $N = 1, \Omega = (-3, 3), \psi(x) = 1 - x^2$, and the bilinear form

$$a(u, v) = \int_{-3}^{3} u'v' \, dx, \qquad u, v \in H_0^1(\Omega).$$

The solution to Problem 6.1 for $f = 0$ is given by a function which is partially the parabola ψ and partially its tangents, because such a function is the smallest supersolution in $\mathbb{K}$. We discuss this problem from a different viewpoint in the next section.

7. The Obstacle Problem in the One Dimensional Case

Before consideration of the general question of the regularity of the solution of the obstacle problem, we shall discuss the one dimensional case. Let us remark, to begin, that when $\Omega = (\alpha, \beta)$ is an open interval of $\mathbb{R}$, the functions of $H^1(\Omega)$ are the absolutely continuous functions $u(x)$ on (α, β) whose derivatives $u'(x)$ are in $L^2(\Omega)$. The functions of $H_0^1(\Omega)$ are those functions of $H^1(\Omega)$ which vanish at α and β. Let

$$\mathbb{K} = \{v \in H_0^1(\Omega) : v \ge \psi \text{ in } \Omega\} \ne \varnothing$$

where

$$\psi \in H^1(\Omega), \qquad \max_{\Omega} \psi > 0, \qquad \text{and} \qquad \psi(\alpha) < 0, \quad \psi(\beta) < 0.$$

Define

$$a(u, v) = \int_{\alpha}^{\beta} u'(x)v'(x)\,dx, \qquad u, v \in H^1(\Omega).$$

Obviously, $a(u, v)$ is a coercive bilinear form on $H_0^1(\Omega)$. Given $f \in H^{-1}(\Omega)$, we know from Theorem 2.1 that there exists a unique solution of the variational inequality

$$u \in \mathbb{K}: \quad a(u, v - u) \geq \langle f, v - u \rangle \qquad \text{for all} \quad v \in \mathbb{K}.$$

From Section 6, there exists a nonnegative measure μ whose support is in the coincidence set $I = \{x \in \Omega : u(x) = \psi(x)\}$ and

$$u \in H_0^1(\Omega): \; -u'' = f + \mu \tag{7.1}$$

in the sense of distributions. Since $\psi(\alpha) < 0$ and $\psi(\beta) < 0$, I is compact in Ω and $\mu(\Omega) = \mu(I) < \infty$ by Theorem 6.11. Now μ is nonnegative so there is a nondecreasing function $\varphi(x)$ such that

$$\varphi' = \mu \qquad \text{(in the sense of distributions).}$$

Of course, it is clear that

$$\varphi(x) = \mu([\alpha, x)).$$

Moreover, since $f \in H^{-1}(\Omega)$, we may write

$$f = F' \qquad \text{where} \quad F \in L^2(\Omega).$$

Therefore Eq. (7.1) may be written

$$-u'' = F' + \varphi',$$

which implies that

$$u'(x) = -(F(x) + \varphi(x) + \text{const}).$$

We remark that if $F(x) \in L^p(\Omega)$, $p > 1$; i.e., if $f \in H^{-1,p}(\Omega)$, then $u \in H^{1,p}(\Omega)$. In fact, $\varphi(x)$, being a nondecreasing function, is bounded in Ω. It then follows that $u(x) \in C^{0,\lambda}(\Omega)$, $\lambda = 1 - (1/p)$.

Henceforth assume that F is continuous. Then it follows that $u'(x)$ is continuous on the set $\Omega - I = \{x \in \Omega : u(x) > \psi(x)\}$, or in other words, the discontinuities of $u'(x)$ can occur only at points of I.

A discontinuity of φ is a jump discontinuity, i.e.,

$$\varphi(x - 0) \leq \varphi(x + 0), \qquad x \in \Omega,$$

inasmuch as φ is nondecreasing. Therefore

$$u'(x - 0) \geq u'(x + 0), \qquad x \in \Omega. \tag{7.2}$$

Let $\xi \in I$, so $u(\xi) = \psi(\xi)$ and

$$u(x) - u(\xi) \geq \psi(x) - \psi(\xi).$$

If $x < \xi$, then

$$[u(x) - u(\xi)]/(x - \xi) \leq [\psi(x) - \psi(\xi)]/(x - \xi);$$

thus

$$u'(\xi - 0) \leq \psi'(\xi - 0),$$

while for $x > \xi$, $\xi \in I$,

$$[u(x) - u(\xi)]/(x - \xi) \geq [\psi(x) - \psi(\xi)]/(x - \xi),$$

so

$$u'(\xi + 0) \geq \psi'(\xi + 0).$$

Theorem 7.1. *In addition to our previous hypotheses, assume now that $\psi'(x)$ has only discontinuities of the form*

$$\psi'(x - 0) \leq \psi'(x + 0).$$

Then $u'(x)$ is continuous.

Indeed,

$$u'(\xi - 0) \leq \psi'(\xi - 0) \leq \psi'(\xi + 0) \leq u'(\xi + 0) \qquad \text{for} \quad \xi \in I.$$

Consequently, by (7.2),

$$u'(\xi - 0) = u'(\xi + 0),$$

and this proves that $u'(x)$ is continuous.

Appendix A. Sobolev Spaces

A systematic investigation of the real variable properties of elements of $H^{m,s}(\Omega)$ may be found in Morrey [1, Chapter 3] but for the convenience of the reader we will provide selected proofs. We first discuss real variable properties giving two proofs of Theorem A.1 below. The first employs the notion of "absolutely continuous representative," Lemma A.2, and the second utilizes an appropriate choice of test function, Lemmas A.3 and A.4. Some generalizations of Theorem A.1 are mentioned. The trace operator and a "matching lemma" are considered.

We then state and prove some Sobolev and Poincaré type lemmas. To conclude this appendix we state the well-known Rellich theorem, whose proof may be easily read in Morrey [1].

Theorem A.1. *Let* $\Omega \subset \mathbb{R}^N$, $u \in H^{1,s}(\Omega)$, $1 \le s \le \infty$. *Then* $\max(u, 0) \in H^{1,s}(\Omega)$ *and for* $1 \le i \le N$

$$[\max(u, 0)]_{x_i} = \begin{cases} u_{x_i} & \textit{in} \quad \{x \in \Omega : u(x) > 0\} \\ 0 & \textit{in} \quad \{x \in \Omega : u(x) \le 0\} \end{cases} \tag{A.1}$$

in the sense of distributions.

We emphasize the meaning of this theorem: the derivatives of $\max(u, 0)$ are determined almost everywhere by the formula (A.1). In particular, for any test function ζ,

$$-\int_\Omega \max(u, 0)\zeta_{x_i}\,dx = \int_\Omega [\max(u, 0)]_{x_i}\zeta\,dx = \int_{\Omega \cap \{u>0\}} u_{x_i}\zeta\,dx.$$

It suffices to prove the theorem in the case $u \in H_0^{1,s}(\Omega)$, $1 \le s < +\infty$. Since ζ has compact support, the case $s = \infty$ corresponds to Rademacher's theorem about the differentiability of Lipschitz functions, which we assume known.

Lemma A.2. *Let* $v \in H_0^{1,s}(\Omega)$, $1 \le s \le \infty$. *Then for each i,* $1 \le i \le N$, *there is a representative* $\tilde{v}_i$ *of v which is absolutely continuous on almost every parallel to the* x_i *axis in* Ω. *Moreover,*

$$(\partial/\partial x_i)\tilde{v}_i = \tilde{h}_i \qquad \textit{a.e. in} \quad \Omega,$$

where $\tilde{h}_i$ *is a representative of* $h_i \in L^s(\Omega)$, *the ith distributional derivative of v.*

Proof. Let $v_m \in C_0^1(\Omega)$ be a sequence such that

$$v_m \to v \qquad \text{in} \quad H_0^{1,s}(\Omega),$$
$$v_m \to v \qquad \text{point-wise a.e. in} \quad \Omega.$$

We extend v_m, v to be zero outside Ω so that

$$v_m(x) = \int_{-\infty}^{x_i} (\partial v_m/\partial x_i)(x_1, \ldots, x_i, t, x_{i+1}, \ldots, x_N)\,dt,$$

that is, the integral is taken over a parallel to the x_i axis passing through x. For $\zeta \in C_0^\infty(\Omega)$,

$$\int_{\mathbb{R}^N} v_m(x)\zeta(x)\,dx = \int_{\mathbb{R}^N}\left(\int_{-\infty}^{x_i} v_{mx_i}\,dt\right)\zeta(x)\,dx$$
$$= \int_{\mathbb{R}^N} v_{mx_i}(x)\left(\int_{x_i}^{\infty}\zeta\,dt\right)dx.$$

Now v_{mx_i} converges in $L^s(\Omega)$ to function h_i, the ith distributional derivative of v. Consequently,

$$\int_{\mathbb{R}^N} v(x)\zeta(x)\,dx = \int_{\mathbb{R}^N} h_i(x)\left(\int_{x_i}^{\infty}\zeta\,dt\right)dx = \int_{\mathbb{R}^N}\left(\int_{-\infty}^{x_i} h_i\,dt\right)\zeta(x)\,dx.$$

This implies that v is equal almost everywhere to a function $\tilde{v}(x)$ which is absolutely continuous on almost every parallel to the x_i axis and whose derivative $\partial\tilde{v}/\partial x_i = \tilde{h}_i$, a representative of $h_i \in L^s(\Omega)$.

In this way, we can understand

$$v_{x_i} = \lim v_{mx_i} = h_i,$$

with equality being understood either in $L^s(\Omega)$ or a.e. in Ω for a subsequence of the v_{mx_i}. Q.E.D.

Proof of Theorem A.1. First observe that $\max(u, 0) \in H^{1,s}(\Omega)$, $1 < s < \infty$. For let $u_n \in C^1(\Omega)$ satisfy $u_n \to u$ in $H^{1,s}(\Omega)$. Then $\max(u_n, 0)$ is Lipschitz and satisfies

$$\|\max(u_n, 0)\|_{H_0^{1,s}(\Omega)} \le \|u_n\|_{H_0^{1,s}(\Omega)} \le C \tag{A.2}$$

for some constant and, recalling that $\max(u, 0) \in L^s(\Omega)$,

$$\|\max(u_n, 0) - \max(u, 0)\|_{L^s(\Omega)} \le \|u_n - u\|_{L^s(\Omega)}. \tag{A.3}$$

From (A.2), it follows that $\max(u_n, 0)$ converges weakly to an element of $H^{1,s}(\Omega)$ and from (A.3) we conclude that this element is equal to $\max(u, 0)$ a.e.

Now we apply Lemma A.2 to $v = \max(u, 0)$. Let $\tilde{v}_i$ and $\tilde{u}_i$ be the representatives of v and u, respectively, with $\tilde{h}_i$ and $\tilde{g}_i$ the representatives of their distributional derivatives determined by the lemma.

First, let $F \subset \Omega$ be any subset such that $v = 0$ on F. We want to show that $\tilde{h}_i = 0$ a.e. on F, or what is the same thing, that

$$\int_F h_i^2\,dx = 0. \tag{A.4}$$

First, we replace F by a subset of equal measure where $\tilde{v}_i = 0$. By Fubini's theorem, viz., (A.4), we need only prove $\tilde{h}_i = 0$ a.e. on F when $F \subset \mathbb{R}$, that is, for functions of a single variable x_i. Now $F \subset \mathbb{R}$ may be written $F = F' \cup \tilde{N}$, where F' is the set of density points of F and meas $\tilde{N} = 0$. On $F' \subset \mathbb{R}$, $\tilde{v}_i$ is differentiable a.e. and its derivative on F' may be computed from its value there. Hence

$$\tilde{h}_i = (\partial/\partial x_i)\tilde{v}_i = 0 \qquad \text{a.e. on} \quad F'.$$

Finally, let $E \subset \Omega$ be any set where $v = u$. We want to show that $\tilde{h}_i - \tilde{g}_i = 0$ on E. We argue as before. After replacement of E by a set of equal measure where $\tilde{u}_i - \tilde{v}_i$, we need only prove that $\tilde{h}_i = \tilde{g}_i$ a.e. on E when $E \subset \mathbb{R}$ by Fubini's theorem. Again, this may be verified by consideration of $\tilde{h}_i$ and $\tilde{g}_i$ on the set of density points E' of E.
The theorem is proved. Q.E.D.

We can deduce Theorem A.1 from the following lemmas:

Lemma A.3. *Let $\theta(t)$ be a $C^1(\mathbb{R})$ function such that θ' is bounded; $|\theta'(t)| \le M$. If $u \in H^{1,s}(\Omega)$, Ω a bounded open set of $\mathbb{R}^N$, $1 \le s < +\infty$, then*

$$\theta(u) \in H^{1,s}(\Omega)$$

and

$$(\partial/\partial x_i)\theta(u) = \theta'(u)(\partial u/\partial x_i) \qquad \text{a.e. in} \quad \Omega, \qquad i = 1, 2, \ldots, N. \tag{A.5}$$

Proof. Since $u \in H^{1,s}(\Omega)$, there exists a sequence $\{u_n\}$ of $C^1(\overline{\Omega})$ functions such that u_n tends to u in $H^{1,s}(\Omega)$ and also u_n tends to u a.e. in Ω. Obviously

$$\theta(u_n) \in C^1(\overline{\Omega})$$

and since

$$|\theta(u_n) - \theta(u)| \le M|u_n - u|,$$

$\theta(u_n)$ converges to $\theta(u)$ in $L^s(\Omega)$.
On the other hand,

$$\theta'(u_n)(\partial u_n/\partial x_i) = (\partial/\partial x_i)\theta(u_n) \to \theta'(u)(\partial u/\partial x_i) \qquad \text{in} \quad L^s(\Omega).$$

In fact,

$$\begin{aligned}\theta'(u_n)(\partial u_n/\partial x_i) - \theta'(u)(\partial u/\partial x_i) = {}& \theta'(u_n)[(\partial u_n/\partial x_i) - (\partial u/\partial x_i)] \\ & + [\theta'(u_n) - \theta'(u)](\partial u/\partial x_i) = A_n + B_n,\end{aligned}$$

and A_n converges to zero in $L^s(\Omega)$ while B_n converges to zero a.e. in Ω and

$$|B_n|^s \le (2M)^s |\partial u/\partial x_i|^s.$$

Therefore, by the Lebesgue theorem, B_n converges to zero in $L^s(\Omega)$.

Since the derivatives in the sense of distribution of $\theta(u)$ are the limit in $L^s(\Omega)$ of $(\partial/\partial x_i)\theta(u_n)$, (A.5) follows.

If $u \in H^{1,s}(\Omega)$ is such that $u:\Omega \supset \Omega' \to (\alpha, \beta)$ and θ is $C^1((\alpha, \beta))$, then (A.5) holds a.e. in Ω'.

Lemma A.4. *Let $u \in H^{1,s}(\Omega)$. Then*

$$\partial u/\partial x_i = 0 \qquad \text{a.e. in} \quad E = \{x \in \Omega : u = 0\}, \quad 1 \le i \le N.$$

Proof. For each $\gamma \in [-1, 1]$ define

$$\sigma_{\gamma\varepsilon}(t) = \begin{cases} -1, & -\infty < t \le -\varepsilon(1+\gamma) \\ \gamma + (t/\varepsilon), & -\varepsilon(1+\gamma) < t < \varepsilon(1-\gamma) \\ 1, & \varepsilon(1-\gamma) \le t < \infty, \end{cases}$$

so $\sigma_{\gamma\varepsilon}(0) = \gamma$ and

$$\lim_{\varepsilon \to 0} \sigma_{\gamma\varepsilon}(t) = \operatorname{sgn}_\gamma(t) = \begin{cases} -1, & t < 0 \\ \gamma, & t = 0 \\ 1, & t > 0. \end{cases}$$

Define

$$\theta_{\gamma\varepsilon}(t) = \int_0^t \sigma_{\gamma\varepsilon}(\tau)\, d\tau, \qquad -\infty < t < \infty.$$

Then $\theta_{\varepsilon\gamma} \in C^1(\mathbb{R})$ and $|\theta'_{\gamma\varepsilon}(t)| \le 1$. Hence $\theta_{\varepsilon\gamma}(u) \in H^{1,s}(\Omega)$ and

$$(\partial/\partial x_i)\theta_{\gamma\varepsilon}(u) = \sigma_{\gamma\varepsilon}(u)(\partial/\partial x_i)u \qquad \text{a.e. in} \quad \Omega, \quad 1 \le i \le N.$$

Now $\theta_{\varepsilon\gamma}(u) \to |u|$ in $L^s(\Omega)$ as $\varepsilon \to 0$ since $\theta_{\varepsilon\gamma}(u) \to |u|$ pointwise a.e. in Ω and $|\theta_{\gamma\varepsilon}(u)| \le |u| \in L^s(\Omega)$. In addition,

$$(\partial/\partial x_i)\theta_{\gamma\varepsilon}(u) \to (\operatorname{sgn}_\gamma u)(\partial/\partial x_i)u \qquad \text{in} \quad \Omega \quad \text{a.e.}$$

and

$$|(\partial/\partial x_i)\theta_{\gamma\varepsilon}(u)| \le |(\partial/\partial x_i)u| \in L^s(\Omega).$$

Hence

$$(\partial/\partial x_i)\theta_{\gamma\varepsilon}(u) \to (\operatorname{sgn}_\gamma u)(\partial/\partial x_i)u \qquad \text{in} \quad L^s(\Omega).$$

Thus $|u| \in H^{1,s}(\Omega)$ and

$$(\partial/\partial x_i)|u| = (\operatorname{sgn}_\gamma u)(\partial/\partial x_i)u \qquad \text{a.e. in} \quad \Omega.$$

In particular,

$$(\partial/\partial x_i)|u| = \gamma(\partial/\partial x_i)u \qquad \text{a.e. on} \quad E$$

for every $\gamma \in [-1, 1]$, so

$$(\partial/\partial x_i)|u| = (\partial/\partial x_i)u = 0 \qquad \text{a.e. on} \quad E. \quad \text{Q.E.D.}$$

A slightly simpler proof may be based on Exercise 15.

A particular consequence of the last lemma is that for any $u \in H^{1,s}(\Omega)$,

$$(\partial/\partial x_i) \max(u, 0) = 0 \qquad \text{a.e. where} \quad u \leq 0$$

and

$$(\partial/\partial x_i)(\max(u, 0) - u) = 0 \qquad \text{a.e. where} \quad u \geq 0.$$

In other words, (A.1) holds, which proves Theorem A.1.

We give two generalizations of Theorem A.1.

Corollary A.5. *Let $\theta(t)$, $-\infty < t < \infty$, be a piecewise linear function whose derivative has discontinuities at $\{a_1, \ldots, a_M\}$ and let $u \in H^{1,s}(\Omega)$, $1 \leq s \leq \infty$. Then $\theta(u) \in H^{1,s}(\Omega)$ and*

$$\theta(u)_{x_i} = \theta'(u)u_{x_i} \qquad \textit{(in the sense of distributions)}$$

with the convention that both sides are zero when $x \in \bigcup_j \{y : u(y) = a_j\}$. In particular, $\min(u, 0)$, $|u| \in H^{1,s}(\Omega)$.

The proof of this easy corollary is omitted. To justify the convention, merely note that by Lemma A.4 applied to $\theta(u)$ and u,

$$(\partial/\partial x_i)\theta(u) = 0 = (\partial/\partial x_i)u \qquad \text{a.e. on} \quad \bigcup_j \{y : u(y) = a_j\}, \qquad 1 \leq i \leq N.$$

A more subtle refinement of Lemma A.4 is

Corollary A.6. *Let $\theta(t)$, $-\infty < t < \infty$, be a Lipschitz function whose derivative $\theta'(t)$ exists except at finitely many points $\{a_1, \ldots, a_M\}$ and let $u \in H^{1,s}(\Omega)$, $1 \leq s \leq \infty$. Then*

$$\theta(u) \in H^{1,s}(\Omega)$$

and

$$\theta(u)_{x_i} = \theta'(u)u_{x_i} \qquad \textit{(in the sense of distributions)} \tag{A.6}$$

with the convention that both sides are zero when $x \in \bigcup_j \{y : u(y) = a_j\}$.

We prove an $H^{1,s}$ "matching lemma" which will be helpful in our study of free boundaries. Its proof depends on the notion of the trace of an $H^{1,s}$ function. Suppose, for example, that

$$G = \{x \in \mathbb{R}^N : |x| < 1 \text{ and } x_N > 0\}$$

and

$$\Sigma = \{x \in \mathbb{R}^N : |x| < 1 \text{ and } x_N = 0\}.$$

The trace operator

$$T: H^{1,s}(G) \to L^s(\Sigma)$$

is a continuous linear mapping with the property

$$Tv(x') = v(x', 0) \qquad \text{for} \quad v \in H^{1,\infty}(G) \subset H^{1,s}(G).$$

Here we denote a point of $\mathbb{R}^N$ by $x = (x', x_N)$. Moreover, denoting by $T_\varepsilon v$ the trace of v on the subset $\Sigma_\varepsilon = G \cap \{x : x_N = \varepsilon\}$, it is easy to check that

$$T_\varepsilon v \to Tv \qquad \text{in} \quad L^s(\Sigma), \qquad \text{as} \quad \varepsilon \to 0,$$

in the obvious way. Assume now that $\zeta \in H^{1,s}(G)$ and $T\zeta = 0$. Identify ζ with one of its representatives absolutely continuous on segments parallel to the x_N axis. Existence of such a representative follows from Lemma A.2. By Lebesgue's theorem, there is a subsequence $\varepsilon_j \to 0$ such that $\zeta(x', \varepsilon_j) \to 0$ for a.e. $x' \in \Sigma$ as $\varepsilon_j \to 0$.

Now extend $\zeta(x)$ to be zero in $G^- = \{x : |x| < 1, x_N < 0\}$. For any x' with $\lim \zeta(x', \varepsilon_j) = 0$, the function of a single variable $\zeta(x', x_N)$, $|x_N| < 1$, is absolutely continuous and

$$\int_{-(1-|x'|^2)^{1/2}}^{(1-|x'|^2)^{1/2}} |\zeta_{x_N}(x', x_N)|^s \, dx_N = \int_0^{(1-|x'|^2)^{1/2}} |\zeta_{x_N}(x', x_N)|^s \, dx_N;$$

by Fubini's theorem, $\zeta_{x_N} \in L^s(B)$, $B = \{x : |x| < 1\}$. It is easy to check that $\zeta_{x_i} \in L^s(B)$ for $i < N$. Hence a function ζ with $T\zeta = 0$ on Σ may be extended as 0 in G^- to become an element of $H^{1,s}(B)$.

From this it is obvious that $\zeta = 0$ on Σ in the sense that $\zeta = \lim \zeta_n$ in $H^{1,s}(G)$, where $\zeta_n \in H^{1,\infty}(G)$ and $\zeta_n = 0$ for $x_N \leq 1/n$.

Finally, observe that the converse is also true: if $\zeta \in H^{1,\infty}(G)$ is the limit of smooth functions ζ_n vanishing near Σ, then $T\zeta = \lim T\zeta_n = 0$. We have shown

Lemma A.7. *Let $\zeta \in H^{1,s}(G)$, $1 \leq s < \infty$. Then $T\zeta = 0$ on Σ if and only if ζ is in the closure in $H^{1,s}(G)$ norm of the smooth functions which vanish near Σ.*

From this we deduce a "matching" lemma.

Lemma A.8. *Let B denote the unit ball in $\mathbb{R}^N$, $G^+ = \{x \in B : x_N > 0\}$, $G^- = \{x \in B : x_N < 0\}$, and $\Sigma = B \cap \{x_N = 0\}$. Let $v^+ \in H^{1,s}(G^+)$ and $v^- \in H^{1,s}(G^-)$ satisfy*

$$Tv^+ = Tv^- \qquad \text{on} \quad \Sigma.$$

Then

$$v(x) = \begin{cases} v^+(x), & x \in G^+ \\ v^-(x), & x \in G^- \end{cases}$$

is in $H^{1,s}(B)$.

Proof. Set $w(x) = v^-(x', -x_n)$ for $x \in G^+$. Then $w \in H^{1,s}(G^+)$ and $Tv^+ = Tw$ on Σ. Consequently,

$$T(w - v^+) = 0 \quad \text{on} \quad \Sigma.$$

Let $\zeta_\varepsilon \in H^{1,\infty}(G)$ be a sequence of Lipschitz functions satisfying

$$\zeta_\varepsilon = 0 \quad \text{on} \quad \Sigma$$

and

$$\zeta_\varepsilon \to w - v^+ \quad \text{in} \quad H^{1,s}(G) \quad \text{as} \quad \varepsilon \to 0.$$

Also let $v_\varepsilon^+ \in H^{1,\infty}(G^+)$ converge to v^+ in $H^{1,s}(G^+)$. Define

$$v_\varepsilon(x) = \begin{cases} v_\varepsilon^+(x), & x \in G^+ \cup \Sigma \\ v_\varepsilon^+(x', -x_N) + \zeta_\varepsilon(x', -x_N), & x \in G^-, \end{cases}$$

which is Lipschitz in B. Obviously

$$\|v_\varepsilon - v\|_{L^s(B)} \leq \|v_\varepsilon^+ - v\|_{L^s(G^+)} + \|\zeta_\varepsilon - (w - v_\varepsilon^+)\|_{L^s(G^+)} \to 0 \quad \text{as} \quad \varepsilon \to 0.$$

Moreover,

$$\|v_x\|_{L^s(B)} \leq \liminf_{\varepsilon \to 0} \|v_{\varepsilon x}\|_{L^s(B)} \leq \|v_x^+\|_{L^s(G)} + \|v_x^-\|_{L^s(G)},$$

so $v \in H^{1,s}(B)$. Q.E.D.

We now state and prove some Sobolev lemmas.

Lemma A.9. (Sobolev). *Let $\partial\Omega$ be Lipschitz and $u \in H^{1,s}(\Omega)$, $1 \leq s < \infty$. Then*

(i) *if $1 \leq s < N$, then $u \in L^{s^*}(\Omega)$, $1/s^* = (1/s) - (1/N)$, and*

$$\|u\|_{L^{s^*}(\Omega)} \leq C\|u\|_{H^{1,s}(\Omega)},$$

whereas

(ii) *if $s > N$, then $u \in C^{0,\lambda}(\overline{\Omega})$, $\lambda = 1 - (N/s)$, and*

$$\|u\|_{C^{0,\lambda}(\overline{\Omega})} \leq C\|u\|_{H^{1,s}(\Omega)},$$

where C depends only on Ω, s, N.

The first assertion of Lemma A.9 is a consequence of the following useful inequality about functions in $H_0^{1,s}(\Omega)$.

Lemma A.10. *There exists a constant C, depending only on N, such that for $s < N$,*

$$\|u\|_{L^{s^*}(\Omega)} \leq C(s^*/1^*)\|u_x\|_{L^s(\Omega)}, \qquad u \in H_0^{1,s}(\Omega).$$

Poincaré's inequality follows from this lemma and Hölder's inequality. We state a form of particular use to us.

Corollary A.11. *Let $u \in H_0^1(B_r)$, $B_r = \{x \in \mathbb{R}^N : |x| < r\}$. Then*

$$\int_{B_r} u(x)^2\, dx \leq Cr^2 \int_{B_r} u_x(x)^2\, dx, \qquad u \in H_0^1(B_r), \tag{A.7}$$

where C depends only on N.

Proof of Lemma A.10. We begin by proving that the inequality is true for $s > 1$ if we assume it true for $s = 1$. Let $u(x)$ be a function of $C_0^1(\overline{\Omega})$; the function $|u|^{s^*/1^*}$ belongs to $H_0^{1,1}(\Omega)$ and therefore we have

$$\left(\int_\Omega |u|^{s^*}\, dx\right)^{1/1^*} \leq C(s^*/1^*) \int_\Omega |u|^{(s^*/1^*)-1}|u_x|\, dx.$$

Using Hölder's inequality and noting that $(1/1^*) - (1/s^*) = 1/s'$, we have

$$\left(\int_\Omega |u|^{s^*} dx\right)^{(1/1^*)-(1/s')} \leq C(s^*/1^*)\left(\int_\Omega |u_x|^s\, dx\right)^{1/s},$$

and thus

$$\left(\int_\Omega |u|^{s^*}\, dx\right)^{1/s^*} \leq C(s^*/1^*)\left(\int |u_x|^s\, dx\right)^{1/s}.$$

It remains to prove the inequality for $s = 1$. To this end let $x = (x_1, \ldots, x_N)$ be any point of $\mathbb{R}^N$. From the inequalities

$$|u(x)| \leq \tfrac{1}{2} \int_i |u_{x_i}(\xi)|\, d\xi_i, \qquad i = 1, 2, \ldots, N,$$

where the right-hand side means that the integral is over the straight line

$$\xi_s = x_s \qquad \text{with} \quad s = i,$$

we infer

$$u(x)^{N/(N-1)} \leq \frac{1}{2^{1^*}} \prod_i \left(\int_i |u_{x_i}(\xi)| \, d\xi_i \right)^{1/(N-1)}$$

Integrating with respect to x_1, then with respect to x_2, and so on and making use of the Hölder inequality,[1] we get

$$\int_1 |u(x)|^{N/(N-1)} \, dx_1 \leq (1/2^{1^*}) \left(\int_1 |u_{x_1}(\xi)| \, d\xi_1 \right)^{1/(N-1)}$$

$$\cdot \left(\int_1 \int_2 |u_{x_2}(\xi)| \, d\xi_1 \, d\xi_2 \right)^{1/(N-1)} \cdots$$

$$\cdot \left(\int_1 \int_N |u_{x_N}(x)| \, dx_1 \, d\xi_N \right)^{1/(N-1)},$$

$$\int_1 \int_2 |u(x)|^{N/(N-1)} \, dx_1 \, dx_2 \leq (1/2^{1^*}) \left(\int_1 \int_2 |u_x(x)| \, d\xi_1 \, dx_2 \right)^{2/(N-1)} \cdots$$

$$\cdot \left(\int_1 \int_2 \int_3 |u_x(x)| \, dx_1 \, dx_2 \, d\xi_3 \right)^{1/(N-1)} \cdots,$$

$$\int_{\mathbb{R}^N} |u(x)|^{N/(N-1)} \, dx_1 \cdots dx_N \leq (1/2^{1^*}) \left(\int_{\mathbb{R}^N} |u_x(x)| \, dx_1 \cdots dx_N \right)^{N/(N-1)}.$$

In such a way the inequality is proved also for $s = 1$ with the constant $C = 1/2^{1^*}$. Q.E.D.

Proof of Lemma A.9(ii). We first show that

$$|u(x)| \leq C(\text{meas } \Omega)^{-1/s}(\|u\|_{L^s(\Omega)} + [1 - (N/s)]^{-1} \operatorname{diam} \Omega \|u_x\|_{L^s(\Omega)}) \quad \text{(A.8)}$$

for $u \in H^{1,s}(\Omega)$. By extending u to a function, still called u, of $H_0^{1,s}(\Omega_0)$, $\overline{\Omega} \subset \Omega_0$, so that

$$\|u\|_{H_0^{1,s}(\Omega_0)} \leq C\|u\|_{H^{1,s}(\Omega)},$$

we see that it is sufficient to consider functions $u \in H_0^{1,s}(\Omega)$. Since $u \in H_0^{1,s}(\Omega)$ is the limit of functions $u_n \in C_0^\infty(\Omega)$, we need only show (A.8) for $u \in C_0^\infty(\Omega)$.

[1] We use the inequality

$$\int f_1 f_2 \cdots f_k \, dx \leq \|f_1\|_{L^{\alpha_1}} \|f_2\|_{L^{\alpha_2}} \cdots \|f_k\|_{L^{\alpha_k}}, (1/\alpha_1) + (1/\alpha_2) + \cdots + (1/\alpha_k) = 1.$$

Given x, we expand u in a Taylor expansion about each $y \in \Omega$ and integrate with respect to y. This gives

$$\text{meas } \Omega\, u(x) = \int_{\Omega} u(y)\, dy + \int_0^1 \int_{\Omega} u_x(x + t(y - x))(y - x)\, dy\, dt. \quad \text{(A.9)}$$

We now apply Hölder's inequality to each term. To estimate the last term

$$I = \int_0^1 \int_{\Omega} u_x(x + t(y - x))(y - x)\, dy\, dt, \quad \text{(A.10)}$$

set $z = x + t(y - x)$ for a fixed t and $\Omega(t) = \{z : x + t(y - x) \in \Omega\}$. Note that meas $\Omega(t) = t^N$ meas Ω. Then

$$\begin{aligned} |I| &\leq \int_0^1 \int_{\Omega(t)} |u_x(z)|\, |y - x| t^{-N}\, dz\, dt \\ &\leq \text{diam } \Omega \int_0^1 t^{-N} (\text{meas}(\Omega(t)))^{1-(1/s)} \|u_x\|_{L^s(\Omega)}\, dt \\ &\leq \text{diam } \Omega\, (\text{meas } \Omega)^{1-(1/s)} \frac{1}{1 - (N/s)} \|u_x\|_{L^s(\Omega)}. \end{aligned}$$

From (A.9) we now deduce

$$\text{meas } \Omega\, |u(x)| \leq (\text{meas } \Omega)^{1-(1/s)} \left\{ \|u\|_{L^s(\Omega)} + \frac{\text{diam } \Omega}{1 - (N/s)} \|u_x\|_{L^s(\Omega)} \right\},$$

which proves (A.8).

A particular consequence of (A.8) is that functions of $H^{1,s}(\Omega)$, $s > N$, are continuous.

Lemma A.12. *Let $v \in H^{1,s}(B_r(x))$ satisfy*

$$\int_{B_r(x)} v(y)\, dy = 0.$$

Then

$$\|v\|_{L^s(B_r(x))} \leq Cr \|v_x\|_{L^s(B_r(x))},$$

where C depends only on N and s.

The demonstration of this lemma is left as an exercise. To complete the proof of Lemma A.9(ii), given $u \in H_0^{1,s}(\Omega)$ and $x, y \in \Omega$, $|x - y| = \delta$, we shall apply Lemma A.12 to

$$u - c = u - (1/\text{meas } B_\delta) \int_{B_\delta(x)} u(\xi)\, d\xi,$$

whose average over $B_\delta(x)$ is zero. Consequently,

$$|u(x) - u(y)| \le |u(x) - c| + |u(y) - c|$$
$$\le 2C\delta^{1-(N/s)}\{1 + 2[1 - (N/s)]^{-1}\}\|u_x\|_{L^s(\Omega)},$$

employing Lemma A.12 to estimate the first term of (A.8). Here $C = C(s, N)$. Q.E.D.

Another useful inequality concerns functions which vanish on a subset of B_r.

Lemma A.13. *Let $E \subset B_r$ satisfy*

$$\text{meas } E \ge \vartheta \text{ meas } B_r, \qquad 0 < \vartheta < 1,$$

and let $u \in H^{1,1}(B_r)$ vanish almost everywhere on E. Then for each $\sigma > 0$

$$\text{meas}\{x \in B_r : |u(x)| > \sigma\} \le \left[(C/\sigma)\int_{B_r} |u_x(x)|\, dx\right]^{N/(N-1)},$$

where C depends only on ϑ.

We shall use

Lemma A.14. *Let $f(x) \in L^1(\mathbb{R}^N)$ and define*

$$Kf(x) = \int_{\mathbb{R}^N} (f(y)/|x - y|^{N-1})\, dy, \qquad x \in \mathbb{R}^N,$$

the potential of order 1 *generated by f. There exists a constant $C > 0$ such that*

$$\text{meas}\{x : |Kf(x)| > \sigma\} \le \left[(C/\sigma)\int_{\mathbb{R}^N} |f(y)|\, dy\right]^{N/(N-1)} \qquad \textit{for all} \quad \sigma > 0.$$

Proof. Now

$$|K^\varepsilon f(x)| \le \varepsilon^{1-N}\|f\|_{L^1(\mathbb{R}^N)} = \varepsilon^{1-N}.$$

Therefore

$$|K^\varepsilon f(x)| \le \sigma/2 \qquad \text{provided} \quad \varepsilon = (2/\sigma)^{1/(N-1)}.$$

Hence

$$\text{meas}\{x : |Kf(x)| > \sigma\} \le \text{meas}\{x : |K_\varepsilon f(x)| > \sigma/2\},$$

where $\varepsilon = (2/\sigma)^{1/(N-1)}$.

On the other hand, it is easy to check that $K_\varepsilon f \in L^1(\mathbb{R}^N)$. Indeed, by Fubini's theorem

$$\int_{\mathbb{R}^N} |K_\varepsilon f(x)|\, dx \le \int_{|y| \le \varepsilon} |y|^{1-N}\, dy = C_1 \varepsilon.$$

Thus

$$(\sigma/2)\, \mathrm{meas}\{x : |K_\varepsilon f(x)| > \sigma/2\} \le \int_{\mathbb{R}^N} |K_\varepsilon f(x)|\, dx \le C_1 \varepsilon.$$

Consequently

$$\begin{aligned} \mathrm{meas}\{x : |Kf(x)| > \sigma\} &\le \mathrm{meas}\{x : |K_\varepsilon f(x)| > \sigma/2\} \\ &\le (2C_1/\sigma)\varepsilon = C(1/\sigma^{N/(N-1)}). \quad \text{Q.E.D.} \end{aligned}$$

Lemma A.15. *Let $v \in H^{1,\infty}(B_r)$ vanish on $E \subset B_r$, where*

$$\mathrm{meas}\, E \ge \vartheta\, \mathrm{meas}\, B_r, \qquad 0 < \vartheta < 1.$$

Then there exists a constant $\beta > 0$ depending only on ϑ and N such that

$$|v(x)| \le \beta \int_{B_r} (|v_x(y)|/|x - y|^{N-1})\, dy, \qquad x \in B_r.$$

Proof. Let $x \in B_r$. If $x \in E$, then the conclusion is valid for any β. For $x \notin E$, consider the function $r \to v(x + r\xi)$ with $\xi \in S^{N-1}$, $S^{N-1} = \partial B_1(0)$. Let $\Sigma = \{\xi \in S^{N-1} :$ there exists $R > 0$ such that $x + R\xi \in E\}$. Then

$$v(x) - v(x + r\xi) = \int_r^0 (dv/d\rho)(x + \rho\xi)\, d\rho$$

and if $x + R\xi \in E$,

$$|v(x)| \le \int_0^R |v_x(x + \rho\xi)|\, d\rho.$$

Let $d\omega$ denote the measure on S^{N-1}, so that

$$dy = |x - y|^{N-1}\, d\rho\, d\omega.$$

We have that

$$\begin{aligned} |\Sigma|\, |v(x)| = \int_\Sigma |v(x)|\, d\omega &\le \int_\Sigma \int_0^{2r} |v_x(y)|\, d\rho\, d\omega \\ &\le \int_{B_r} (|v_x(y)|/|x - y|^{N-1})\, dy. \end{aligned}$$

In order to prove the lemma, it is sufficient to show that the measure $|\Sigma|$ of Σ is bounded below.

Now

$$\vartheta \text{ meas } B_r \leq \text{meas } E \leq \int_\Sigma d\omega \int_0^{2r} \rho^{N-1}\, d\rho = [(2r)^N/N]|\Sigma|.$$

Hence

$$|\Sigma| \geq \vartheta N[\text{meas } B_r/(2r)^N] = \vartheta\gamma(N) = 1/\beta(\vartheta, N)$$

and thus

$$|v(x)| \leq \beta(\vartheta, N) \int_\Omega (|v_x(y)|/|x-y|^{N-1})\, dy. \quad \text{Q.E.D.}$$

Proof of Lemma A.13. It suffices to verify the conclusion for functions $u \in H^{1,\infty}(B_r)$ which vanish on E, where

$$\text{meas } E \geq \vartheta \text{ meas } B_r.$$

For such u,

$$\{x : |u(x)| \geq \sigma\} \subset \{x : K(|u_x|)(x) \geq \sigma/\beta\}$$

by the preceding lemma. Therefore, by Lemma A.14,

$$\begin{aligned} \text{meas}\{x : |u(x)| \geq \sigma\} &\leq \text{meas}\{x : K(|u_x|) \geq \sigma/\beta\} \\ &\leq \left[(C\beta/\sigma) \int_{\mathbb{R}^N} |u_x(x)|\, dx\right]^{N/(N-1)}. \quad \text{Q.E.D.} \end{aligned}$$

We state without proof the well-known Rellich compactness theorem (Morrey [1]).

Theorem A.16. *Suppose that Ω is a bounded domain with smooth boundary $\partial\Omega$, $m \geq 1$, and $1 \leq s < \infty$. If $u_n \to u$ weakly in $H^{m,s}(\Omega)$, then $u_n \to u$ in $H^{m-1,s}(\Omega)$.*

Appendix B. Solutions to Equations with Bounded Measurable Coefficients

We shall now prove the estimate for solutions of the Dirichlet problem,

$$\begin{aligned} Lu &= -(a_{ij}u_{x_j})_{x_i} = f \quad &&\text{in} \quad \Omega, \\ u &= 0 &&\text{on} \quad \partial\Omega, \end{aligned}$$

that was stated in Section 5. This estimate will be useful to us in Chapter IV. We need this lemma:

Lemma B.1. *Let $\varphi(t)$, $k_0 \le t < \infty$, be nonnegative and nonincreasing such that*

$$\varphi(h) \le [C/(h-k)^\alpha]\,|\varphi(k)|^\beta, \qquad h > k > k_0, \tag{B.1}$$

where C, α, β are positive constants with $\beta > 1$. Then

$$\varphi(k_0 + d) = 0, \tag{B.2}$$

where

$$d^\alpha = C\,|\varphi(k_0)|^{\beta-1}\,2^{\alpha\beta/(\beta-1)}. \tag{B.3}$$

Proof. Consider the sequence

$$k_r = k_0 + d - (d/2^r), \qquad r = 0, 1, 2, \ldots .$$

By assumption, we have

$$\varphi(k_{r+1}) \le C[2^{(r+1)\alpha}/d^\alpha]\,|\varphi(k_r)|^\beta, \qquad r = 0, 1, 2, \ldots . \tag{B.4}$$

We now prove by induction that

$$\varphi(k_r) \le \varphi(k_0)/2^{-r\mu}, \qquad \text{where} \quad \mu = \alpha/(1-\beta) < 0. \tag{B.5}$$

For $r = 0$, (B.5) is trivial. Assuming it holds up to r, we shall show that it holds for $r + 1$. Indeed, by (B.4) we obtain

$$\varphi(k_{r+1}) \le C[2^{(r+1)\alpha}/d^\alpha][\,|\varphi(k_0)|^\beta/2^{-r\beta\mu}].$$

Using (B.3),

$$\varphi(k_{r+1}) \le \varphi(k_0)/2^{-(r+1)\mu}.$$

As $r \to +\infty$, the right-hand side tends to zero, and so

$$0 \le \varphi(k_0 + d) \le \varphi(k_r) \to 0. \quad \text{Q.E.D.}$$

Theorem B.2. *Let $a_{ij}(x) \in L^\infty(\Omega)$ satisfy*

$$\nu|\xi|^2 \le a_{ij}(x)\xi_i\xi_j \qquad \text{for} \quad \xi \in \mathbb{R}^N \quad \text{a.e.} \quad x \in \Omega,$$

where $\nu > 0$. Let $f_0, f_1, \ldots, f_N \in L^s(\Omega)$ for $s > N$ and let

$$u \in H_0^1(\Omega)\colon \quad \int_\Omega a_{ij}u_{x_j}\zeta_{x_i}\,dx = \int_\Omega \{f_0\zeta + f_i\zeta_{x_i}\}\,dx, \qquad \zeta \in H_0^1(\Omega). \tag{B.6}$$

Then

$$\max_{\Omega} |u| \leq \frac{K}{\nu} \sum_{0}^{N} \| f_i \|_{L^s(\Omega)} (\text{meas } \Omega)^{(1/N)-(1/s)},$$

where K is a constant independent of ν.

Proof. For $k > 0$ define

$$\zeta = (\text{sgn } u) \max(|u| - k, 0) = \begin{cases} u - k & \text{for } u \geq k \\ 0 & \text{for } |u| \leq k \\ u + k & \text{for } u \leq -k. \end{cases}$$

According to Corollary A.5 and Proposition 5.3, $\zeta \in H_0^1(\Omega)$. Note that $\zeta_{x_i} = u_{x_i}$ on the set $A(k) = \{x \in \Omega : |u(x)| \geq k\}$ whereas $\zeta_{x_i} = 0$ elsewhere. Thus

$$\begin{aligned} \nu \| \zeta_{x_i} \|_{L^2(\Omega)}^2 &\leq \int_{\Omega} a_{ij} \zeta_{x_j} \zeta_{x_i} \, dx = \int_{\Omega} a_{ij} u_{x_j} \zeta_{x_i} \, dx \\ &= \int_{A(k)} \{ f_0 \zeta + f_i \zeta_{x_i} \} \, dx \\ &\leq b \left(\sum_{0}^{N} \int_{A(k)} f_i^2 \, dx \right)^{1/2} \| \zeta_x \|_{L^2(\Omega)}, \end{aligned}$$

where we have used the Poincaré inequality to control the first term, allowing the constant $b \geq 1$ as a factor in the inequality. Therefore,

$$\nu \| \zeta \|_{H_0^1(\Omega)}^2 \leq \frac{\nu}{2} \| \zeta \|_{H_0^1(\Omega)}^2 + \frac{K_1}{\nu} \sum_{i=0}^{N} \int_{A(k)} f_i^2 \, dx.$$

By Hölder's inequality,

$$\int_{A(k)} f_i^2 \, dx \leq \left(\int_{A(k)} |f_i|^s \, dx \right)^{2/s} (\text{meas } A(k))^{1-(2/s)}$$

and hence

$$\| \zeta \|_{H_0^1(\Omega)}^2 \leq \frac{2K_1}{\nu^2} \sum_{i=0}^{N} \| f_i \|_{L^s(\Omega)}^2 [\text{meas } A(k)]^{1-(2/s)}.$$

From Sobolev's inequality, Lemma A.9,

$$\begin{aligned}\left(\int_{A(k)} (|u| - k)^{2^*}\,dx\right)^{2/2^*} &= \left(\int_\Omega |\zeta|^{2^*}\,dx\right)^{2/2^*}\\ &\le C\|\zeta\|^2_{H_0^1(\Omega)}\\ &\le \frac{KC_1}{\nu^2}\sum_{i=0}^N \|f_i\|^2_{L^s(\Omega)}[\text{meas } A(k)]^{1-(2/s)},\end{aligned}$$

where $1/2^* = (1/2) - (1/N)$.

If $h > k > 0$, $A(h) \subset A(k)$ and we have that

$$\begin{aligned}(h-k)^2[\text{meas } A(h)]^{2/2^*} &\le \left(\int_{A(h)} (|u| - k)^{2^*}\,dx\right)^{2/2^*}\\ &\le \left(\int_{A(k)} (|u| - k)^{2^*}\,dx\right)^{2/2^*}.\end{aligned}$$

Consequently

$$(h-k)^2[\text{meas } A(h)]^{2/2^*} \le \frac{K_1 C}{\nu^2}\left(\sum_{i=0}^N \|f_i\|^2_{L^s(\Omega)}\right)[\text{meas } A(k)]^{1-(2/s)}.$$

That is, we have obtained that

$$\text{meas } A(h) \le \frac{C}{\nu^{2^*}(h-k)^{2^*}}\left(\sum_{i=0}^N \|f_i\|_{L^s(\Omega)}\right)^{2^*/2}[\text{meas } A(k)]^\beta,$$

where $\beta = [1 - (2/s)](2^*/2) = [1 - (2/s)][1 - (2/N)]^{-1} > 1$, since $s > N$.

Now we apply the Lemma B.1 to derive that

$$\begin{aligned}&\text{meas } A(d) = 0,\\ &d = \frac{C}{\nu}\left(\sum_{i=0}^N \|f_i\|^2_{L^s(\Omega)}\right)^{1/2}(\text{meas } \Omega)^{(1/N)-(1/s)}.\end{aligned}$$

This implies that

$$u(x) \le d \qquad \text{a.e. in} \quad \Omega. \quad \text{Q.E.D.}$$

Definition B.3. A bounded open set $\Omega \subset \mathbb{R}^N$ is of *class S* if there exist two constants α, with $0 < \alpha < 1$, and ρ_0 such that, for all $x_0 \in \partial\Omega$ and for all $\rho < \rho_0$,

$$\text{meas}(B_\rho(x_0) - \Omega \cap B_\rho(x_0)) \ge \alpha \text{ meas } B_\rho(x_0).$$

We have the following theorem:

Theorem B.4. *Let $u(x) \in H_0^1(\Omega)$ be a solution to the equation*

$$Lu = -\frac{\partial}{\partial x_i}(a_{ij}(x)u_{x_j}) = \sum_{i=1}^{N} (f_i)_{x_i},$$

where $a_{ij}(x)\xi_i\xi_j > \nu|\xi|^2$ $(\nu > 0)$, $|a_{ij}(x)| \leq M$, and $f_i \in L^s(\Omega)$ with $s > N$; let Ω be of class S. Then $u(x)$ is Hölder continuous in $\bar{\Omega}$; indeed, there exist two constants k and λ with $0 < \lambda < 1$, depending on ν, M, such that

$$\operatorname*{osc}_{B_\rho(x)\cap\Omega} u \leq k\left(\sum_{i=1}^{N} \|f_i\|_{L^s(\Omega)}\right)\rho^\lambda.$$

Observe that the right-hand side may be of the form

$$f_0 + \sum_{i=1}^{N} (f_i)_{x_i} \qquad \text{where} \quad f_0 \in L^r(\Omega) \quad \text{with} \quad r > N/2.$$

In fact let V be the solution of the equation

$$V \in H_0^1(D); \qquad -\Delta V = f_0 \quad \text{in} \quad D \supset \bar{\Omega};$$

then $f_0 = -\sum (\partial/\partial x_i)V_{x_i}$. Since from the Calderon–Zygmund inequality (recalled in Section 2 of Chapter IV) $V_{x_i x_j} \in L^r(D)$, it follows, by the Sobolev inequality, that $V_{x_i} \in L^{r^*}(\Omega)$ with $r^* > N$.

The proof of the theorem will be given in Appendix C.

Appendix C. Local Estimates of Solutions

We recall

Definition C.1. An open set Ω is said *of class S* if there exist two numbers α $(0 < \alpha \leq 1)$ and $\rho_0 > 0$ such that

$$\text{meas}\{B_\rho(x_0) - \Omega(x_0, \rho)\} \geq \alpha \text{ meas } B_\rho(x_0) \tag{C.1}$$

for all $x_0 \in \bar{\Omega}$. Here $\Omega(x, \rho)$ denotes the set $\Omega \cap B_\rho(x)$.

Consider the operator

$$Lu = -(a_{ij}(x)u_{x_j})_{x_i} \tag{C.2}$$

and assume that

$$\nu\xi^2 \le a_{ij}(x)\xi_i\xi_j, \qquad \nu > 0, \quad |a_{ij}(x)| < M.$$

The aim of this and of the next sections is to prove the following theorem:

Theorem C.2. *Let $u \in H_0^1(\Omega)$ be a solution of the equation*

$$Lu = \sum_{i=1}^{N} (f_i)_{x_i} \tag{C.3}$$

where $f_i \in L^s$, $s > N$, and Ω is of class S. Then u is Hölder continuous in $\overline{\Omega}$, i.e., there exist two constants $K > 0$ and λ $(0 < \lambda < 1)$, depending only on ν, M, Ω, N, such that

$$\operatorname*{osc}_{\Omega(x,\rho)} u \le K\left\{\sum_{i=1}^{N} \|f_i\|_{L^s(\Omega)}\right\}\rho^{\lambda}. \tag{C.4}$$

In Appendix B we have already proved that the solution u is bounded. We need to establish the inequality (C.4). This will be done in several steps. First of all we need some local estimates for solutions, supersolutions, and subsolutions relative to the operator L.

Definition C.3. A function $u \in H_{\text{loc}}^1(\Omega)$ is said to be a *local L subsolution* if

$$\int_\Omega a_{ij}(x) u_{x_j} \varphi_{x_i}\, dx \le 0 \qquad \text{for all} \quad \varphi \in C_0^\infty(\Omega), \quad \varphi \ge 0. \tag{C.5}$$

A function $u \in H_{\text{loc}}^1(\Omega)$ is said to be a *local L supersolution* if the function $-u$ is a local L subsolution, i.e., if the inequality in (C.5) is reversed. Therefore a local solution is at the same time a local L supersolution and a local L subsolution.

The following theorem holds, and it will be proved in the sequel.

Theorem C.4. *Let u be a local L subsolution; then there exists a constant $K = K(\nu, M, N)$ such that, for $x \in \Omega$ and $\Omega(x, 2\rho) \subset \Omega$,*

$$\max_{\Omega(x,\rho)} u(x) \le K\left\{(1/\rho^N)\int_{\Omega(x,2\rho)} u^2\, dx\right\}^{1/2}. \tag{C.6}$$

Corollary C.5. *Let $u(x)$ be a local L supersolution; then there exists a constant $K = K(\nu, M, N)$ such that, for $x \in \Omega$ and $\Omega(x, 2\rho) \subset \Omega$,*

$$\min_{\Omega(x,\rho)} u(x) \ge K\left\{(1/\rho^N)\int_{\Omega(x,2\rho)} u^2\, dx\right\}^{1/2}. \tag{C.7}$$

In fact $-u$ is a subsolution and thus (C.7) follows from (C.6).

Theorem C.6. *If $u(x)$ is a solution of $Lu = 0$ which belongs to $H^1(\Omega(x, 2\rho))$ and vanishes on $\partial\Omega \cap B_{2\rho}(x)$, then* (C.6) *and* (C.7) *are valid.*

If $x \in \Omega$ and $2\rho < \operatorname{dist}(x, \partial\Omega)$, the statement is a consequence of Theorem C.4 and corollary C.5. If $x \in \partial\Omega$, this will follow from the proof of Theorem C.4.

In order to prove Theorem C.4 and Theorem C.6 we need some lemmas. First of all we prove the following one, which is analogous to Lemma B.1.

Lemma C.7. *If $\varphi(h, \rho)$ is a real, nonnegative function defined for $h > k_0$ and $\rho < R_0$ which is nonincreasing with respect to h and nondecreasing with respect to ρ, such that*

$$\varphi(h, \rho) \le \frac{C}{(h-k)^\alpha (R-\rho)^\gamma} [\varphi(k, R)]^\beta \tag{C.8}$$

for $h > k > k_0$ and $\rho < R < R_0$, where C, α, β, γ are positive constants with $\beta > 1$, then if $0 < \sigma < 1$,

$$\varphi(k_0 + d, R_0 - \sigma R_0) = 0, \tag{C.9}$$

where

$$d^\alpha = \frac{2^{(\alpha+\gamma)\beta/(\beta-1)} C [\varphi(k_0, R_0)]^{\beta-1}}{\sigma^\gamma R_0^\gamma}. \tag{C.10}$$

Proof. We consider the sequences

$$k_r = k_0 + d - (d/2^r) \qquad \text{and} \qquad \rho_r = R_0 - \sigma R_0 + (\sigma R_0/2^r) \tag{C.11}$$

and we prove by induction that

$$\varphi(k_r, \rho_r) \le \varphi(k_0, R_0)/2^{\mu r}, \qquad \text{where} \quad \mu = (\alpha + \gamma)/(\beta - 1). \tag{C.12}$$

(C.12) is true for $r = 0$; assume that it holds true for r and we shall prove that it is true for $r + 1$. Indeed, from (C.8), (C.10), and (C.12) we get

$$\varphi(k_{r+1}, \rho_{r+1}) \le C[2^{(\alpha+\gamma)(r+1)}/(d^\alpha \sigma^\gamma R_0^\gamma)][\varphi(k_r, \rho_r)]^\beta \le \varphi(k_0, R_0)/2^{\mu(r+1)}.$$

Passing to the limit as $r \to +\infty$, we get (C.9). Q.E.D.

Lemma C.8. *If v is a local nonnegative subsolution relative to the operator L and if $\alpha \in C_0^1(\Omega)$, then there exists a constant $K = K(v, M, N)$ such that*

$$\int_\Omega \alpha^2 v_x^2 \, dx \le K \int_\Omega \alpha_x^2 v^2 \, dx. \tag{C.13}$$

Proof. By assumption we have

$$\int_\Omega a_{ij}(x) v_{x_j} \varphi_{x_i}\, dx \le 0 \qquad \text{for all} \quad \varphi \in H_0^1(\Omega')$$

with

$$\bar{\Omega}' \subset \Omega \quad \text{and} \quad \varphi \ge 0 \quad \text{in} \quad \Omega'.$$

Setting $\varphi = \alpha^2 v$, by standard computation we have (C.13). Q.E.D.

Corollary C.9. *Under the same assumptions of Lemma* C.8 *we have*

$$\left(\int_\Omega (\alpha v)^{2^*}\, dx\right)^{2/2^*} \le K \int_\Omega \alpha_x^2 v^2\, dx, \qquad \frac{1}{2^*} = \frac{1}{2} - \frac{1}{N}. \tag{C.14}$$

It is enough to make use of the Sobolev inequality for functions of $H_0^1(\Omega)$.

Corollary C.10. *Under the same assumptions of Lemma C.8 we have*

$$\int_\Omega \alpha^2 v^2\, dx \le K \int_\Omega \alpha_x^2 v^2\, dx\, [\text{meas}\{x \in \Omega : \alpha v \ne 0\}]^{2/N}. \tag{C.15}$$

Use Corollary C.8 and the Hölder inequality.

Remark C.11. *If v is a subsolution in $H^1(\Omega(x, R))$, vanishing on $\partial\Omega \cap B_R(x)$, then we have, for $x \in \partial\Omega$,*

$$\int_{\Omega(x,\rho)} v_x^2\, dx \le \frac{K}{(R-\rho)^2} \int_{\Omega(x,R)} v^2\, dx, \tag{C.16}$$

$$\int_{\Omega(x,\rho)} v^2\, dx \le \frac{K}{(R-\rho)^2} \int_{\Omega(x,R)} v^2\, dx\, [\text{meas}\{x \in \Omega(x,R) : v > 0\}]^{2/N}. \tag{C.17}$$

It is enough in Lemma C.8 and in the Corollaries C.9 and C.10 to choose $\alpha \in C^1(\Omega)$ such that $\alpha = 1$ in $B_\rho(x)$, $\alpha = 0$ outside of $B_{2\rho}(x)$ and in such a way that $|\alpha_x| \le 2/\rho$.

Proof of Theorem C.4. Since u is a local L subsolution the function

$$\max(u, k) - k = u - \min(u, k) \tag{C.18}$$

is a local L subsolution for any constant k (see Theorem 6.6). Notice that the same function is a subsolution in $H^1(\Omega(x, 2\rho))$ vanishing on $\partial\Omega \cap B_{2\rho}(x)$ if $k \ge 0$.

Let $\alpha \in C_0^1(\Omega)$ with $\alpha = 1$ in $\Omega(x, \rho)$ and $\alpha = 0$ outside $B_R(x)$, where $\rho < R$ with $|\alpha_x| \leq 2/(R - \rho)$. Making use of Corollary C.10, we have

$$\int_{A(k,\rho)} (u - k)^2\, dx \leq \frac{K}{(R - \rho)^2} \int_{A(k,R)} (u - k)^2\, dx\, [\text{meas}\, A(x, R)]^{2/N}, \quad \text{(C.19)}$$

where $A(k, \rho) = \{y : u(y) \geq k\} \cap \Omega(x, \rho)$. For $h > k$ we have

$$(h - k)^2 \text{meas}\, A(h, \rho) \leq \int_{A(h,\rho)} (u - k)^2\, dx \leq \int_{A(k,R)} (u - k)^2\, dx. \quad \text{(C.20)}$$

If we set

$$\begin{aligned} a(h, \rho) &= \text{meas}\, A(h, \rho), \\ u(h, \rho) &= \int_{A(h,\rho)} (u - h)^2\, dx, \end{aligned} \quad \text{(C.21)}$$

we have

$$\begin{aligned} u(h, \rho) &\leq [K/(R - \rho)^2] u(k, R) [a(k, R)]^{2/N}, \\ a(h, \rho) &\leq [1/(h - k)^2] u(k, R). \end{aligned} \quad \text{(C.22)}$$

Let ξ, η be two positive numbers that will be fixed later. From (C.22) we get

$$|u(h, \rho)|^{\xi} |a(h, \rho)|^{\eta} \leq \{K^{\xi}/[(R - \rho)^{2\xi}(h - k)^{2\eta}]\} |u(k, R)|^{\xi + \eta} |a(k, R)|^{2\xi/N}. \quad \text{(C.23)}$$

Choose ξ and η in such a way that

$$\xi + \eta = \theta\xi, \qquad 2\xi/N = \theta\eta;$$

then θ needs to be a positive solution of the equation

$$\theta^2 - \theta - (2/N) = 0.$$

So

$$\theta = (1/2) + \sqrt{(1/4) + (2/N)} > 1$$

and we can fix $\eta = 1$, $\xi = N\theta/2$. Setting

$$\varphi(h, \rho) = |u(h, \rho)|^{\xi} |a(h, \rho)|^{\eta}.$$

(C.23) may be written

$$\varphi(h, \rho) \leq \frac{K^{\xi}}{(R - \rho)^{2\xi}(h - k)^{2\eta}} [\varphi(k, R)]^{\theta} \quad \text{(C.24)}$$

where $h > k$, $\rho < R$ and $\theta > 1$. We are now in condition to use Lemma C.7 for $\sigma = \frac{1}{2}$, $k_0 = 0$, $R_0 = 2\rho$; then

$$\varphi(d, \rho) = 0,$$

where

$$d = K \frac{[\varphi(0, 2\rho)]^{(\theta-1)/2}}{(2\rho)^{N\theta/2}}.$$

(C.6) follows from an easy computation. Theorem C.6 follows with the same proof taking into account Remark C.11. Q.E.D.

With the same proof of Theorem C.4 we can establish the following:

Theorem C.12. *Under the same assumption of Theorem C.4 (or of Theorem C.5) the following inequality is true:*

$$\max_{\Omega(x,\rho)} u \leq k_0 + K\left(\frac{1}{\rho^N}\int_{A(k_0,R)} (u - k_0)^2\, dx\right)^{1/2}\left(\frac{\text{meas } A(k_0, R)}{R^N}\right)^{(\theta-1)/2} \tag{C.25}$$

where

$$\theta = (1/2) + \sqrt{(1/4) + (2/N)} > 1, \qquad \rho < R,$$

$$A(k, R) = \{y \in \Omega(x, R) : u(y) \geq k\},$$

and k_0 is any real number if $x \in \Omega$ (or k_0 is any nonnegative number if $x \in \partial\Omega$).

The proof proceeds exactly as in the proof in Theorem C.4 in order to obtain (C.24); then we make use of Lemma C.7 with k_0 instead of 0 and we obtain (C.25).

Consider now the solution of the elliptic equation

$$Lu = \sum_{i=1}^{N} (f_i)_{x_i} \tag{C.26}$$

with $f_i \in L^s(\Omega)$, $s \geq 2$. Then we have the following:

Theorem C.13. *If $u(x)$ is a local solution of (C.26), then there exists a constant $K = K(v, M, N)$ such that for $B_R(x) = \Omega(x, R) \subset \Omega$ and $s > N$*

$$\max_{\Omega(x,R/2)} |u| \leq K\left[\left\{\frac{1}{R^N}\int_{\Omega(x,R)} u^2\, dx\right\}^{1/2} + \sum_{i=1}^{N} \|f_i\|_{L^s(\Omega(x,R))} R^{1-(N/s)}\right]. \tag{C.27}$$

Theorem C.14. *Let $x \in \partial\Omega$ and let $u(x)$ be a solution of (C.26), vanishing on $\partial\Omega \cap B_R(x)$; then the inequality (C.27) holds.*

Proof of Theorems C.13 *and* C.14. Let v be the solution in $H_0^1(\Omega(x, R))$ of Eqn. (C.26) and set

$$u = v + w;$$

then w is a solution in $H^1(\Omega(x, R))$ of the equation $Lu = 0$, which vanishes on $\partial\Omega \cap B_R(x)$ in the case of Theorem C.14.

By Theorem C.4, Corollary C.5, and Theorem C.6 we have

$$\begin{aligned} \max_{\Omega(x, R/2)} |w| &\leq K\left\{\frac{1}{R^N}\int_{\Omega(x, R)} w^2\, dx\right\}^{1/2} \\ &\leq K\left[\left\{\frac{1}{R^N}\int_{\Omega(x, R)} u^2\, dx\right\}^{1/2} + \left\{\frac{1}{R^N}\int_{\Omega(x, R)} v^2\, dx\right\}^{1/2}\right]. \end{aligned}$$

But, from Theorem B.2

$$\begin{aligned} \left(\int_{\Omega(x, R)} v^2\, dx\right)^{1/2} &\leq \max_{\Omega(x, R)} |v| [\text{meas}\, \Omega(x, R)]^{1/2} \\ &\leq K \sum_{i=1}^{N} \|f_i\|_{L^s(\Omega(x, R))} R^{1-(N/S)} [\text{meas}\, \Omega(x, R)]^{1/2}. \end{aligned}$$

Therefore

$$\begin{aligned} \max_{\Omega(x, R/2)} |u| &\leq \max_{\Omega(x, R/2)} |w| + \max_{\Omega(x, R/2)} |v| \\ &\leq K\left[\left(\frac{1}{R^N}\int_{\Omega(x, R)} u^2\, dx\right)^{1/2} + \sum_{i=1}^{N} \|f_i\|_{L^s(\Omega(x, R))} R^{1-(N/s)}\right]. \end{aligned}$$

Theorem C.13 and Theorem C.14 are proved. Q.E.D.

Appendix D. Hölder Continuity of the Solutions

In this section we assume that the open set Ω is "of class S" (see Definition C.1).

We prove the following lemmas.

Lemma D.1. *Let* $u(x) \in H^1(\Omega(x, R))$ *and denote by* $A(k, R)$ *the set*

$A(k, R) = \{y \in \Omega(x, R) : u(y) \geq k\}$; *if there exist two constants* k_0 *and* θ, *with* $0 \leq \theta < 1$, *such that*

$$\text{meas } A(k_0, R) < \theta \text{ meas } \Omega(x, R)$$

then, for $h > k > k_0$, *we have*

$$(h - k)[\text{meas } A(h, R)]^{(N-1)/N} \leq K \int_{A(k,R)-A(h,R)} |u_x(t)| \, dt \tag{D.1}$$

where $K = K(\theta, N)$.

Moreover, making use of Cauchy's inequality, we get

$$(h - k)^2[\text{meas } A(h, R)]^{(2N-2)/N}$$
$$\leq \int_{A(k,R)} |u_x(t)|^2 \, dt \text{ meas}\{A(k, R) - A(h, R)\}. \tag{D.2}$$

Proof. If $k > k_0$, then $v = \min\{u, h\} - \min\{u, k\} \in H^1(B_R(x))$ and $\text{meas}\{x \in B_R(x) : v(x) = 0\} = \text{meas}\{B_R(x) - A(k, R)\} > (1 - \theta) \text{ meas } B_R(x)$. From the formula of Lemma A.13 we have

$$\text{meas } A(h, R) \leq \left[K/(h - k) \int_{A(k,R)-A(h,R)} |u_x(t)| \, dt \right]^{N/(N-1)}$$

and (D.1) and (D.2) follow. Q.E.D.

We need the following:

Lemma D.2. *Let* $h \to \varphi(h)$ *be a nonnegative and nonincreasing function defined for* $h \in [k_0, M]$ *such that the inequality*

$$(h - k)^\alpha |\varphi(h)|^\beta \leq C[M - k]^\alpha[\varphi(k) - \varphi(h)] \tag{D.3}$$

for $k < h < M$ *holds, where* α, β, *and* C *are positive constants. Then*

$$\lim_{h \to M} \varphi(h) = 0. \tag{D.4}$$

Moreover, if we set $k_n = M - (M - k_0)/2^n$, *we have*

$$\varphi(k_n) \leq 2^\alpha C \left(\frac{\varphi(k_0)}{n} \right)^{1/\beta}. \tag{D.5}$$

Proof. Since $k_n - k_{n-1} = (M - k_0)/2^n$, $M - k_{n-1} = (M - k_0)/2^{n-1}$, we have from (D.3)

$$|\varphi(k_n)|^\beta \leq C2^\alpha[\varphi(k_{n-1}) - \varphi(k_n)].$$

Adding for $n = 1, \ldots, \nu$ and noting that $\varphi(k_n) \geq \varphi(k_\nu)$, we find

$$\nu|\varphi(k_\nu)|^\beta \leq 2^\alpha C[\varphi(k_0) - \varphi(k_1)]$$

and thus

$$\varphi(k_\nu) \leq (2^\alpha C\varphi(k_0)/\nu)^{1/\beta},$$

which implies (D.4). Q.E.D.

If u is a solution of the equation $Lu = 0$, we set

$$m(r) = \inf_{\Omega(x,r)} u, \qquad M(r) = \sup_{\Omega(x,r)} u.$$

Theorem D.3. *Let $u(x)$ be a local solution of the equation $Lu = 0$. Assume that for $x_0 \in \Omega$*

$$\text{meas } A(k_0, R) \leq \tfrac{1}{2} \text{ meas } \Omega(k_0, R), \tag{D.6}$$

where $k_0 = \frac{1}{2}(M(2R) + m(2R))$; then

$$\lim_{h \to M(2R)} \text{meas } A(h, R) = 0. \tag{D.7}$$

Theorem D.4. *Let $u(x)$ be a solution of the equation $Lu = 0$, vanishing on $\partial\Omega \cap B_R(x_0)$, where Ω is of class S and $x_0 \in \partial\Omega$. Then the same conclusion as Theorem D.3 holds true.*

Proof of Theorems D.3 and D.4. Let $\alpha \in C_0^1(\Omega(x_0, 2R))$, where $\alpha = 1$ in $\Omega(x_0, R)$; then from Lemma C.8 and Remark C.11 with $v = \max(u, k) - k$, we have

$$\int_{A(k,R)} u_x^2 \, dx \leq \frac{C}{R^2} \int_{A(k,2R)} (u - k)^2 \, dx;$$

here k is any number if $x_0 \in \Omega$ and $k \geq 0$ if $x_0 \in \partial\Omega$.

From Lemma D.1 we have that when $h > k$,

$$(h - k)^2[\text{meas } A(h, R)]^{(2N-r)/N}$$
$$\leq C[M(2R) - k]^2 R^{N-2}[\text{meas } A(k, R) - \text{meas } A(h, R)],$$

and thus Lemma D.2 implies (D.7). Q.E.D.

Theorem D.5. *Under the same assumption of Theorems D.3 and D.4, there exist η with $0 < \eta < 1$ and ρ_0 such that for $\rho < \rho_0$*

$$\omega(\rho) \leq \eta\omega(4\rho), \tag{D.8}$$

where $\omega(\rho)$ denotes the oscillation $M(\rho) - m(\rho)$ of the solution u in $\Omega(x_0, \rho)$.

Proof. We can assume that (D.6) is satisfied, indeed, for $x_0 \in \partial\Omega$ if (D.6) does not hold, the solution $v = -u$ satisfies (D.6) since

$$\operatorname{meas}\{x \in \Omega(x_0, R): u(x) \geq k\} + \operatorname{meas}\{x \in \Omega(x_0, R): \\ -u(x) \geq k\} = \operatorname{meas} \Omega(x_0, R).$$

Otherwise, if $x_0 \in \partial\Omega$, we can assume that (D.6) is satisfied since for one of the two solutions u or $-u$ we have $k_0 = \frac{1}{2}(M(2R) + m(2R)) \geq 0$. If $k_0 \geq 0$ for u, the function $v = \max(u, k) - k$ vanishes outside $\partial\Omega \cap B_\rho(x_0)$ and therefore, extended by zero in $B_R(x_0) - \Omega(x_0, R)$, Remark C.11 holds. Making use of Theorem C.12 with k_0 there given by

$$k_\nu = M(2R) - [M(2R) - m(2R)]/2^{\nu+1}$$

with ν large enough so that $k_\nu > k_0$, we have •

$$M(R/2) \leq k_\nu + K[M(2R) - k_\nu][\operatorname{meas} A(k_\nu, R)/R^N]^{(\theta-1)/2}$$

with $\theta > 1$. Because of Theorems D.3 and D.4, we can fix ν in such a way that

$$K[\operatorname{meas} A(k_\nu, R)/R_N]^{(\theta-1)/2} < \tfrac{1}{2},$$

and thus

$$M(R/2) \leq M(2R) - [M(2R) - m(2R)]/2^{\nu+2}.$$

Moreover, noticing that $m(R/2) \geq m(2R)$, we have

$$\omega(R/2) \leq \omega(2R)[1 - (1/2^{\nu+2})]$$

and Theorem D.5 is proved. Q.E.D.

It is known that from Theorem D.5 it follows that the solution is Hölder continuous (Exercise 18).

At this point we may conclude in particular that a local solution of the equation $Lu = 0$ satisfies a Hölder condition in each compact C of Ω; in other words, there exist two constants K and λ with $0 \leq \lambda < 1$, depending on ν, M, N and the compact, such that

$$\operatorname*{osc}_{\Omega(x,\rho)} u(y) \leq K\left(\left[\int_{\Omega(x,R)} u^2\,dx\right]/R^N\right)^{1/2}(\rho/R)^\lambda, \qquad 0 < \rho < R < \operatorname{dist}(C, \partial\Omega). \tag{D.9}$$

Let us now consider the solution of the equation $Lu = \sum_{i=1}^N (f_i)_{x_i}$, vanishing on $\partial\Omega$.

Let $v \in H_0^1(\Omega(x, 8\rho))$ be the solution of the equation $Lv = \sum_{i=1}^N (f_i)_{x_i}$ and set $u = v + w$; w is a solution of the equation $Lw = 0$ vanishing on $\partial\Omega \cap B_{8\rho}(x)$. Since

$$\underset{\Omega(x,\rho)}{\operatorname{osc}} u \leq 2 \max_{\Omega(x,\rho)} |v| + \operatorname{osc} w,$$

we deduce from Theorems D.5 and B.2 that for $f_i \in L^s(\Omega)$ with $s > N$ there exist constants $K > 0$, $\rho_0 > 0$, and η, $0 < \eta < 1$, such that

$$\omega(\rho) \leq \eta\omega(4\rho) + K\rho^{1-(N/s)}, \qquad \rho < \rho_0. \tag{D.10}$$

And from this estimate we may deduce analogously that u is Hölder continuous in $\bar{\Omega}$. Thus we deduce Theorem C.2.

Remark D.6. *Under the same assumptions of Theorem* C.2 *there exist constants* $K > 0$, $\rho_0 > 0$, *and* λ *with* $0 \leq \lambda < 1$ *such that*

$$\int_{\Omega(x,\rho)} u_x^2\, dx \leq K\rho^{n-2+2\lambda}, \qquad \rho < \rho_0. \tag{D.11}$$

Proof. It is enough to remark that $u - u(x_0)$ is still a solution of the same equation in $\Omega(x_0, \rho_0)$ vanishing on $\partial\Omega \cap B_{\rho_0}(x_0)$ and that $|u(x) - u(x_0)| \leq K\rho^\lambda$ for $x \in \Omega(x_0, 2\rho)$ because of Theorem C.2.

Making use of an inequality similar to (C.16) for the inhomogeneous equation and the solution $u(x) - u(x_0)$, we get (D.11). Q.E.D.

The inequality (D.11) implies, because of a theorem of Morrey, that u is Hölder continuous. It is worth noticing that, while in two variables it is possible to prove (D.11) and from this deduce Hölder continuity of u, in more than two variables we had to prove first Hölder continuity of u and from this deduce the validity of (D.11).

Comments and Bibliographical Notes

Theorem 2.1 was first proved in Stampacchia [3]. A different proof was given in Lions and Stampacchia [1]. The special case of a symmetric form had been treated previously. We also mention the papers of Littman *et al.* [1] and Fichera [1] cited in the bibliography.

The first example of a variational inequality given in Section 3 introduces the truncation of an L^2 function. This basic procedure connects the variational approach of boundary value problems to classical potential theory because it leaves the Sobolev spaces $H^{1,s}(\Omega)$ invariant. In addition, the distributional derivatives of the truncated function may be computed explicitly (cf. Theorem A.1 and Corollaries A.5 and A.6).

This fact was well known for the functions which are absolutely continuous in the sense of Tonelli and/or of Morrey but requires a careful proof when the spaces $H^{1,s}(\Omega)$ are defined as completion of smooth functions in a suitable topology. Many results of second order elliptic differential equations in divergence form depend on this result, for instance, the weak form of the maximum principle, proved in Section 5, and those connected with the celebrated theorem of E. De Giorgi on the Hölder continuity of solutions of a second order elliptic differential equation in divergence form.

The results of Section 6 on the obstacle problem make use of the paper of Lions and Stampacchia [1], mentioned above and of that of Lewy and Stampacchia [1]. The theorems of Section 6 depend on this paper. We do not treat the problem of regularity of the solutions, which is postponed to Chapter IV. But in Section 7 we treat the obstacle problem in the one dimensional case.

The appendices to this chapter contain some facts relevant to the content of the book. Besides the results about Sobolev spaces and inequalities, the L^∞ estimates and Hölder continuity of solutions of second order equations with bounded measurable coefficients are proved. The presentation of the latter material is based on Stampacchia [2]. The proof of Lemma A.4 is due to M. Crandall and that of Lemma A.13 to E. Fabes.

Exercises

1. Let $a(u, v)$ be a bilinear coercive form on a real Hilbert space H, i.e., $|a(u, v)| \leq C\|u\| \cdot \|v\|$ and $a(u, v) \geq \alpha\|v\|^2$, for all $u, v \in H$ (see Definition 1.1). Let ρ be a real number such that $0 < \rho < 2\alpha/C^2$. Show that there exists a ϑ, $0 < \vartheta < 1$, such that

$$|(u, v) - \rho a(u, v)| < \vartheta\|u\| \cdot \|v\| \qquad \text{for} \quad u, v \in H.$$

2. Prove Theorem 2.1 by making use of the result of the previous problem (Lions and Stampacchia [1]).

3. Let F be a convex function from H in $\mathbb{R} \cup \{+\infty\}$ which is proper, i.e., not merely the constant $+\infty$ and lower semicontinuous. Then there exists a unique $u \in H$ such that

$$a(u, v - u) + F(v) - F(u) \geq 0 \qquad \text{for all} \quad v \in H.$$

Deduce from this statement Theorem 2.1 by choosing

$$F(v) = \begin{cases} -(f, v) & \text{for} \quad v \in \mathbb{K} \\ +\infty & \text{for} \quad v \notin \mathbb{K}, \end{cases}$$

where $f \in H'$ and $\mathbb{K}$ is a closed convex subset of H.

4. Let $E \subset B_r$ be a subset of the ball B_r of $\mathbb{R}^N$. The points

$$\xi = x + (y - x)/\|y - x\|$$

belong to the unit sphere Σ: $\|\xi - x\| = 1$. The $N - 1$ dimensional measure of the set $\{\xi \in \Sigma : y \in E\}$ is called the angle under which the set E is seen from x in B_r and is denoted by $\operatorname{ang}\{x, E\}$. Assume that there exists a constant $m > 0$ such that

$$\operatorname{ang}\{x, E\} \geq m \qquad \text{for all} \quad x \in B_r.$$

Let $u \in H^{1,1}(B_r)$ vanish on E. Show that, for each $\sigma > 0$,

$$\operatorname{meas}\{x \in B_r : |u(x)| > \sigma\} \leq \left[(C/\sigma) \int_{B_r} |u_x(t)|\, dt \right]^{N/(N-1)},$$

where C depends only on E. (*Hint*: Adapt the proof of Lemma A.7 to this situation. Let T be a bi-Lipschitz mapping of Ω onto B_r, i.e., $T : \Omega \to B_r$ and

$$0 < \alpha \leq \|T_x - T_y\|/\|x - y\| < \beta < +\infty \qquad \text{for all} \quad x, y \in \Omega.)$$

Show that if $u \in H^{1,1}(\Omega)$ vanishes on $T(E)$ with E satisfying the condition above, then the inequality

$$\int_\Omega u^2\, dx \leq C \int_\Omega u_x^2\, dx,$$

C constant, holds true.

5. Let Ω be an open bounded set of $\mathbb{R}^N$ and E a subset of Ω. Let u be a function of $H^{1,1}(\Omega)$ vanishing on E. How must the set E be chosen in order that there exist a constant β such that

$$\operatorname{meas}\{x \in \Omega : |u(x)| > \sigma\} \leq \left[\frac{\beta}{\sigma} \int_\Omega |u_x(t)|\, dt \right]^{N/(N-1)} ?$$

(see Lemma A.15.) Is the condition of Lemma A.15 necessary?

6. Use Exercise 4 in order to study in the framework of Section 4 the existence of the solution of the mixed problem

$$\begin{aligned} -\Delta u &= f, \\ u &= 0 \qquad \text{on} \quad \partial_1 \Omega, \\ du/dn &= 0 \qquad \text{on} \quad \partial_2 \Omega, \end{aligned}$$

where $f \in H^{-1}(\Omega)$ and $\partial\Omega = \partial_1\Omega \cup \partial_2\Omega$ with $\partial_1\Omega \cap \partial_2\Omega = \varnothing$.

7. Prove Lemma A.12.

8. Study the variational inequality

$$u \in \mathbb{K}: \quad \int_\Omega u_{x_i}(v-u)_{x_i}\,dx \geq \langle f, v-u \rangle \qquad \text{for all} \quad v \in \mathbb{K},$$

where

$$\mathbb{K} = \{v \in H_0^1(\Omega) : \psi_1 \leq u \leq \psi_2 \text{ in } \Omega\},$$

$\psi_1, \psi_2 \in H^1(\Omega)$, and $\psi_1 \leq 0 \leq \psi_2$ on $\partial\Omega$.

9. Let $\Omega \subset R^N$ be a domain, $f \in H^{-1}(\Omega)$, and $\psi \in H^1(\Omega)$. Set $\mathbb{K} = \{v \in H_0^1(\Omega) : v \geq \psi \text{ on } \Omega\}$ and consider

$$u \in \mathbb{K}: \quad \int_\Omega u_{x_i}(v-u)_{x_i} \geq \langle f, v-u \rangle \qquad \text{for} \quad v \in \mathbb{K}. \tag{$*$}$$

Suppose that $f \in L^2(\Omega)$ and $u \in H^2(\Omega)$. Show that u is a solution of $(*)$ if and only if

$$-\Delta u - f \geq 0, \qquad u - \psi \geq 0, \qquad \text{and} \qquad -(\Delta u + f)(u - \psi) = 0 \quad \text{a.e. in} \quad \Omega. \tag{$**$}$$

The form $(**)$ is called the "complementarity form" by analogy with Complementarity Problem I.5.4.

10. Let Ω be a bounded domain with smooth boundary $\partial\Omega$ and suppose given $f \in H^{-1}(\Omega)$ and $g \in L^2(\partial\Omega)$. For $\lambda \in \mathbb{R}$, study the existence and uniqueness of solutions of the variational inequality

$$u \in \mathbb{K}: \quad \int_\Omega [u_{x_i}(v-u)_{x_i} + \lambda(v-u)]\,dx \geq \langle f, v-u \rangle + \int_{\partial\Omega} g(v-u)\,dx \qquad \text{for all} \quad v \in \mathbb{K},$$

where $\mathbb{K} = \{v \in H^1(\Omega) : v \geq 0 \text{ on } \partial\Omega\}$. Write the solution in the "complementarity form."

11. Again let Ω be a bounded domain with smooth boundary $\partial\Omega$ and suppose that $f \in H^{-1}(\Omega)$. Let

$$V = \{v \in H_0^1(\Omega) : \Delta v \in H_0^1(\Omega)\} \subset H^3(\Omega),$$

$$\mathbb{K} = \{v \in V : -\Delta v \geq 0 \text{ a.e. in } \Omega\},$$

and

$$a(u, v) = \int_\Omega (\Delta u_{x_i}\,\Delta v_{x_i} + u_{x_i} v_{x_i})\,dx.$$

Show there exists a unique solution

$$u \in \mathbb{K}: \quad a(u, v - u) \geq \langle f, v - u \rangle \qquad \text{for} \quad v \in \mathbb{K}.$$

Write this problem in "complementarity form." Show that there is a constant $c > 0$ such that

$$\|u_1 - u_2\|_{H^3(\Omega)} \leq c\|f_1 - f_2\|_{H^{-1}(\Omega)},$$

where u_i is the solution of the variational inequality corresponding to f_i.

12. Show that a solution to the Neumann problem

$$-\Delta u = f \quad \text{in} \quad \Omega, \qquad \partial u/\partial \nu = g \quad \text{on} \quad \partial\Omega$$

for $f \in L^2(\Omega)$ and $g \in L^2(\partial\Omega)$ exists provided the compatibility condition

$$\int_\Omega f\, dx + \int_{\partial\Omega} g\, d\sigma = 0$$

is fulfilled. Find a compatibility condition for this problem when $f \in H^{-1}(\Omega)$ and $g \in L^2(\Omega)$.

13. Let Ω be a bounded domain of $\mathbb{R}^N$ with smooth boundary $\partial\Omega$ and let $\varphi^1, \varphi^2 \in H^1(\Omega)$, satisfy $\varphi^1 \geq \varphi^2$ on $\partial\Omega$. Let $\mathbb{K}$ be the closed convex set of pairs $v = (v^1, v^2) \in H^1(\Omega) \times H^1(\Omega)$ such that $v^1 \geq v^2$ in Ω and $v^1 = \varphi^1$ and $v^2 = \varphi^2$ on $\partial\Omega$. Let $(f_1, f_2) \in H^{-1}(\Omega) \times H^{-1}(\Omega)$.

Consider the bilinear form

$$a(v, \zeta) = \int_\Omega (v^1_{x_i}\zeta^1_{x_i} + v^2_{x_i}\zeta^2_{x_i})\, dx + \int_\Omega (\lambda v^1\zeta^1 + \mu v^2\zeta^2)\, dx$$

where $\lambda, \mu \in \mathbb{R}$. Study the variational inequality

$$u \in \mathbb{K}: \quad a(u, v - u) \geq \langle f_1, v^1 - u^1 \rangle + \langle f_2, v^2 - u^2 \rangle \qquad \text{for all} \quad v \in \mathbb{K},$$

imposing restrictions on λ, μ if necessary.

14. Let Ω be a bounded domain of $\mathbb{R}^N$ with smooth boundary $\partial\Omega$ and let Γ, Γ' be open nonempty submanifolds of $\partial\Omega$ satisfying (i) $\Gamma \cap \Gamma' = \emptyset$, (ii) $\Gamma \cup \Gamma' = \partial\Omega$, (iii) $\overline{\Gamma} \cap \overline{\Gamma'}$ is a smooth $(N - 2)$ dimensional manifold.

Given $f_1, f_2 \in H^{-1}(\Omega)$, consider the problem: To find $u_1, u_2 \in H^1(\Omega)$ such that

$$\begin{aligned} -\Delta u_i &= f_i, & i &= 1, 2, & &\text{in} \quad \Omega, \\ u_1 - u_2 &= 0, & (du_1/dn) &+ (du_2/dn) = 0, & &\text{on} \quad \Gamma, \\ u_1 &= u_2 = 0 & & & &\text{on} \quad \Gamma', \end{aligned}$$

where n is the normal to Γ.

Find a weak formulation for this problem and show that a solution exists.

[*Hint*: Formulate the problem for a bilinear form defined on the Hilbert space

$$H = \{v = (v_1, v_2) : v_i \in H^1(\Omega),\ i = 1, 2;\ v_1 - v_2 = 0 \text{ on } \Gamma,\quad v_1 = v_2 = 0 \text{ on } \Gamma'\}.]$$

15. Let $u \in H^{1,p}(B_r)$, $1 < p < +\infty$, where B_r is the ball of $\mathbb{R}^N$ of radius r. Show that there exists a constant γ such that

$$\int_{B_r} |u - \gamma|^p \, dx \leq C \int_{B_r} u_x^p \, dx,$$

where C is a suitable constant. [*Hint*: We can write

$$u - \gamma = [\max(u, \gamma) - \gamma] + [\min(u, \gamma) - \gamma]$$

for any constant γ. Choose γ in such a way that the measures, where the first and the second term on the right are zero, are $\geq \frac{1}{2}$ meas B_r. Then make use of Lemmas A.15 and A.14.]

16. Prove Corollary A.6 [*Hint*: Assume that $|u(x)| \leq M$ and choose $f_n \in C(-M, M)$ such that $f_n(t) \to \theta'(t)$ for $|t| \leq M$, $t \neq a_1, \ldots, a_M$. Prove first that $f_n(u(x))u_{x_i}(x) \to \theta'(u(x))u_{x_i}(x)$ a.e. in Ω. Now write that (assuming that $\theta(0) = 0$)

$$-\int_\Omega \theta(u(x))\zeta_{x_i}(x)\, dx = -\int_\Omega \lim_{n\to\infty} \int_0^{u(x)} f_n(t)\, dt\, \zeta_{x_i}(x)\, dx.$$

For $f_n(t)$ it is always possible to choose

$$f_n(t) = \frac{1}{2n} \int_{|t-\tau|<1/n} \theta'(\tau)\, d\tau.]$$

17. Suppose that $u_n \in L^s(\Omega)$ and $0 \leq w_n \in L^s(\Omega)$ satisfy

$$u_n \to u \quad \text{a.e. in } \Omega,$$

$$w_n \to w \quad \text{a.e. in } \Omega \text{ and in } L^s(\Omega),$$

and

$$|u_n| \leq w_n \quad \text{a.e. in } \Omega.$$

Then $u_n \to u$ in $L^s(\Omega)$. Use this fact to show that if $u_n \in H^{1,s}(\Omega)$ and $u_n \to u$ in $H^{1,s}(\Omega)$, then $|u_n| \to |u|$ in $H^{1,s}(\Omega)$. As a corollary, $\max(u_n, 0) \to \max(u, 0)$ in $H^{1,s}(\Omega)$. Can you also use this in Exercise 14 or Lemma A.4?

18. Let $\omega(\rho)$, $0 \leq \rho \leq R$, be a continuous function which satisfies

$$0 \leq \omega(\rho) \leq \eta\omega(4\rho) + C(\rho/R)^\alpha, \qquad 0 \leq \rho \leq R,$$

where $C \geq 0$, $0 \leq \eta < 1$, and $\alpha > 0$. Then

$$\omega(\rho) \leq 2 \max(\omega(R), CK)(\rho/R)^{\lambda}, \qquad 0 \leq \rho \leq R/4,$$

where K is a universal constant and

$$\lambda = \min\left(\left|\frac{\log \eta}{\log 4}\right|, \alpha\right).$$

[*Hint*: By induction there is a constant K such that

$$\omega(4^{-n}R) \leq \eta^n \omega(R) + CK\rho^{\alpha}, \qquad n = 1, 2, 3, \ldots$$

if $4^n \leq R/\rho \leq 4^{n+1}$, then $\eta^n \leq \eta^q \leq \eta^{n+1}$ where $q = (\log R/\rho)/\log 4$ and $\omega(\rho) \leq \omega(4^{-n}R)$.]

CHAPTER **III**

Variational Inequalities for Monotone Operators

1. An Abstract Existence Theorem

We have already encountered monotone operators in connection with the uniqueness of the solution to a *variational inequality* in a finite dimensional space. In this chapter, dedicated to the study of variational inequalities in general spaces, the property of monotonicity becomes important for the existence of a solution. In the last chapter we saw that a variational inequality associated to a bilinear form could be solved provided the form was coercive. In fact, the property of coerciveness implies strict monotonicity for a bilinear form, but certainly not vice versa. In general if the convex $\mathbb{K}$ is unbounded, it will be necessary to add hypotheses of coerciveness to achieve the existence of a solution; cf., for example, Corollary 1.8.

Let X be a reflexive Banach space with dual X'. Let $\langle \cdot, \cdot \rangle$ denote a pairing between X' and X. Let $\mathbb{K} \subset X$ be a closed convex set.

Definition 1.1. A mapping $A\colon \mathbb{K} \to X'$ is called *monotone* if

$$\langle Au - Av, u - v \rangle \geq 0 \qquad \text{for all} \quad u, v \in \mathbb{K}. \tag{1.1}$$

The monotone mapping A is called *strictly monotone* if

$$\langle Au - Av, u - v \rangle = 0 \qquad \text{implies} \quad u = v. \tag{1.2}$$

Definition 1.2. The mapping A from $\mathbb{K} \to X'$ *is continuous on finite dimensional subspaces* if for any finite dimensional subspace $M \subset X$ the restriction of A to $\mathbb{K} \cap M$ is weakly continuous, namely, if

$$A : \mathbb{K} \cap M \to X'$$

is weakly continuous.

Definition 1.3. Finally, A is *coercive* on $\mathbb{K}$ if there exists an element $\varphi \in \mathbb{K}$ such that

$$\langle Au - A\varphi, u - \varphi\rangle / \|u - \varphi\| \to +\infty \quad \text{as} \quad \|u\| \to +\infty \qquad \text{for any} \quad u \in \mathbb{K}. \tag{1.3}$$

We now prove

Theorem 1.4. *Let $\mathbb{K}$ be a closed bounded convex subset of $X (\neq \varnothing)$ and let $A : \mathbb{K} \to X'$ be monotone and continuous on finite dimensional subspaces. Then there exists a*

$$u \in \mathbb{K}: \quad \langle Au, v - u\rangle \geq 0 \qquad \textit{for all} \quad v \in \mathbb{K}. \tag{1.4}$$

Note that if A is strictly monotone the solution u to the variational inequality (1.4) is unique.

We first prove a lemma due to Minty [1].

Lemma 1.5. *Let $\mathbb{K}$ be a closed convex set of X and let $A : \mathbb{K} \to X'$ be monotone and continuous on finite dimensional subspaces. Then u satisfies*

$$u \in \mathbb{K}: \quad \langle Au, v - u\rangle \geq 0 \qquad \textit{for all} \quad v \in \mathbb{K} \tag{1.5}$$

if and only if it satisfies

$$u \in \mathbb{K}: \quad \langle Av, v - u\rangle \geq 0 \qquad \textit{for all} \quad v \in \mathbb{K}. \tag{1.6}$$

Proof. First we show that (1.5) implies (1.6). By monotonicity of A,

$$0 \leq \langle Av - Au, v - u\rangle = \langle Av, v - u\rangle - \langle Au, v - u\rangle \qquad \text{for} \quad u, v \in \mathbb{K}.$$

Thus

$$u \in \mathbb{K}: \quad 0 \leq \langle Au, v - u\rangle \leq \langle Av, v - u\rangle \qquad \text{for all} \quad v \in \mathbb{K}.$$

Now we show that (1.6) implies (1.5). Let $w \in \mathbb{K}$, and set, for $0 \leq t \leq 1$, $v = u + t(w - u) \in \mathbb{K}$ since $\mathbb{K}$ is convex. Hence by (1.6) for $t > 0$,

$$\langle A(u + t(w - u)), t(w - u)\rangle \geq 0,$$

or

$$\langle A(u + t(w - u)), w - u \rangle \geq 0 \qquad \text{for all} \quad w \in \mathbb{K}.$$

Since A is weakly continuous on the intersection of $\mathbb{K}$ with the finite dimensional subspace spanned by u and w, we may allow $t \to 0$ to obtain

$$u \in \mathbb{K}: \quad \langle Au, w - u \rangle \geq 0 \qquad \text{for any} \quad w \in \mathbb{K}. \quad \text{Q.E.D.}$$

Proof of Theorem 1.4. To begin, let $M \subset X$ be a finite dimensional subspace of X of dimension, say $N < \infty$. We may assume without loss of generality that $0 \in \mathbb{K}$. Define

$$j: M \to X$$

to be the injection map and

$$j': X' \to M'$$

to be its dual. The pairing between M' and M, $\langle \cdot, \cdot \rangle_M$, is chosen, of course, so that

$$\langle f, jx \rangle = \langle j'f, x \rangle_M \qquad \text{whenever} \quad x \in M, \quad f \in X'.$$

We set $\mathbb{K}_M = \mathbb{K} \cap M \equiv \mathbb{K} \cap jM$ and consider the mapping $j'Aj: \mathbb{K}_M \to M'$. Now $\mathbb{K}_M$ is a compact convex set of M and $j'Aj$ is continuous by hypothesis from $\mathbb{K}$ into M'. Hence there exists an element $u_M \in \mathbb{K}_M$ such that

$$\langle j'Aju_M, v - u_M \rangle_M \geq 0 \qquad \text{for all} \quad v \in \mathbb{K}_M,$$

or what is the same, since $ju_M = u_M$ and $jv_M = v$,

$$\langle Au_M, v - u_M \rangle \geq 0 \qquad \text{for all} \quad v \in \mathbb{K}_M.$$

By Minty's lemma 1.5,

$$\langle Av, v - u_M \rangle \geq 0 \qquad \text{for all} \quad v \in \mathbb{K}_M. \tag{1.7}$$

In view of this we define

$$S(v) = \{u \in \mathbb{K} : \langle Av, v - u \rangle \geq 0\}.$$

Obviously, $S(v)$ is weakly closed for each $v \in \mathbb{K}$. Moreover, since $\mathbb{K}$ is bounded, $\mathbb{K}$ is weakly compact. Consequently $\bigcap_{v \in \mathbb{K}} S(v)$, a closed subset of $\mathbb{K}$, is weakly compact. To conclude that it is nonempty, we employ the finite intersection property. Let $\{v_1, \ldots, v_m\} \subset \mathbb{K}$. We claim that

$$S(v_1) \cap S(v_2) \cap \cdots \cap S(v_m) \neq \varnothing. \tag{1.8}$$

Let M be the finite dimensional subspace of X spanned by $\{v_1, \ldots, v_m\}$ and define $\mathbb{K}_M = \mathbb{K} \cap M$ as before. According to the argument given earlier, there is an element $u_M \in \mathbb{K}_M$ such that

$$\langle Av, v - u_M \rangle \geq 0 \qquad \text{for all} \quad v \in \mathbb{K}_M,$$

viz., (1.7). In particular,

$$\langle Av_i, v_i - u_M \rangle \geq 0 \qquad \text{for} \quad i = 1, \ldots, m,$$

so $u_M \in S(v_i)$, $i = 1, \ldots, m$. Hence for any finite collection $v_1, \ldots, v_m$, (1.8) holds. Therefore there exists an element

$$u \in \bigcap_{v \in \mathbb{K}} S(v),$$

which means that

$$u \in \mathbb{K}: \quad \langle Av, v - u \rangle \geq 0 \qquad \text{for all} \quad v \in \mathbb{K}.$$

By Lemma 1.5, once again

$$u \in \mathbb{K}: \quad \langle Au, v - u \rangle \geq 0 \qquad \text{for all} \quad v \in \mathbb{K}. \quad \text{Q.E.D.}$$

The solution to (1.4) need not be unique, but it is easy to verify, using Lemma 1.5, that its set of solutions is a closed convex subset of $\mathbb{K}$.

Corollary 1.6. *Let H be a Hilbert space and $\mathbb{K} \subset H$ a closed bounded convex set ($\neq \emptyset$). Suppose that $F: \mathbb{K} \to \mathbb{K}$ is nonexpansive. Then F posseses a nonempty closed convex subset $K. \subset \mathbb{K}$ of fixed points.*

The definition of nonexpansive mapping is given in Definition 1.1. of Chapter I.

Proof. The proof of this Corollary is very simple. It suffices to observe that we may take $H' = H$ and pairing $(\cdot, \cdot)$ with the scalar product of H. Now if F is nonexpansive, $I - F$ is monotone, so we may apply Theorem 1.4. Any solution to the variational inequality for $I - F$ is a fixed point for F. Q.E.D.

Theorem 4.2 and Corollary 4.3 of Chapter I have analogs here.

Theorem 1.7. *Let $\mathbb{K}$ be a closed convex subset of X and let $A: \mathbb{K} \to X'$ be monotone and continuous on finite dimensional subspaces. A necessary and sufficient condition that there exists a solution to the variational inequality*

$$u \in \mathbb{K}: \quad \langle Au, v - u \rangle \geq 0 \qquad \textit{for any} \quad v \in \mathbb{K}$$

is that there exist an $R > 0$ such that at least one solution of the variational inequality

$$u_R \in \mathbb{K}_R: \quad \langle Au_R, v - u_R \rangle \geq 0 \qquad \textit{for any} \quad v \in \mathbb{K}_R,$$
$$\mathbb{K}_R = \mathbb{K} \cap \{v : \|v\| \leq R\},$$

satisfies the inequality

$$\|u_R\| < R.$$

We also state

Corollary 1.8. *Let $\mathbb{K} \subset X$ be a closed convex set ($\neq \emptyset$) and $A: \mathbb{K} \to X'$ be monotone, coercive and, continuous on finite dimensional subspaces. Then there exists*

$$u \in \mathbb{K}: \quad \langle Au, v - u \rangle \geq 0 \qquad \textit{for any} \quad v \in \mathbb{K}.$$

The proofs of Theorem 1.7 and Corollary 1.8 are identical to those for the case of finite dimension, so are omitted.

2. Noncoercive Operators

In this section we shall consider some cases when the assumption of coerciveness is relaxed. We confine ourselves to the cases of a linear operator in a Hilbert space, for simplicity. The bilinear form associated to it is assumed only to be nonnegative. This leads to a method which is called elliptic regularization. We explain it in the next two theorems.

Then we consider the case when the form is semicoercive, for which we prove an existence theorem provided the right-hand side, i.e., the inhomogeneous term, of the variational inequality satisfies certain conditions.

Consider the following case: when $X = H$, a Hilbert space, $(\cdot, \cdot)$ denotes the inner product in H, and $\langle \cdot, \cdot \rangle$ denotes the pairing between H and H', let $f \in H'$. Let $\mathbb{K}$ be a closed cone in H and let $f \in H'$.

Let $a(u, v)$ be a bilinear form on $H \times H$ such that $a(v, v) \geq 0$. Define the operator $A: H \to H'$ by

$$\langle Au, v \rangle = a(u, v).$$

From Theorem 1.4 it follows that there is a solution to the variational inequality

$$u_0 \in \mathbb{K}: \quad \langle Au_0, v - u_0 \rangle \geq \langle f, v - u_0 \rangle \qquad \text{for all} \quad v \in \mathbb{K},$$

or equivalently

$$u_0 \in \mathbb{K}: \quad a(u_0, v - u_0) \geq \langle f, v - u_0 \rangle \qquad \text{for all} \quad v \in \mathbb{K},$$

provided $\mathbb{K}$ is bounded or $\mathbb{K}$ in unbounded but $A - f$ satisfies the condition of Theorem 1.7.

However, if neither of these conditions is satisfied, such a solution $u_0 \in \mathbb{K}$ may not exist. In order to recognize whether such a solution exists the following theorem is useful. Its proof makes use of a procedure that is called elliptic regularization.

Let us define

$$\chi = \{u_0 \in \mathbb{K} : \langle Au_0, v - u_0 \rangle \geq \langle f, v - u_0 \rangle \text{ for all } v \in \mathbb{K}\}.$$

We know that χ is a closed, convex set. It should be noted, however, that χ may be empty.

Now let $\beta(u, v)$ be a coercive bilinear form on $H \times H$ and let $g \in H'$. For each $\varepsilon > 0$ we consider the form

$$a(u, v) + \varepsilon\beta(u, v).$$

Clearly this form is bilinear and coercive on H, so in view of Theorem 2.1 of Chapter II, there exists a unique solution u_ε of the variational inequality

$$u_\varepsilon \in \mathbb{K}: \quad a(u_\varepsilon, v - u_\varepsilon) + \varepsilon\beta(u_\varepsilon, v - u_\varepsilon) \geq \langle f + \varepsilon g, v - u_\varepsilon \rangle \qquad \text{for all} \quad v \in \mathbb{K}.$$

We have the following theorem.

Theorem 2.1. *Using the notation given above, $\chi \neq \varnothing$ if and only if there exists a constant L, independent of ε, such that*

$$\|u_\varepsilon\| \leq L.$$

Proof. Assume first that $\chi \neq \varnothing$. Since χ is a closed convex subset of $\mathbb{K} \subset H$, there exists a unique solution of the variational inequality

$$u_0 \in \chi: \quad \beta(u_0, v - u_0) \geq \langle g, v - u_0 \rangle \qquad \text{for all} \quad v \in \chi.$$

Moreover, since $u_0 \in \chi$, we have

$$a(u_0, v - u_0) \geq \langle f, v - u_0 \rangle \qquad \text{for all} \quad v \in \mathbb{K}.$$

Setting $v = u_\varepsilon$ in this variational inequality and $v = u_0$ in the variational inequality defining u_ε and adding the two inequalities, we get

$$a(u_0, u_\varepsilon - u_0) + a(u_\varepsilon, u_0 - u_\varepsilon) + \varepsilon\beta(u_\varepsilon, u_0 - u_\varepsilon) \geq \varepsilon\langle g, u_0 - u_\varepsilon \rangle.$$

But $a(u_0, u_\varepsilon - u_0) + a(u_\varepsilon, u_0 - u_\varepsilon) = -a(u_\varepsilon - u_0, u_\varepsilon - u_0) \le 0$. Therefore

$$\beta(u_\varepsilon, u_0 - u_\varepsilon) \ge \langle g, u_0 - u_\varepsilon \rangle$$

and by coerciveness of $\beta(\cdot, \cdot)$,

$$\begin{aligned}\beta_0 \|u_\varepsilon\|^2 &\le \beta(u_\varepsilon, u_\varepsilon) \le \beta(u_0, u_\varepsilon) + \langle g, u_0 - u_\varepsilon \rangle \\ &\le C_1 \|u_\varepsilon\| \cdot \|u_0\| + C_2(\|u_\varepsilon\| + \|u_0\|) \le C(1 + \|u_\varepsilon\|),\end{aligned}$$

where $C = C(\|u_0\|, g, \beta)$. Hence, since $C\|u_\varepsilon\| \le (\beta_0/2)\|u_\varepsilon\|^2 + [C^2/(2\beta_0)]$,

$$\|u_\varepsilon\| \le L = 2 + (C/\beta_0),$$

where L is independent of ε.

Assume now that $\|u_\varepsilon\| \le L$, independently of ε. Then there is a subsequence u_η of u_ε such that $u_\eta \to w$ weakly in H. Thus $w \in \mathbb{K}$ and we wish to show that

$$w \in \chi.$$

Using Lemma 1.5 we have

$$a(v, v - u_\eta) + \eta\beta(v, v - u_\eta) \ge \langle f, v - u_\eta \rangle + \eta\langle g, v - u_\eta \rangle.$$

Letting $\eta \to 0$, since $u_\eta \to w$ weakly, we have

$$a(v, v - w) \ge \langle f, v - w \rangle \qquad \text{for all} \quad v \in \mathbb{K}$$

and using Lemma 1.5 again, we get

$$a(w, v - w) \ge \langle f, v - w \rangle \qquad \text{for all} \quad v \in \mathbb{K}$$

and therefore $w \in \chi$. Q.E.D.

We have an additional theorem.

Theorem 2.2. *With the same assumption of Theorem 2.1, if the sequence u_ε is uniformly bounded, then $u_\varepsilon \to u_0$ as $\varepsilon \to 0$ strongly in H and*

$$u_0 \in \chi: \quad \beta(u_0, v - u_0) \ge \langle g, v - u_0 \rangle \qquad \textit{for all} \quad v \in \chi.$$

Proof. Since $\|u_\varepsilon\| \le L$ (independently of ε), then $\chi = \varnothing$ from the previous theorem (and vice versa). Thus there exists a unique solution of the variational inequality

$$u_0 \in \chi: \quad \beta(u_0, v - u_0) \ge \langle g, v - u_0 \rangle \qquad \text{for all} \quad v \in \chi.$$

Since $\|u_\varepsilon\| < L$ there exists a subsequence u_η of u_ε such that $u_\eta \to w$ weakly in H.

In the first part of the proof of previous theorem we have proved that

$$\beta(u_\eta, u_0 - u_\eta) \geq \langle g, u_0 - u_\eta \rangle.$$

Since $\beta(v, v)$ is lower semicontinuous we have

$$\beta(w, w) \leq \liminf_{\eta \to 0} \beta(u_\eta, u_\eta)$$

and therefore

$$\beta(w, u_0 - w) \geq \langle g, u_0 - w \rangle.$$

Next we show that $u_\varepsilon \to u_0$ strongly in H, proving in succession that

(i) $w = u_0$,
(ii) $u_\varepsilon \to u_0$ weakly in H, and
(iii) $u_\varepsilon \to u_0$ strongly in H.

Since $u_0 \in \chi$ and $\beta(u_0, v - u_0) \geq \langle g, v - u_0 \rangle$ for all $v \in \chi$ and $w \in \chi$, we have

$$\beta(u_0, w - u_0) \geq \langle g, w - u_0 \rangle.$$

Then from

$$\beta(w - u_0, w - u_0) \leq 0$$

and from coerciveness of β

$$\|w - u_0\| = 0;$$

hence $w = u_0$.

Since u_0 is the unique limit of any weakly convergent subsequence of u_ε, we have $u_\varepsilon \to u_0$ weakly in H.

Therefore it remain to be shown that the convergence is strong.

But

$$\beta_0 \|u_\varepsilon - u_0\|^2 \leq \beta(u_\varepsilon - u_0, u_\varepsilon - u_0) = \beta(u_\varepsilon, u_\varepsilon - u_0) - \beta(u_0, u_\varepsilon - u_0).$$

Now as $\varepsilon \to 0$, $\beta(u_0, u_\varepsilon - u_0) \to 0$. On the other hand,

$$\beta(u_\varepsilon, u_\varepsilon - u_0) \leq \langle g, u_\varepsilon - u_0 \rangle \to 0 \qquad \text{as} \quad \varepsilon \to 0.$$

Hence $u_\varepsilon \to u_0$ strongly in H. Q.E.D.

In the remainder of this section, we apply Theorem 1.7 to show the existence of a solution to a variational inequality for a certain type of non-coercive bilinear form on a Hilbert space H. The result may be used to solve an obstacle problem with Neumann boundary conditions (Lions and Stampacchia [1]) (cf. Exercise 12) or the problem in Fichera [1].

We consider a Hilbert space H with norm $\|\cdot\|$ and pairing $\langle\cdot, \cdot\rangle$ and assume that $\|\cdot\|$ is equivalent to $p_0(\cdot) + p_1(\cdot)$, where

$$p_0(\cdot) \text{ is a norm under which } H \text{ is a pre-Hilbert space} \tag{2.1,i}$$

and

$$p_1(\cdot) \text{ is a seminorm on } H. \tag{2.1,ii}$$

Suppose that

$$M = \{v \in H : p_1(v) = 0\} \tag{2.1,iii}$$

is finite dimensional and that there exists $c_1 > 0$ such that

$$\inf_{\zeta \in M} p_0(v - \zeta) \leq c_1 p_1(v) \qquad \text{for any} \quad v \in M. \tag{2.1,iv}$$

Suppose that $a(u, v)$ is a continuous bilinear form on $H \times H$ such that

$$a(v, v) \geq \alpha p_1(v)^2 \qquad \text{for} \quad v \in H \quad \text{and some fixed} \quad \alpha > 0. \tag{2.1,v}$$

In particular, note that $a(v, v) = 0$ implies that $v \in M$ but not that $v = 0$. Let $\mathbb{K} \subset H$ be a closed convex set with $0 \in \mathbb{K}$. Finally, let $f \in H'$ admit the representation $f = f_0 + f_1$ with

$$|\langle f_1, v\rangle| \leq c_2 p_1(v) \qquad \text{for all} \quad v \in H \tag{2.1,vi}$$

and

$$\langle f_0, \zeta\rangle < 0 \qquad \text{for} \quad \zeta \in M \cap \mathbb{K}, \quad \zeta \neq 0. \tag{2.1,vii}$$

Theorem 2.3. *Assume the hypotheses* (2.1). *Then there exists*

$$u \in \mathbb{K}: \quad a(u, v - u) \geq \langle f, v - u\rangle \qquad \textit{for all} \quad v \in \mathbb{K}. \tag{2.2}$$

Proof. Let $\mathbb{K}_R = \mathbb{K} \cap \{v : \|u\| \leq R\}$. Since $\mathbb{K}_R$ is bounded, we know from Theorem 1.4 that there exists a $u_R \in \mathbb{K}_R$ such that

$$a(u_R, v - u_R) \geq \langle f, v - u_R\rangle \qquad \text{for} \qquad v \in \mathbb{K}_R. \tag{2.3}$$

By Theorem 1.4, there exists a solution to (2.2) if and only if $\|u_R\| < R$ for some R. Assume that (2.2) does not admit a solution. We shall show that this leads to a contradiction.

In this circumstance the solution u_R to (2.3) satisfies $\|u_R\| = R$. Let $w_R = (1/R)u_R$, $R \geq 1$. Then $\|w_R\| = 1$. Moreover, since $\mathbb{K}$ is convex, $u_R \in \mathbb{K}$

and $0 \in \mathbb{K}$, we see that $w_R \in \mathbb{K}$. Once again, since $0 \in \mathbb{K}_R \subset \mathbb{K}$, we may take $v = 0$ in (2.3) so by (2.1,v)

$$\alpha p_1(u_R)^2 \leq a(u_R, u_R) \leq \langle f, u_R \rangle \leq \|f\|_{H'} \|u_R\| = \|f\|_{H'} R. \tag{2.4}$$

Therefore

$$p_1(u_R) = O(\sqrt{R}),$$

so

$$p_1(w_R) = O(1/\sqrt{R}), \qquad \text{or} \qquad p_1(w_R) \to 0 \qquad \text{as} \quad R \to +\infty.$$

Since $\|w_R\| = 1$, $R \geq 1$, there is a subsequence, again called w_R, and an element $w \in H$ such that $w_R \to w$ weakly in H. Because $\mathbb{K}$ is closed, $w \in \mathbb{K}$. In addition, p_1 is lower semi-continuous, so $p_1(w) = 0$. Therefore $w \in \mathbb{K} \cap M$.

Consider the projection $P: H \to M$ given by $Pv = \eta$, i.e.,

$$p_0(v - \eta) = \inf_{\zeta \in M} p_0(v - \zeta).$$

We know η exists because M is finite dimensional. We prove at this point that $v_n \to v$ weakly in H implies $Pv_n \to Pv$ (strongly) in M.

At this point, let $(\cdot, \cdot)_M$ and $(\cdot, \cdot)_{M^\perp}$ denote the inner products on M and its perpendicular complement $M^\perp$, respectively. Observe that we may take

$$(u, v)_M = \tfrac{1}{4}(p_0(u + v) - p_0(u - v))$$

and the inner product $(\cdot, \cdot)$ on all of H to be

$$(u, v) = (u, v)_M + (u, v)_{M^\perp}.$$

In particular, if $\zeta \in M$, then $(v, \zeta) = (v, \zeta)_M$. Since

$$(Pv, \zeta) = (v, \zeta) \qquad \text{for all} \quad \zeta \in M$$

and

$$(Pv_n, \zeta) = (v_n, \zeta) \qquad \text{for all} \quad \zeta \in M,$$

$$\|Pv_n\| \leq \|v_n\| \leq C \qquad \text{for all} \quad n$$

for some constant $C > 0$. Now since M is finite dimensional there is a subsequence of v_n, still called v_n, for which $Pv_n \to \vartheta \in M$ for some element ϑ. But by the weak convergence of v_n to v in H, $(Pv, \zeta) = (\vartheta, \zeta)$ for all $\zeta \in M$. Therefore $\vartheta = Pv$.

Note that there is a constant $c_3 > 0$ such that $p_0(Pw_R) \geq c_3 > 0$ for R

sufficiently large. Otherwise there is a sequence of R for which $p_0(Pw_R) \to 0$ as $R \to \infty$. But

$$p_0(w_R) \le p_0(w_R - Pw_R) + p_0(Pw_R) \le c_1 p_1(w_R) + p_0(Pw_R) \to 0$$

since $p_1(w_R) = O(1/\sqrt{R})$. Since the norm $\|\cdot\|$ on H is equivalent to $p_0(\cdot) + p_1(\cdot)$, $\|w_R\| \to 0$. This contradicts $\|w_R\| = 1$. Finally, since $w \in M$, we have $Pw = w$ and hence

$$p_0(w) \ge c_3 > 0. \tag{2.5}$$

If $M \cap \mathbb{K} = \{0\}$, then $w = 0$, which contradicts (2.5), so $M \cap \mathbb{K} \ne \{0\}$. Now $\langle f_0, w \rangle < 0$ since $0 \ne w \in M \cap \mathbb{K}$. Thus

$$-\langle f_0, w \rangle = 2\beta > 0 \qquad \text{for some} \quad \beta > 0.$$

Now $Pw_R \to Pw = w$ strongly in M by the foregoing and hence there is an $R_0 > 0$ such that

$$-\langle f_0, Pw_R \rangle \ge \beta > 0 \qquad \text{for} \quad R \ge R_0.$$

Employing (2.4),

$$\begin{aligned} \alpha p_1(u_R)^2 &\le \langle f, u_R \rangle = \langle f_0, u_R - Pu_R \rangle + \langle f_0, Pu_R \rangle + \langle f_1, u_R \rangle \\ &\le \|f_0\|_{H'} \cdot \|u_R - Pu_R\| + R\langle f_0, Pw_R \rangle + c_2 p_1(u_R). \end{aligned}$$

Therefore

$$\alpha p_1(u_R)^2 - R\langle f_0, Pw_R \rangle \le c(p_0(u_R - Pu_R) + p_1(u_R)) \le c' p_1(u_R),$$

and, since $\alpha p_1(u_R)^2 \ge 0$,

$$-R\langle f_0, Pw_R \rangle \le c' p_1(u_R) = O(\sqrt{R}).$$

But $-\langle f_0, Pw_R \rangle \ge \beta > 0$ for $R \ge R_0$, which implies that $\beta R \le O(\sqrt{R})$. This is impossible. The proof of the theorem is complete. Q.E.D.

3. Semilinear Equations

As an example of the application of Theorems 1.4 and 1.7 and Corollary 1.8, we solve a problem for a semilinear equation. Let $\Omega \subset \mathbb{R}^N$ be a bounded connected domain with smooth boundary $\partial\Omega$. Set

$$\mathbb{K} = \{v \in H_0^1(\Omega) : \psi_1(x) \le v(x) \le \psi_2(x) \text{ in } \Omega\},$$

where

$$\psi_1, \psi_2 \in H^1(\Omega) \cap L^\infty(\Omega) \qquad \text{and} \qquad \psi_1(x) \le 0 \le \psi_2(x) \qquad \text{for} \quad x \in \partial\Omega.$$

Let $F(x, u)$, $x \in \Omega$, $-\infty < u < \infty$, be bounded and measurable on compact subsets of $\overline{\Omega} \times \mathbb{R}$ and continuous and nondecreasing in u. Define

$$L\colon \mathbb{K} \to H^1(\Omega)$$

by

$$\begin{aligned}\langle Lu, \varphi\rangle &= \langle -\Delta u + F(x, u), \varphi\rangle \\ &= \int_\Omega \{u_{x_i}\varphi_{x_i} + F(x, u)\varphi\}\, dx, \qquad \varphi \in H_0^1(\Omega).\end{aligned}$$

Then

$$\begin{aligned}\langle Lu - Lv, u - v\rangle &= \int_\Omega (u - v)_x^2\, dx + \int_\Omega (F(x, u) - F(x, v))(u - v)\, dx \\ &\geq \int_\Omega (u - v)_x^2\, dx\end{aligned}$$

because F is nondecreasing in u. So L is a coercive and strictly monotone operator on $\mathbb{K}$. Note that it is not defined on all of $H_0^1(\Omega)$, but that it is defined on the closed convex $\mathbb{K}$. It is easy to see that L is continuous on finite dimensional subspaces of $\mathrm{H}_0^1(\Omega)$ intersected with $\mathbb{K}$. Hence there exists a unique

$$u \in \mathbb{K}\colon \quad \int_\Omega \{u_{x_i}(v - u)_{x_i} + F(x, u)(v - u)\}\, dx \geq 0 \qquad \text{for} \quad v \in \mathbb{K}.$$

4. Quasi-Linear Operators

Let $a(\xi) = (a_1(\xi), \ldots, a_N(\xi))$ be a continuous monotone mapping from $\mathbb{R}^N$ to $(\mathbb{R}^N)'$, which we identify with $\mathbb{R}^N$, as defined in Chapter I, Section 4. In this section we shall be concerned with some special vector fields satisfying a condition about their growth at infinity and their monotonicity. More precisely, we introduce

Definition 4.1. $a(\xi)$ is a *strongly coercive monotone vector field* if for $|\xi|$ sufficiently large, i.e., $|\xi| > R_0$,

$$|a(\xi)| \leq K|\xi| \qquad \text{for} \quad |\xi| > R_0, \quad \text{some} \quad K > 0, \tag{4.1}$$

and

$$(a_i(\xi) - a_i(\eta))(\xi_i - \eta_i) \geq c_0 |\xi - \eta|^2 \qquad \text{for all} \quad \xi, \eta \in \mathbb{R}^N \tag{4.2}$$

for some $c_0 > 0$.

When $a(\xi)$ is a strongly coercive monotone vector field

$$-\frac{\partial}{\partial x_i} a_i(u_x)$$

is well defined in $H_0^1(\Omega)$ and it maps $H_0^1(\Omega)$ into $H^{-1}(\Omega)$ since $a_i(u_x) \in L^2(\Omega)$, $1 \leq i \leq N$, for $u \in H_0^1(\Omega)$ due to (4.1). It is strictly monotone (Definition 1.1), coercive in the sense of Definition 1.3, and continuous on finite dimensional spaces (see Definition 1.2). An example of such an operator is obtained by choosing $a(\xi) = \xi$ with the associated operator

$$-\frac{\partial}{\partial x_i} a_i(u_x) = -\Delta u.$$

It satisfies all the assumptions above.

Let $f = f_0 - \sum_1^N (\partial/\partial x_i) f_i \in H^{-1}(\Omega)$ be given and let $\mathbb{K}$ be a closed convex set in $H_0^1(\Omega)$. Then by Theorem 1.7 and Corollary 1.8, there exists a unique solution u to the variational inequality

$$u \in \mathbb{K}: \quad \int_\Omega a_i(u_x)(v - u)_{x_i}\, dx \geq \int_\Omega \{f_0(v - u) + f_i(v - u)_{x_i}\}\, dx, \qquad v \in \mathbb{K}. \tag{4.3}$$

As a second example, we consider the vector field

$$a_i(\xi) = (1 + \xi^2)^{(\alpha - 2)/2} \xi_i, \qquad \xi \in \mathbb{R}^N, \quad 1 \leq \alpha < 2, \tag{4.4}$$

and the operator

$$Au = -\frac{\partial}{\partial x_i} a_i(u_x) = -\frac{\partial}{\partial x_i} \{[1 + u_x^2]^{(\alpha - 2)/2} u_{x_i}\}. \tag{4.5}$$

Using again Theorem 1.7 and Corollary 1.8 we may solve a variational inequality relative to A for $1 < \alpha \leq 2$ and the convex set

$$\mathbb{K} = \{v \in H_0^{1,\alpha}(\Omega) : v \geq \psi \text{ in } \Omega\}, \qquad \Omega \subset \mathbb{R}^N \text{ bounded.}$$

Indeed, since $|a(\xi)| \leq 1 + |\xi|^{\alpha - 1}$, the operator A maps $H_0^{1,\alpha}(\Omega)$ into $H^{-1,\alpha}(\Omega)$. It is obviously monotone since $a(\xi)$ is the gradient of the convex function $(1/\alpha)(1 + \xi^2)^{\alpha/2}$, continuous on finite dimensional subspaces, and coercive in the sense of Definition 13; that is, for $\varphi_0 \in \mathbb{K}$,

$$\frac{\langle Av - A\varphi_0, v - \varphi_0\rangle}{\|v - \varphi_0\|_{H_0^{1,\alpha}(\Omega)}} \to +\infty \qquad \text{as} \quad \|v\|_{H_0^{1,\alpha}(\Omega)} \to +\infty.$$

In fact, we first note that

$$\begin{aligned}
\langle Au, u\rangle/\|u\|_{H_0^{1,\alpha}(\Omega)} &= \int_\Omega (1 + u_x^2)^{(\alpha-2)/2} u_x^2 \, dx \Big/ \left(\int_\Omega |u_x|^\alpha \, dx\right)^{1/\alpha} \\
&\geq 2^{(\alpha-2)/2} \int_\Omega (|u_x|^\alpha - 1) \, dx \Big/ \left(\int_\Omega |u_x|^\alpha \, dx\right)^{1/\alpha} \\
&\geq 2^{(\alpha-2)/2} \left(\int_\Omega |u_x|^\alpha \, dx\right)^{1-(1/\alpha)} \\
&\quad - \left[2^{(\alpha-2)/2} \operatorname{meas} \Omega \Big/ \left(\int |u_x|^\alpha \, dx\right)^{1/\alpha}\right] \to +\infty
\end{aligned}$$

as $\|u\|_{H_0^{1,\alpha}(\Omega)} \to +\infty$. Second,

$$\begin{aligned}
\frac{\langle Au - A\varphi_0, u - \varphi_0\rangle}{\|u\|_{H_0^{1,\alpha}(\Omega)}} &\geq \frac{\langle Au, u\rangle}{\|u\|_{H_0^{1,\alpha}(\Omega)}} - \frac{\|a(u_x)\|_{L^{\alpha'}(\Omega)} \|\varphi_0\|_{H_0^{1,\alpha}(\Omega)}}{\|u\|_{H_0^{1,\alpha}(\Omega)}} \\
&\quad - \|a(\varphi_{0x})\|_{L^{\alpha'}(\Omega)} \frac{\|u - \varphi_0\|_{H_0^{1,\alpha}(\Omega)}}{\|u\|_{H_0^{1,\alpha}(\Omega)}},
\end{aligned}$$

where the first term on the right tends to infinity as $\|u\|_{H_0^{1,\alpha}(\Omega)}$ tends to infinity while the second and the third are bounded.

Now applying Theorem 1.7, given $f \in H^{-1,\alpha'}(\Omega)$, there exists a unique

$$u \in \mathbb{K}: \quad \int_\Omega (1 + u_x^2)^{(\alpha-2)/2} u_{x_i} (v - u)_{x_i} \, dx \geq \langle f, v - u\rangle \qquad \text{for} \quad v \in \mathbb{K}. \tag{4.6}$$

For $\alpha = 1$, the operator above is the minimal surface operator, or the operator of prescribed mean curvature. It is worthwhile to note that this operator is not included in the abstract theorems of Section 1. This fact will be clarified later.

The operator of prescribed mean curvature is related to the vector field

$$a_i(\xi) = (1 + \xi^2)^{-1/2} \xi_i, \qquad 1 \leq i \leq N, \quad \xi \in \mathbb{R}^N. \tag{4.7}$$

It is strictly monotone because it is the gradient of the strictly convex function

$$(1 + \xi^2)^{1/2}.$$

However, it is not a strongly coercive vector field, that is, condition (4.2) does not hold and, in fact, the variational inequality (4.6) need not admit a solution even when $f = 0$ (c.f. Serrin [1]). The case $\alpha = 1$ suggests the introduction of the notion of a locally coercive vector field.

Definition 4.2. A continuous vector field $a(\xi) = (a_1(\xi), \ldots, a_N(\xi))$ is *locally coercive* if for every compact $C \subset \mathbb{R}^N$ there exists a constant $\nu = \nu(C)$ such that

$$(a(\xi) - a(\eta)) \cdot (\xi - \eta) \geq \nu|\xi - \eta|^2 \qquad \text{for} \quad \xi, \eta \in C.$$

We remark that a locally coercive vector field is monotone and that the vector field (4.7) is locally coercive.

In the sequel this lemma will be useful:

Lemma 4.3. *Let $a(\xi)$ be a locally coercive vector field on $\mathbb{R}^N$ and let $M > 0$ be given. Then there exists a strongly coercive vector field $\tilde{a}(\xi)$ such that*

$$\tilde{a}(\xi) = a(\xi) \qquad \textit{for} \quad |\xi| < M.$$

More precisely, for $|\xi| > 3M$

$$|\tilde{a}(\xi)| \leq K|\xi|$$

for a suitable constant $K > 0$.

In other words the lemma affirms that we may modify a locally coercive vector field outside a given ball to obtain a strongly coercive field which satisfies (4.1).

Proof of the Lemma. Let $a(\xi)$ be the given vector field which we suppose locally coercive. Thus there exists $\nu > 0$ such that

$$(a(\xi) - a(\eta))(\xi - \eta) \geq \nu|\xi - \eta|^2 \qquad \text{for} \quad |\xi|, |\eta| \leq 2M.$$

Let ψ be defined on $\mathbb{R}^1$ such that

$$\psi(t) = \begin{cases} 1 & \text{for} \quad 0 \leq t \leq 2M \\ 0 & \text{for} \quad t \geq 3M \end{cases}$$

nonnegative and smooth for $0 \leq t < +\infty$.

Let $g(t)$ be a smooth and increasing function such that

$$g(t) = \begin{cases} 0 & \text{for} \quad 0 \leq t \leq M \\ 1 & \text{for} \quad t \geq 3M. \end{cases}$$

Assume that $g(t) \geq c_0 > 0$ for $t \geq 2M$. Define

$$\tilde{a}(\xi) = \psi(|\xi|)a(\xi) + kg(|\xi|)\xi,$$

where k is a positive constant we will fix later.

We wish to show that $\tilde{a}$ is strongly coercive and satisfies the statement of the lemma.

Let $\xi, \eta \in \mathbb{R}^N$ and, without loss of generality, assume $|\xi| \leq |\eta|$.

Observe that, since $|\xi| \leq |\eta|$,

$$0 \geq |\xi| \cdot |\eta| - |\eta|^2 \geq (\xi, \eta) - |\eta|^2 = (\eta, \xi - \eta) = \eta \cdot (\xi - \eta).$$

Consider first the following cases.

(1) $|\xi| \leq |\eta| \leq M$; then $\tilde{a}(\xi) = a(\xi)$ and $\tilde{a}(\eta) = a(\eta)$ and obviously

$$(\tilde{a}(\xi) - \tilde{a}(\eta))(\xi - \eta) \geq \nu|\xi - \eta|^2.$$

(2) $3M \leq |\xi| \leq |\eta|$; then $\tilde{a}(\xi) = kg(|\xi|)\xi$ and $\tilde{a}(\eta) = kg(|\eta|)\eta$ and it is obvious that

$$(\tilde{a}(\xi) - \tilde{a}(\eta))(\xi - \eta) = k|\xi - \eta|^2 \geq \nu|\xi - \eta|^2$$

if $k > \nu$.

(3) $M \leq |\xi| \leq |\eta| \leq 2M$; then $\tilde{a}(\xi) = a(\xi) + kg(|\xi|)\xi$, $\tilde{a}(\eta) = a(\eta) + kg(|\eta|)\eta$. Thus

$$\begin{aligned}(\tilde{a}(\xi) - \tilde{a}(\eta))(\xi - \eta) &= (a(\xi) - a(\eta))(\xi - \eta) + k(g(|\xi|)\xi - g(|\eta|)\eta)(\xi - \eta) \\ &\geq \nu|\xi - \eta|^2 + k(g(|\xi|)\xi - g(|\eta|)\eta)(\xi - \eta).\end{aligned}$$

We show that

$$k(g(|\xi|)\xi - g(|\eta|)\eta)(\xi - \eta) \geq 0.$$

In fact

$$(g(|\xi|)\xi - g(|\eta|)\eta)(\xi - \eta) = g(|\xi|)|\xi - \eta|^2 + (g(|\xi|) - g(|\eta|))\eta(\xi - \eta).$$

But $|\xi| \leq |\eta|$ implies $g(|\xi|) \leq g(|\eta|)$ and, as we have seen above, $\eta \cdot (\xi - \eta) \leq 0$. Therefore the left-hand side is nonnegative.

(4) $2M \leq |\xi| \leq |\eta| \leq 3M$; then

$$\tilde{a}(\xi) = \psi(|\xi|)a(\xi) + kg(|\xi|)\xi$$

and

$$\tilde{a}(\eta) = \psi(|\eta|)a(\eta) + kg(|\eta|)\eta.$$

Hence

$$\begin{aligned}(\tilde{a}(\xi) - \tilde{a}(\eta))(\xi - \eta) &= [\psi(|\xi|)a(\xi) - \psi(|\eta|)a(\eta)](\xi - \eta)\\ &\quad + k[g(|\xi|)\xi - g(|\eta|)\eta](\xi - \eta)\\ &= \psi(|\xi|)(a(\xi) - a(\eta))(\xi - \eta)\\ &\quad + [\psi(|\xi|) - \psi(|\eta|)]a(\eta)(\xi - \eta)\\ &\quad + kg(|\xi|)|\xi - \eta|^2 + k[g(|\xi|) - g(|\eta|)]\eta(\xi - \eta)\\ &\geq [\psi(|\xi|) - \psi(|\eta|)]a(\eta)(\xi - \eta) + kg(|\xi|)|\xi - \eta|^2\end{aligned}$$

since, as in case (3), $[g(|\xi|) - g(|\eta|)]\eta(\zeta - \eta) \geq 0$. Also

$$\sup_{2M \leq |\xi| \leq |\eta| \leq 3M} \left| \frac{[\psi(|\xi|) - \psi(|\eta|)]a(\eta)(\xi - \eta)}{|\xi - \eta|^2} \right| = K < +\infty$$

since ψ is smooth. Hence

$$[\psi(|\xi|) - \psi(|\eta|)]a(\eta)(\xi - \eta) \geq -K|\xi - \eta|^2,$$

and therefore

$$\begin{aligned}[\tilde{a}(\xi) - \tilde{a}(\eta)](\xi - \eta) &\geq [kg(|\xi|) - K]|\xi - \eta|^2\\ &\geq [kc_0 - K]|\xi - \eta|^2 \geq \nu|\xi - \eta|^2.\end{aligned}$$

Choosing now $k > \max(\nu, (K + \nu)/c_0)$ also in case (4), we have

$$[\tilde{a}(\xi) - \tilde{a}(\eta)](\xi - \eta) \geq \nu|\xi - \eta|^2.$$

Now let ξ and η be any two points in $\mathbb{R}^N$. If $|\xi| = |\eta|$, then the previous arguments apply. Assume $|\xi| < |\eta|$.

There exist at most six points ξ_i on the line joining ξ and η intersecting the spheres of radius M, $2M$, and $3M$. Denote them by ξ_i, $i = 1, 2, \ldots, j \leq 6$. Then

$$\tilde{a}(\xi) - \tilde{a}(\eta) = \tilde{a}(\xi) - \tilde{a}(\xi_1) + \tilde{a}(\xi_1) - \tilde{a}(\xi_2) + \cdots + \tilde{a}(\xi_j) - \tilde{a}(\eta),$$

and

$$\xi - \eta = (|\xi - \eta|/|\xi - \xi_i|)(\xi - \xi_i).$$

Therefore, since two consecutive ξ_i belong to same ball, we have

$$\begin{aligned}(\tilde{a}(\xi) - \tilde{a}(\eta))(\xi - \eta) &= (\tilde{a}(\xi) - \tilde{a}(\xi_1))(\xi - \xi_1)(|\xi - \eta|/|\xi - \xi_1|)\\ &\quad + (\tilde{a}(\xi_1) - \tilde{a}(\xi_2))(\xi_1 - \xi_2)(|\xi_1 - \eta|/|\xi_1 - \xi_2|) + \cdots\\ &\geq \nu|\xi - \xi_1|^2(|\xi - \eta|/|\xi - \xi_1|)\\ &\quad + \nu|\xi_1 - \xi_2|^2(|\xi - \eta|/|\xi_1 - \xi_2|) + \cdots\\ &= \nu|\xi - \eta|(|\xi - \xi_1| + |\xi_1 - \xi_2| + \cdots)\\ &= \nu|\xi - \eta|^2. \quad \text{Q.E.D.}\end{aligned}$$

We remark in connection with Lemma 4.3 that if $a(\xi)$ is C^1 and a locally coercive vector field on $\mathbb{R}^N$, then the vector field $\tilde{a}(\xi)$ may be chosen to be in $C^1(\mathbb{R}^N)$.

By the way, we remark that a C^1 vector field $a(\xi)$ in $\mathbb{R}^N$ is coercive if and only if

$$\sum \frac{\partial a_i}{\partial \xi_j}(\xi)\lambda_i\lambda_j \geq v|\lambda|^2 \qquad \text{for} \quad \xi \in C, \quad \lambda \in \mathbb{R}^N. \tag{4.8}$$

In fact, if $a(\xi)$ is locally coercive, for any compact C there is a constant $v = v(C)$ such that

$$[a(\xi) - a(\eta)](\xi - \eta) \geq v|\xi - \eta|^2 \qquad \text{for all} \quad \xi, \eta \in C$$

and vice versa. This implies that for each $\xi \in \text{int } C$ and $\eta \in \mathbb{R}^N$, there is an $\varepsilon > 0$ such that

$$\sum_i [a_i(\xi + t\eta) - a_i(\xi)]t\eta_i \geq vt^2|\eta|^2, \qquad |t| < \varepsilon,$$

or

$$\sum_i \frac{[a_i(\xi + t\eta) - a_i(\xi)]\eta_i}{t} \geq v|\eta|^2, \qquad |t| < \varepsilon.$$

Allowing $t \to 0$, we obtain

$$\sum_{i,j} \frac{\partial a_i}{\partial p_j}(\xi)\eta_j\eta_i \geq v|\eta|^2 \qquad \text{for all} \quad \eta \in \mathbb{R}^N.$$

The above clearly holds for $\xi \in C$.

On the other hand, assuming (4.8),

$$\begin{aligned}(a(\xi) - a(\eta))(\xi - \eta) &= \sum \left(\int_0^1 \frac{\partial a_i}{\partial p_j}(\xi + t(\eta - \xi))\,dt\right)(\xi_i - \eta_i)(\xi_j - \eta_j) \\ &\geq v|\xi - \eta|^2 \qquad \text{for} \quad \xi, \eta \in C,\end{aligned}$$

where v is the constant of the convex hull of C.

Moreover, if $|\partial a_i/\partial p_j(\xi)|$ are bounded, then (4.1) is satisfied but (4.1) does not imply boundedness of $(\partial a_i/\partial p_j)(\xi)$.

Comments and Bibliographical Notes

Although it is difficult to ascertain the first person to consider monotone operators, among the many modern initiators of the field are F. Browder

[1, 2], Goulomb, Minty [1, 2], Vishik [1], and Zarantonello [1]. Certainly it was Browder who systematically employed the monotonicity of operators in the study of nonlinear elliptic problems. Studying variational inequalities in this framework is due to Hartman and Stampacchia and, independently, to Browder.

The proofs of Section 1 follow Hartman and Stampacchia [1]. The material of Section 2 is from Lions and Stampacchia [1]. A special case of Theorem 2.3 had been previously proved by Fichera [1]. The presentation in Section 4 employs the results of the paper of Brezis and Stampacchia [4]. The special case when the monotone vector field is the gradient of a convex function was considered in Stampacchia [1].

The theory of monotone operators also has interesting applications to the theory of ordinary differential equations and systems (Schiaffino and Troianello [1]; Vergara-Caffarelli [1]).

Exercises

1. Let $a(u, v)$ be a bilinear from on a Hilbert space H and let Au defined by

$$\langle Au, v\rangle = a(u, v)$$

be the associated linear transformation from H to H'. Show that Au is coercive on H in the sense of Definition 1.3 of Chapter III if and only if $a(u, v)$ is coercive in the sense of Definition 1.1 of Chapter II.

2. Let H be a Hilbert space and $a(u, v)$ a nonnegative bilinear form on H. Prove that $a(u, v)$ is lower semicontinuous with respect to weak convergence in H, i.e.,

$$a(u, u) \leq \liminf a(u_\varepsilon, u_\varepsilon) \qquad \text{as} \quad u_\varepsilon \to u \quad \text{weakly in} \quad H.$$

3. Prove that the set of solutions to the variational inequality (1.4) is a closed convex subset of the given convex $\mathbb{K}$.

4. Let Ω be a bounded domain in $\mathbb{R}^N$ and let $\mathscr{K}$ be a closed linear subspace of $L^2(\Omega)$. Let $\gamma(t)$ be a continuous increasing function satisfying

$$|\gamma(t)| \leq A + B|t|, \qquad A, B \quad \text{constants.}$$

Finally, let $F \in L^2(\Omega)$. Show that there exists a unique

$$u \in \mathscr{K}: \quad u + \gamma(u) - F \in \mathscr{K}^{\perp},$$

where $\mathscr{K}^{\perp}$ is the orthogonal complement of $\mathscr{K}$. (*Hint*: Solve the variational inequality

$$u \in \mathscr{K}: \quad (u + \gamma(u) - F, v - u) \geq 0 \qquad \text{for all} \quad v \in \mathscr{K}.)$$

5. Let $\mathbb{K}$ be a closed bounded subset of a Hilbert space H and let $F: \mathbb{K} \to \mathbb{K}$ be a nonexpansive mapping. Let W be a contraction on H such that $W(\mathbb{K}) \subset \mathbb{K}$. Show that for $0 < \varepsilon < 1$,

$$F_{\varepsilon} = (1 - \varepsilon)F - \varepsilon W$$

is a contraction mapping from $\mathbb{K}$ to $\mathbb{K}$. Let x_{ε} denote the unique fixed point of F_{ε} and show that x_{ε} converges to a fixed point of F as ε tends to 0.

6. Let Ω be a bounded open domain in $\mathbb{R}^N$ and let E and F be closed nonempty subsets of Ω. Let $\psi \in H^1(\Omega)$ and denote by $\mathbb{K}$ the convex set

$$\mathbb{K} = \{v \in H^1(\Omega): v \geq \psi \text{ on } E \text{ and } v \leq \psi \text{ on } F\}.$$

Let

$$a(u, v) = \int_{\Omega} u_{x_i} v_{x_i}\, dx \qquad \text{for} \quad u, v \in H^1(\Omega).$$

Show that there exists a solution to the variational inequality

$$u \in \mathbb{K}: \quad a(u, v - u) \geq 0 \qquad \text{for all} \quad v \in \mathbb{K}$$

(cf. Iordanov [1]).

7. Let $T: L^2(\Omega) \to L^2(\Omega)$ be a coercive monotone operator, where $\Omega \subset \mathbb{R}^N$ is a bounded open set. Show that the problem: *Find* $u \in L^2(\Omega)$ *such that*

$$u \geq 0, \qquad Tu \geq 0, \qquad \textit{and} \quad uTu = 0 \quad \textit{a.e. in} \quad \Omega,$$

or equivalently,

$$\min(u, Tu) = 0 \qquad \textit{a.e. in} \quad \Omega,$$

can be formulated as the variational inequality: *Find*

$$u \in \mathbb{K}: \quad (Tu, v - u) \geq 0 \qquad \textit{for all} \quad v \in \mathbb{K}.$$

where

$$\mathbb{K} = \{v \in L^2(\Omega): v \geq 0 \textit{ a.e. in } \Omega\}$$

(cf. H. Brezis and L. Evans [1]).

8. Show that the form

$$a(u, v) = \int_{\Omega} \Delta u\, \Delta v\, dx$$

is coercive on $H_0^1(\Omega) \cap H^2(\Omega)$, $\Omega \subset \mathbb{R}^N$ bounded. Generalize this result for the form

$$a(u, v) = \int_\Omega A^1 u A^2 u \, dx,$$

where A^1 and A^2 are linear elliptic operators of the form

$$A^i u = -(a^i_{jk} u_{x_k})_{x_j} + \mu^i u$$

(cf. Brezis and Evans [1], Ladyzhenskaya and Ural'tseva [1]).

9. Let $\Omega \subset \mathbb{R}^N$ be a bounded domain and let A^1 and A^2 be two elliptic operators

$$A^i : H_0^1(\Omega) \cap H^2(\Omega) \to L^2(\Omega)$$

given by

$$A^i u = -(a^i_{jk} u_{x_k})_{x_j}, \qquad i = 1, 2.$$

Show that a solution of the problem of Bellman [1]: *find*

$$u \in H_0^1(\Omega) \cap H^2(\Omega) : \min(A^1 u - f, A^2 u) = 0,$$

where $f \in L^2(\Omega)$ is given, is a solution of the variational inequality: *Find*

$$u \in \mathbb{K}: \quad \int_\Omega (A^1 u - f) A^2(v - u) \, dx \geq 0 \qquad \textit{for all} \quad v \in \mathbb{K},$$

where

$$\mathbb{K} = \{v \in H_0^1(\Omega) \cap H^2(\Omega) : A^2 v \geq 0\}.$$

Establish conditions which insure the existence of a solution to this variational inequality (Brezis and Evans [1]).

10. Let $\mathbb{K}$ be a closed convex subset of a reflexive Banach space X with dual X' and let $A: \mathbb{K} \to X'$ be monotone and continuous on finite dimensional subspaces. Then a necessary and sufficient condition for the existence of a solution of the variational inequality

$$u \in \mathbb{K}: \quad \langle Au, v - u \rangle \geq 0 \qquad \textit{for all} \quad v \in \mathbb{K} \tag{I}$$

is that there exists a nonempty bounded subset C of $\mathbb{K}$ such that for each $w \in \mathbb{K} - C$, *there exists* $v \in C$ *such that*

$$\langle Aw, v - w \rangle \leq 0$$

Moreover, assume that $\mathbb{K}$ is strictly convex in the sense that for all $u, v \in \mathbb{K}$, the points

$$\lambda u + (1 - \lambda) v, \qquad 0 < \lambda < 1,$$

are in the interior of $\mathbb{K}$, and also that $A^{-1}(0)$ consists of no more than one element. Then the solution of (I) is unique.

This condition says that uniqueness for the equation $Av = 0$ implies uniqueness for the variational inequality (I) provided $\mathbb{K}$ is strictly convex (Coppoletta [1]).

11. Show that the operator of mean curvature

$$Au = -(\partial/\partial x_i)[u_{x_i}/(1 + u_x^2)^{1/2}]$$

is not strongly coercive, but that the vector field $p/(1 + p^2)^{1/2}$ is locally coercive.

12 Let $\Omega \subset \mathbb{R}^N$ be a smooth domain with boundary $\partial\Omega$. Let

$$\mathbb{K} = \{v \in H^1(\Omega) : v \geq 0 \text{ on } \partial\Omega\}$$

and

$$f = f_0 - \sum_1^N (f_i)_{x_i} \in H^{-1}(\Omega)$$

with

$$\int_\Omega f_0 \, dx < 0.$$

Show that there exists a solution of the variational inequality

$$u \in \mathbb{K} : \int_\Omega u_{x_i}(v - u)_{x_i} \, dx \geq \langle f, v - u \rangle \qquad \text{for all} \quad v \in \mathbb{K}.$$

Write this problem in its complementarity form and discuss the uniqueness of its solutions.

CHAPTER **IV**

Problems of Regularity

1. Penalization

We were able to give an abstract existence theorem for the solution of a variational inequality in part because our notion of solution was very weak. In this chapter we study the smoothness of the solution in some detail. The solution of an elliptic boundary value problem enjoys a degree of smoothness depending on that of its data. However, we have already seen, for instance, that the solution of an obstacle problem associated to a second order equation cannot in general be of class C^2 regardless of the smoothness of the data. This illustrates that our regularity theory will have to be especially invented for variational inequalities.

To initiate our discussion, we shall briefly describe a method called *penalization*. We want to emphasize that this method, often used to demonstrate the regularity of solutions, may sometimes be used to attain their existence when the assumptions of general theorems do not apply. This we shall see later. The method of penalization consists in substituting the variational inequality by a family of nonlinear boundary value problems and demonstrating that their solutions converge to the solution of the variational inequality. The principal difficulty lies in obtaining suitable a priori estimates. We stress that there are different choices of penalization.

2. Dirichlet Integral

Let $\Omega \subset \mathbb{R}^N$ be a bounded connected domain with smooth boundary $\partial\Omega$. Let $\psi \in H^1(\Omega)$ with $\psi < 0$ on $\partial\Omega$ and set

$$\mathbb{K} = \mathbb{K}_\psi = \{v \in H_0^1(\Omega) : v \geq \psi \text{ on } \Omega\}.$$

A method of penalization will be adopted to study the following:

Problem 2.1. *Given $f \in L^2(\Omega)$, find*

$$u \in \mathbb{K}: \quad \int_\Omega u_{x_i}(v - u)_{x_i}\, dx \geq \int_\Omega f(v - u)\, dx \qquad \textit{for all} \quad v \in \mathbb{K}. \tag{2.1}$$

We shall suppose that f and $-\Delta\psi$ are measures with the properties

$$f \in L^s(\Omega) \qquad \textit{and} \qquad \max(-\Delta\psi - f, 0) \in L^s(\Omega) \qquad \textit{for some} \qquad N < s < \infty. \tag{2.2}$$

Under these assumptions it will be shown that the solution u of (2.1) is in $H^{2,s}(\Omega) \cap C^{1,\lambda}(\overline{\Omega})$, $\lambda = 1 - (N/s)$. A major tool will be the following estimate, which is based on a theorem of Calderon–Zygmund.
Suppose that

$$v \in H_0^1(\Omega): \quad -\Delta v = F \qquad \textit{in} \quad \Omega, \qquad \textit{where} \quad F \in L^s(\Omega).$$

Then

$$\|v\|_{H^{2,s}(\Omega)} \leq C_0(s, \Omega)\|F\|_{L^s(\Omega)}. \tag{2.3}$$

Formally, it is evident that the solution u of Problem 2.1 satisfies, with $I = \{x \in \Omega : u(x) = \psi(x)\}$,

$$\begin{aligned} -\Delta u &= f & &\textit{in} \quad \Omega - I, \\ -\Delta u &= -\Delta\psi & &\textit{in} \quad I, \end{aligned}$$

so we would like to obtain u as a solution to the Dirichlet problem

$$\begin{aligned} -\Delta u &= \begin{cases} f & \textit{in} \quad \Omega - I \\ -\Delta\psi & \textit{in} \quad I, \end{cases} \\ u &= 0 \qquad\qquad\quad \textit{on} \quad \partial\Omega, \end{aligned} \tag{2.4}$$

Recalling Section 6 of Chapter II, however, we know that

$$-\Delta u = f + \mu \qquad \textit{in} \quad \Omega$$

(in the sense of distributions), where μ is a certain nonnegative measure whose support is in I. Combining the two representations leads to the formal description

$$d\mu = (-\Delta\psi - f)\,dx \qquad in \quad I,$$
$$d\mu = 0 \qquad in \quad \Omega - I.$$

Denoting by ϑ the function

$$\vartheta(t) = \begin{cases} 1, & t \le 0 \\ 0, & t > 0, \end{cases} \tag{2.5}$$

the formula above may be rewritten

$$d\mu = (-\Delta\psi - f)\vartheta(u - \psi)\,dx \qquad in \quad \Omega. \tag{2.6}$$

Unfortunately, we have no way to ensure a priori that μ is absolutely continuous, which is certainly implied by (2.6), nor can we determine that a solution v to (2.4), where $I = I(u)$, u the known solution of (2.1), is an element of $\mathbb{K}$. This notwithstanding, (2.6) leads to consideration of the problems

$$-\Delta u_\varepsilon = (-\Delta\psi - f)\vartheta_\varepsilon(u_\varepsilon - \psi) + f \qquad in \quad \Omega$$
$$u_\varepsilon = 0 \qquad on \quad \partial\Omega$$

where ϑ_ε is a sequence of Lipschitz functions which tend to ϑ defined in (2.5) almost everywhere on $\mathbb{R}$. Note that since u is a supersolution of $-\Delta - f$, we know that

$$I \subset \{x : -\Delta\psi(x) - f \ge 0\},$$

so we replace $-\Delta\psi - f$ by $\max(-\Delta\psi - f, 0)$. The boundary value problem

$$\begin{aligned} -\Delta u_\varepsilon &= \max(-\Delta\psi - f, 0)\vartheta_\varepsilon(u_\varepsilon - \psi) + f \qquad in \quad \Omega \\ u_\varepsilon &= 0 \qquad on \quad \partial\Omega \end{aligned} \tag{2.7}$$

is called the *penalized problem.*

Lemma 2.2. *Let $\vartheta(t)$, $-\infty < t < \infty$, be uniformly Lipschitz, nonincreasing, and satisfy $0 \le \vartheta(t) \le 1$. Assume (2.2). Then there exists a unique*

$$w \in H_0^1(\Omega): \quad -\Delta w = \max(-\Delta\psi - f, 0)\vartheta(w - \psi) + f \qquad in \quad \Omega.$$

Moreover,

$$\|w\|_{H^{2,s}(\Omega)} \le C_0(s, \Omega)(2\|f\|_{L^s(\Omega)} + \|\max(-\Delta\psi - f, 0)\|_{L^s(\Omega)}),$$

where $C_0(s, \Omega)$ is the constant of (2.3).

Proof. First note that $\vartheta(w - \psi) \in L^\infty(\Omega)$ for any $w \in L^2(\Omega)$ since ϑ is bounded. Therefore the distribution Lw defined by

$$\langle Lw, \zeta \rangle = \int_\Omega \{w_{x_i}\zeta_{x_i} - [\max(-\Delta\psi - f, 0)\vartheta(w - \psi) + f]\zeta\}\, dx$$

is in $H^{-1}(\Omega)$. We claim that it is strictly monotone and coercive on $H_0^1(\Omega)$. Since $\vartheta(t)$ is nonincreasing,

$$-[\max(-\Delta\psi - f, 0)\vartheta(w - \psi) - \max(-\Delta\psi - f, 0)\vartheta(v - \psi)](w - v) \geq 0.$$

We compute

$$\begin{aligned}\langle Lw - Lv, w - v \rangle &= \int_\Omega (w - v)_{x_i}^2\, dx \\ &\quad - \int_\Omega \max(-\Delta\psi - f, 0)(\vartheta(w - \psi) - \vartheta(v - \psi))(w - v)\, dx \\ &\geq \int_\Omega (w - v)_{x_i}^2\, dx \\ &= \|w - v\|_{H_0^1(\Omega)}^2.\end{aligned}$$

This proves that L is strictly monotone and coercive. Furthermore, $w_n \to w$ in $H_0^1(\Omega)$ implies $Lw_n \to Lw$ weakly in $H^{-1}(\Omega)$, which certainly implies that L is continuous on finite dimensional subspaces of $H_0^1(\Omega)$. We may apply Corollary 1.8 of Chapter III to obtain the existence of w. The estimate follows from (2.3). Q.E.D.

We now specify

$$\vartheta_\varepsilon(t) = \begin{cases} 1, & t \leq 0 \\ 1 - (t/\varepsilon), & 0 \leq t \leq \varepsilon, \quad \varepsilon > 0 \\ 0, & t \geq \varepsilon \end{cases} \tag{2.8}$$

and consider the Dirichlet problems, for $\varepsilon > 0$,

$$w \in H_0^1(\Omega): \quad -\Delta w = \max(-\Delta\psi - f, 0)\vartheta_\varepsilon(w - \psi) + f \quad \text{in } \Omega. \tag{2.9}$$

We turn to the main result of this section.

Theorem 2.3. *Assume* (2.2) *and let u denote the solution to Problem* 2.1. *Then $u \in H^{2,s}(\Omega) \cap C^{1,\lambda}(\overline{\Omega})$, $\lambda = 1 - (N/s)$. In addition, let u_ε, $\varepsilon > 0$, denote the solution to* (2.9). *Then $u_\varepsilon \to u$ weakly in $H^{2,s}(\Omega)$.*

Proof. We first claim that $u_\varepsilon \in \mathbb{K}$. Set $\zeta = u_\varepsilon - \max(u_\varepsilon, \psi) \le 0$, an element of $H_0^1(\Omega)$. We want to show that $\zeta = 0$. By (2.9)

$$\int_\Omega \{u_{\varepsilon x_i}\zeta_{x_i} - [\max(-\Delta\psi - f, 0)\vartheta(u_\varepsilon - \psi) + f]\zeta\}\, dx = 0,$$

and by Green's theorem

$$\int_\Omega \{\psi_{x_i}\zeta_{x_i} + \Delta\psi\zeta\}\, dx = 0.$$

Subtracting these two equations,

$$\int_\Omega (u_\varepsilon - \psi)_{x_i}\zeta_{x_i}\, dx = \int_\Omega [\max(-\Delta\psi - f, 0)\vartheta_\varepsilon(u_\varepsilon - \psi) + \Delta\psi + f]\zeta\, dx.$$

Now

$$\zeta_{x_i} = \begin{cases} (u_\varepsilon - \psi)_{x_i} & \text{if } \zeta < 0 \\ 0 & \text{if } \zeta = 0, \end{cases}$$

so

$$\int_\Omega \zeta_{x_i}^2\, dx = \int_{\{\zeta<0\}} [\max(-\Delta\psi - f, 0)\vartheta_\varepsilon(u_\varepsilon - \psi) + \Delta\psi + f]\zeta\, dx.$$

At this point we observe that $\zeta(x) < 0$, which is the same as $u_\varepsilon(x) - \psi(x) < 0$, implies that $\vartheta_\varepsilon(u_\varepsilon(x) - \psi(x)) = 1$; hence

$$\int_\Omega \zeta_{x_i}^2\, dx = \int_{\{\zeta<0\}} [\max(-\Delta\psi - f, 0) + \Delta\psi + f]\zeta\, dx \le 0.$$

Hence $\zeta = 0$.

By Lemma 2.2, the family $\{u_\varepsilon\}$ is bounded in $H^{2,s}(\Omega)$ and hence contains a subsequence $u_{\varepsilon'}$ which converges weakly in $H^{2,s}(\Omega)$ and uniformly in $\overline{\Omega}$ to a $\tilde{u} \in H^{2,s}(\Omega) \subset C^{1,\lambda}(\overline{\Omega})$, $\lambda = 1 - (N/s)$. Since $u_\varepsilon \in \mathbb{K}$, $\tilde{u} \in \mathbb{K}$.

To show that $\tilde{u}$ is a solution to Problem 2.1 we use Minty's Lemma 1.5 of Chapter III, applied to the monotone operator defined by (2.9). Let $v \in \mathbb{K}$ and suppose $v \ge \psi + \delta$ for some positive δ. Then

$$\int_\Omega v_{x_i}(v - u_{\varepsilon'})_{x_i}\, dx - \int_\Omega [\max(-\Delta\psi - f, 0)\vartheta_{\varepsilon'}(v - \psi) + f](v - u_{\varepsilon'})\, dx \ge 0.$$

For $\varepsilon < \delta$, $\vartheta_\varepsilon(v - \psi) = 0$. Now allowing $\varepsilon \to 0$ we obtain, by weak convergence,

$$\int_\Omega v_{x_i}(v - \tilde{u})_{x_i}\, dx - \int_\Omega f(v - \tilde{u})\, dx \ge 0 \qquad \text{for} \quad v \in \mathbb{K}, \quad v \ge \psi + \delta.$$

We now let $\delta \to 0$, whence it follows that

$$\int_\Omega v_{x_i}(v - \tilde{u})_{x_i}\, dx \geq \int_\Omega f(v - \tilde{u})\, dx \qquad \text{for any} \quad v \in \mathbb{K}.$$

Employing Minty's lemma once again, we conclude that $\tilde{u} = u$, the unique solution to Problem 2.1. The solution being unique, the entire family u_ε converges weakly to u in $H^{2,s}(\Omega)$. Q.E.D.

The approximations we have employed possess some interesting properties.

Theorem 2.4. *Assume* (2.2). *Let u denote the solution to the variational inequality* (2.1) *and u_ε, $\varepsilon > 0$, denote the solution to the penalized problem* (2.7) *with ϑ_ε defined by* (2.8). *Then $\{u_\varepsilon\}$ is a nondecreasing sequence and*

$$u(x) \leq u_\varepsilon(x) \leq u(x) + \varepsilon, \qquad x \in \Omega \qquad \textit{for} \quad \varepsilon > 0.$$

Proof. To prove that $\{u_\varepsilon\}$ is a nonincreasing sequence as ε decreases observe, first of all, that the sequence $\vartheta_\varepsilon(t)$ decreases as ε decreases. So given $\varepsilon < \varepsilon'$ we write, for $\zeta \in H_0^1(\Omega)$,

$$\begin{aligned}
&\int_\Omega (u_{\varepsilon'} - u_\varepsilon)_{x_i} \zeta_{x_i}\, dx \\
&\quad = \int_\Omega \max(-\Delta\psi - f, 0)[\vartheta_{\varepsilon'}(u_{\varepsilon'} - \psi) - \vartheta_\varepsilon(u_\varepsilon - \psi)]\zeta\, dx \\
&\quad = \int_\Omega \max(-\Delta\psi - f, 0)[\vartheta_{\varepsilon'}(u_{\varepsilon'} - \psi) - \vartheta_{\varepsilon'}(u_\varepsilon - \psi)]\zeta\, dx \\
&\qquad + \int_\Omega \max(-\Delta\psi - f, 0)[\vartheta_{\varepsilon'}(u_\varepsilon - \psi) - \vartheta_\varepsilon(u_\varepsilon - \psi)]\zeta\, dx.
\end{aligned}$$

Now choose $\zeta = \min(u_{\varepsilon'} - u_\varepsilon, 0)$; $\zeta \in H_0^1(\Omega)$ and satisfies $\zeta \leq 0$. Then

$$\begin{aligned}
\int_\Omega \zeta_{x_i}^2\, dx = &\int_{\{\zeta < 0\}} \max(-\Delta\psi - f, 0)[\vartheta_{\varepsilon'}(u_{\varepsilon'} - \psi) - \vartheta_{\varepsilon'}(u_\varepsilon - \psi)](u_{\varepsilon'} - u_\varepsilon)\, dx \\
&+ \int_\Omega \max(-\Delta\psi - f, 0)[\vartheta_{\varepsilon'}(u_\varepsilon - \psi) - \vartheta_\varepsilon(u_\varepsilon - \psi)]\zeta\, dx.
\end{aligned}$$

Since the function $\vartheta_{\varepsilon'}$ is decreasing,

$$[\vartheta_{\varepsilon'}(u_{\varepsilon'} - \psi) - \vartheta_{\varepsilon'}(u_\varepsilon - \psi)](u_{\varepsilon'} - u_\varepsilon) \leq 0,$$

and since $\vartheta_\varepsilon(t) < \vartheta_{\varepsilon'}(t)$ for $\varepsilon < \varepsilon'$ and $\zeta \leq 0$,

$$[\vartheta_{\varepsilon'}(u_\varepsilon - \psi) - \vartheta_\varepsilon(u_\varepsilon - \psi)]\zeta \leq 0.$$

Therefore

$$\int_\Omega \zeta_{x_i}^2 \, dx = 0, \qquad \text{so} \quad \zeta = 0 \qquad \text{and} \qquad u_\varepsilon \leq u_{\varepsilon'}.$$

Since $u_\varepsilon \to u$, it follows that $u_\varepsilon \geq u$ for $\varepsilon > 0$. Now observe that $u + \varepsilon$ is a supersolution to the approximating equation. Namely,

$$\int_\Omega (u + \varepsilon)_{x_i} \zeta_{x_i} \, dx - \int_\Omega [\max(-\Delta\psi - f, 0)\vartheta_\varepsilon(u + \varepsilon - \psi) + f]\zeta \, dx$$

$$= \int_\Omega u_{x_i} \zeta_{x_i} \, dx - \int_\Omega f\zeta \, dx \geq 0 \qquad \text{for} \quad \zeta \geq 0 \tag{2.10}$$

because u is a solution to the variational inequality. Moreover,

$$u + \varepsilon = \varepsilon \geq u_\varepsilon \qquad \text{on} \quad \partial\Omega.$$

From (2.7) and (2.10), choosing $\zeta = \min(u + \varepsilon - u_\varepsilon, 0) \in H_0^1(\Omega)$, ≤ 0, we conclude that

$$u_\varepsilon \leq u + \varepsilon. \quad \text{Q.E.D.}$$

Similarly, one may choose a sequence of approximating functions $\tilde{u}_\varepsilon$ to be solutions to the problem

$$\begin{aligned} -\Delta \tilde{u}_\varepsilon &= \max(-\Delta\psi - f, 0)\tilde{\vartheta}_\varepsilon(\tilde{u}_\varepsilon - \psi) + f \qquad && in \quad \Omega, \\ \tilde{u}_\varepsilon &= 0 && on \quad \partial\Omega, \end{aligned}$$

with

$$\tilde{\vartheta}_\varepsilon(t) = \begin{cases} 1, & t \leq -\varepsilon \\ -\dfrac{t}{\varepsilon}, & -\varepsilon \leq t \leq 0 \\ 0, & t \geq 0. \end{cases}$$

The family $\tilde{u}_\varepsilon$ has the property that it increases as ε decreases.

For ease of exposition we chose a convex set associated to homogeneous boundary values to illustrate our method of penalization. A general guideline is that the solution of the variational inequality is of class $H^{2,s}(\Omega)$ whenever the solution of the associated boundary value problem enjoys this property. It is easy to check that the conclusion of Theorem 2.3 is valid for the convex $\mathbb{K} = \{v \in H^1(\Omega) : v \geq \psi \text{ in } \Omega, v = g \text{ on } \partial\Omega\}$, where $g \in H^{2,s}(\Omega)$ and $\psi \leq g$ on

$\partial\Omega$. We offer a further example useful in the study of filtration (cf. Chapter VII.)

Let $\Omega \subset \mathbb{R}^N$ be a domain with Lipschitz boundary $\partial\Omega$ which may be written $\partial\Omega = \overline{\partial_1\Omega \cup \partial_2\Omega}$, where $\partial_1\Omega$ and $\partial_2\Omega$ are disjoint open Lipschitz hypersurfaces. In particular, the trace from $H^1(\Omega)$ to $L^2(\partial_1\Omega)$ is well defined. Set

$$V = \{v \in H^1(\Omega) : v = 0 \text{ on } \partial_1\Omega\}$$

and assume that the $H_0^1(\Omega)$ norm is also a norm on V, that is, that the form

$$\int_\Omega v_{x_i}\zeta_{x_i}\, dx, \qquad v, \zeta \in V,$$

is coercive on V. Let $f \in L^2(\Omega)$, g, $\psi \in H^1(\Omega)$ with $\psi \leq g$ on $\partial_1\Omega$, and set

$$\mathbb{K} = \{v \in H^1(\Omega) : v \geq \psi \text{ in } \Omega \text{ and } v - g \in V\}. \tag{2.11}$$

From Theorem 2.1 in Chapter II we conclude that there exists a unique solution u of the variational inequality

$$u \in \mathbb{K}: \quad \int_\Omega u_{x_i}(v - u)_{x_i}\, dx \geq \int_\Omega f(v - u)\, dx \qquad \text{for} \quad v \in \mathbb{K}. \tag{2.12}$$

Written in the complementarity form, which we do not justify, u is a solution of

$$\begin{aligned} (-\Delta u - f)(u - \psi) &= 0 && \text{a.e. in} \quad \Omega, \\ -\Delta u - f \geq 0,\ u - \psi &\geq 0 && \text{a.e. in} \quad \Omega, \\ u &= g && \text{on} \quad \partial_1\Omega, \\ \partial u/\partial \nu &= 0 && \text{on} \quad \partial_2\Omega, \end{aligned}$$

where ν is the exterior normal to $\partial_2\Omega$. Thus (2.12) is related to a mixed boundary value problem (cf. Chapter II, Mixed Problem 4.9). Our object is to study the smoothness of u for which we impose the hypothesis (2.14) below. We stress that (2.14) does not always hold: its validity depends on $\partial_1\Omega$, $\partial_2\Omega$, and also g.

Let $f \in L^s(\Omega)$ and $g \in H^{2,s}(\Omega)$, some s, $N < s < \infty$, and determine w by

$$\begin{aligned} w \in H^1(\Omega): \quad -\Delta w &= f && \text{in} \quad \Omega, \\ w &= g && \text{on} \quad \partial_1\Omega, \\ \partial w/\partial \nu &= 0 && \text{on} \quad \partial_2\Omega. \end{aligned} \tag{2.13}$$

Assume that for a constant $C_0 > 0$ independent of f

$$\|w\|_{H^{2,s}(\Omega)} \leq C_0[\|f\|_{L^s(\Omega)} + \|g\|_{H^{2,s}(\Omega)}]. \tag{2.14}$$

In the formulation of (2.13) we intend of course that

$$w - g \in V: \quad \int_\Omega w_{x_i} \zeta_{x_i}\, dx = \int_\Omega f\zeta\, dx \qquad \text{for all} \quad \zeta \in V.$$

Theorem 2.5. *Let* $f \in L^s(\Omega)$ *and* $\psi, g \in H^{2,s}(\Omega)$, $\psi \le g$ *on* $\partial_1\Omega$, *for some* s, $N < s < \infty$. *Suppose that* (2.14) *holds. Then the solution* u *of* (2.12) *satisfies* $u \in H^{2,s}(\Omega) \cap C^{1,\lambda}(\bar{\Omega})$, $\lambda = 1 - (N/s)$.

To prove the theorem introduce the penalized equations

$$\begin{aligned} -\Delta u_\varepsilon &= \max(-\Delta\psi - f, 0)\vartheta_\varepsilon(u_\varepsilon - \psi) + f && \text{in} \quad \Omega, \\ u_\varepsilon &= g && \text{on} \quad \partial_1\Omega, \\ \partial u_\varepsilon/\partial v &= 0 && \text{on} \quad \partial_2\Omega, \end{aligned}$$

where ϑ_ε is defined by (2.8). The estimate (2.14) applies to u_ε. Details are left to the reader.

3. Coercive Vector Fields

Coercive vector fields were discussed in Section 4 of Chapter III. In order to obtain regularity of the solution of a variational inequality, we must restrict our attention to vector fields which are at least C^1. In this situation Definition 4.1 of Chapter III becomes

Definition 3.1. A vector field $a(p)$ is C^1 *strongly coercive* if $a_i(p) \in C^1(\mathbb{R}^N)$, $1 \le i \le N$,

$$\begin{aligned} (a(p) - a(q))(p - q) &\ge v|p - q|^2, && p, q \in \mathbb{R}^N, \\ |(\partial a_i/\partial p_j)(p)| &\le M \qquad \text{for} \quad p \in \mathbb{R}^N, && 1 \le i, j \le N, \end{aligned} \tag{3.1}$$

for some constants $0 < v \le M < \infty$.

Let Ω be a bounded open subset of $\mathbb{R}^N$ with smooth boundary $\partial\Omega$. Using (3.1) it is easy to verify that the quasi-linear operator

$$A: H_0^1(\Omega) \to H^{-1}(\Omega), \qquad Av = -(\partial/\partial x_i)a_i(v_x)$$

is well defined, monotone, and coercive. If $v_n \to v$ in $H_0^1(\Omega)$, then, by the smoothness of $a_i(p)$, $Av_n \to Av$ weakly in $H^{-1}(\Omega)$, so it is clear that A is continuous on finite dimensional subspaces of $H_0^1(\Omega)$. In addition we pointed out at the end of Section 4 of Chapter III that

$$(\partial a_i/\partial p_j)(p)\xi_i\xi_j \geq \nu|\xi|^2 \qquad \text{for all} \quad p, \xi \in \mathbb{R}^N,$$

so the operator A is uniformly elliptic in the ordinary sense.

Consider a function $\psi \in C^2(\overline{\Omega})$ which satisfies

$$\max_{\Omega} \psi > 0 \qquad \text{and} \qquad \psi < 0 \qquad \text{on} \quad \partial\Omega$$

and define

$$\mathbb{K} = \mathbb{K}_\psi = \{v \in H_0^1(\Omega) : v \geq \psi \text{ in } \Omega\}.$$

We consider again

Problem 3.2. *Given $f \in L^\infty(\Omega)$, find*

$$u \in \mathbb{K}: \quad \int_\Omega a_i(u_x)(v - u)_{x_i}\, dx \geq \int_\Omega f(v - u)\, dx \qquad \textit{for} \quad v \in \mathbb{K}. \tag{3.2}$$

The existence of a solution to (3.2) has already been shown in Section 4 of Chapter III, indeed, under much weaker assumptions about ψ and f. Here we discuss its smoothness.

As before, we define

$$\vartheta_\varepsilon(t) = \begin{cases} 1, & t \leq 0 \\ 1 - (t/\varepsilon), & 0 < t < \varepsilon \\ 0, & t \geq \varepsilon, \quad \varepsilon > 0 \end{cases}$$

and consider the Dirichlet problem

$$w \in H_0^1(\Omega): \quad Aw = \max(A\psi - f, 0)\vartheta_\varepsilon(w - \psi) + f \qquad in \quad \Omega. \tag{3.3}$$

As in the previous section, one checks that the operator

$$A_\varepsilon w = Aw - \max(A\psi - f, 0)\vartheta_\varepsilon(w - \psi) + f, \qquad w \in H_0^1(\Omega),$$

is strictly monotone, coercive, and continuous on finite dimensional subspaces of $H_0^1(\Omega)$. Therefore by Corollary 1.8 of Chapter III there exists a unique solution to (3.3) for each $\varepsilon > 0$. We now state a theorem which will serve us in our proof of regularity. Its proof is discussed in the appendix.

Theorem 3.3. *Let $a(p)$ be a C^1 vector field and let $g \in L^\infty(\Omega)$. Suppose that*

$$\begin{aligned} u \in H^1(\Omega): \quad Au = -(\partial/\partial x_i)a_i(u_x) &= g \qquad in \quad \Omega, \\ u &= 0 \qquad on \quad \partial\Omega \end{aligned} \tag{3.4}$$

and that

$$(\partial a_i/\partial p_j)(u_x(x))\xi_i\xi_j \geq \nu|\xi|^2 \qquad \text{for} \quad \xi \in \mathbb{R}^N \quad \text{a.e.} \quad x \in \Omega,$$
$$|(\partial a_i/\partial p_j)(u_x(x))| \leq M, \qquad 1 \leq i, j \leq N$$

for some numbers $0 < \nu \leq M < \infty$. *Then for each* s *there is a* $c_0 = c_0(s, \nu, M, \Omega)$ *such that*

$$\|u\|_{H^{2,s}(\Omega)} \leq c_0 \|g\|_{L^\infty(\Omega)}.$$

As an immediate corollary we have

Corollary 3.4. *If* $a(p)$ *is a* C^1 *strongly coercive vector field (see Definition* 3.1) *and* $g \in L^\infty(\Omega)$, *then the solution* u *of* (3.4) *satisfies*

$$\|u\|_{H^{2,s}(\Omega)} \leq c_0 \|g\|_{L^\infty(\Omega)}.$$

In particular we state

Corollary 3.5. *Let* $f \in L^\infty(\Omega)$ *and let* u_ε, $\varepsilon > 0$, *denote the solution to* (3.3). *Then* $u_\varepsilon \in H^{2,s}(\Omega)$, $1 \leq s < \infty$, *and*

$$\|u_\varepsilon\|_{H^{2,s}(\Omega)} \leq c_0(2\|f\|_{L^\infty(\Omega)} + \|A\psi\|_{L^\infty(\Omega)}).$$

These considerations lead us to

Theorem 3.6. *Let* u *denote the solution to Problem* 3.2. *Then* $u \in H^{2,s}(\Omega) \cap C^{1,\lambda}(\bar{\Omega})$ *for* $1 \leq s < \infty$ *and* $0 < \lambda < 1$. *In addition,*

$$\|u\|_{H^{2,s}(\Omega)} \leq c_0(2\|f\|_{L^\infty(\Omega)} + \|A\psi\|_{L^\infty(\Omega)}),$$

where $c_0 = c_0(s, \Omega, \nu, M)$. *If* u_ε *denotes the solution to* (3.3), *then* $u_\varepsilon \to u$ *weakly in* $H^{2,s}(\Omega)$ *as* $\varepsilon \to 0$.

Proof. Of course the spirit of this theorem is the same as that of Theorem 2.3. Because of its utility, we explain again the highlights of the arguments. The first step is to show that $u_\varepsilon \in \mathbb{K}$. This is accomplished by truncation. The function $\zeta = u_\varepsilon - \max(u_\varepsilon, \psi) \in H_0^1(\Omega)$ and satisfies $\zeta \leq 0$. Our object is to show that $\zeta = 0$. From (3.3)

$$\langle Au_\varepsilon, \zeta \rangle = \int_\Omega [\max(A\psi - f, 0)\vartheta_\varepsilon(u_\varepsilon - \psi) + f]\zeta \, dx \qquad \text{for} \quad \zeta \in H_0^1(\Omega),$$

and by definition, since $\zeta \in H_0^1(\Omega)$ and ψ is smooth,

$$\langle A\psi, \zeta \rangle = \int_\Omega A\psi\zeta \, dx.$$

Now if $\zeta(x) < 0$, then $\vartheta_\varepsilon(u_\varepsilon(x) - \psi(x)) = 1$; so subtracting the two equations above we obtain

$$\langle Au_\varepsilon - A\psi, \zeta\rangle = \int_{\{\zeta<0\}} [\max(A\psi - f, 0) - (A\psi - f)]\zeta \, dx \geq 0.$$

On the other hand,

$$\zeta_{x_i} = \begin{cases} u_{\varepsilon x_i} - \psi_{x_i}, & \zeta < 0 \\ 0, & \zeta \geq 0 \end{cases}$$

(in the sense of distributions), as we recall from Appendix A of Chapter II, Theorem A.1. Consequently

$$\begin{aligned} 0 &\geq \langle Au_\varepsilon - A\psi, \zeta\rangle \\ &= \int_{\{\zeta<0\}} [a_i(u_{\varepsilon x}) - a_i(\psi_x)](u_\varepsilon - \psi)_{x_i} \, dx \\ &\geq \nu \int_{\{\zeta<0\}} (u_\varepsilon - \psi)_{x_i}^2 \, dx = \nu \int_\Omega \zeta_{x_i}^2 \, dx, \end{aligned}$$

so $\zeta = 0$.

From Corollary 3.5, $\{u_\varepsilon\}$ form a bounded set of $H^{2,s}(\Omega)$, from which we may extract a weakly convergent subsequence u_ε. The limit $\tilde{u}$ of this subsequence is shown to be a solution of (3.2) in a manner identical to that of the proof of Theorem 2.3. Namely, one applies *Minty's Lemma* 1.5 of Chapter III twice. We leave the details to the reader.

Since the solution u of (3.2) is unique, the only limit point of the $\{u_\varepsilon\}$ is u. Q.E.D.

4. Locally Coercive Vector Fields

The existence theory of Chapter III cannot be applied directly, for example, to a variational inequality associated with

$$a_i(p) = p_i/(1 + p^2)^{1/2}, \qquad p \in \mathbb{R}^N,$$

the first variation of the area integrand, because this vector field is not strongly coercive; cf. Chapter III, Section 4. To achieve the existence and the regularity of the solution to such a problem, we shall prove an a priori estimate for the gradient of the solution. We then modify the vector field and avail ourselves of the previously developed theory.

This will require hypotheses about the nature of Ω. The reader will recall that the mere existence of a solution to the boundary value problem

$$-\frac{\partial}{\partial x_i}\frac{u_{x_i}}{\sqrt{1+u_x^2}} = f \quad \text{in} \quad \Omega,$$

$$u = g \quad \text{on} \quad \partial\Omega$$

rests on certain geometrical conditions between f and $\partial\Omega$. Recall that a vector field $a(p) = (a_1(p), \ldots, a_N(p)) \in C^1(\mathbb{R}^N)$ is called *locally coercive* (see Definition 4.2 of Chapter III) if for each compact $C \subset \mathbb{R}^N$ there exists a $v = v(C) > 0$ such that

$$[a(p) - a(q)](p - q) \geq v|p - q|^2 \quad \text{for} \quad p, q \in C. \tag{4.1}$$

Let $\psi \in C^2(\bar{\Omega})$ satisfy

$$\max_{\Omega} \psi > 0 \quad \text{and} \quad \psi < 0 \quad \text{on} \quad \partial\Omega$$

and set

$$\mathbb{K} = \mathbb{K}_\psi = \{v \in H_0^{1,\infty}(\Omega) : v \geq \psi \text{ in } \Omega\}. \tag{4.2}$$

Note that the convex is now of Lipschitz functions, not a reflexive Banach space. We shall assume that Ω is *convex*. Consider

Problem 4.1. *To find*

$$u \in \mathbb{K}: \quad \int_\Omega a_i(u_x)(v - u)_{x_i}\, dx \geq 0 \quad \textit{for} \quad v \in \mathbb{K}.$$

It is easy to see, by (4.1), that a solution of Problem 4.1 is unique provided it exists. We now state our a priori estimate.

Lemma 4.2. *Let Ω be convex and let $a(p)$ be a C^1 locally coercive vector field. Suppose that $u \in C^1(\bar{\Omega}) \cap H^2(\Omega)$ is a solution to Problem* 4.1. *Then*

$$\|u_x\|_{L^\infty(\Omega)} \leq \|\psi_x\|_{L^\infty(\Omega)}.$$

Proof. First we note that the continuously differentiable function $u - \psi$ attains its minimum at any point $x \in I$. Hence

$$u_x(x) = \psi_x(x) \quad \text{for} \quad x \in I = \{x \in \Omega : u(x) = \psi(x)\}. \tag{4.3}$$

Now $u > \psi$ in $\Omega - I$, from which it follows that

$$Au = -(\partial/\partial x_i)a_i(u_x) = 0 \quad \text{in} \quad \Omega - I$$

in the sense of distributions. For any direction ξ, we choose $\zeta \in C_0^\infty(\Omega - I)$ and compute

$$0 = \int_{\Omega - I} a_i(u_x)(\partial\zeta/\partial\xi)_{x_i}\,dx$$

$$= -\int_{\Omega - I} \alpha_{ij}(x)(\partial u/\partial\xi)_{x_j}\zeta_{x_i}\,dx,$$

where

$$\alpha_{ij}(x) = (\partial a_i/\partial p_j)(u_x), \qquad 1 \le i, j \le N.$$

Recalling that $a(p)$ is locally coercive and $|u_x| \in L^\infty(\Omega)$, we have

$$\alpha_{ij}(x)\xi_i\xi_j \ge \nu|\xi|^2 \qquad \text{for} \quad \xi \in \mathbb{R}^N \quad \text{a.e.} \quad x \in \Omega - I$$

and

$$|\alpha_{ij}(x)| \le M \qquad \text{a.e.} \quad x \in \Omega - I,$$

for some numbers $0 < \nu \le M < \infty$. Hence we may apply the weak maximum principle to $(\partial u/\partial\xi)(x)$, Theorem 5.5 of Chapter II, which yields

$$|(\partial u/\partial\xi)(x)| \le \sup_{\partial(\Omega - I)} |(\partial u/\partial\xi)(y)|$$

$$\le \max\left(\sup_I |(\partial\psi/\partial\xi)(y)|, \sup_{\partial\Omega} |(\partial u/\partial\xi)(y)|\right), \qquad x \in \Omega - I,$$

in view of (4.3). Again using (4.3), we infer that

$$|(\partial u/\partial\xi)(x)| \le \max\left(\sup_I |(\partial\psi/\partial\xi)(y)|, \sup_{\partial\Omega} |(\partial u/\partial\xi)(y)|\right) \qquad \text{for} \quad x \in \Omega. \tag{4.4}$$

By convexity of Ω, given $x_0 \in \partial\Omega$ there is an affine function $\pi(x)$ with the properties

$$\pi(x_0) = 0,$$

$$\pi(x) \ge \max(\psi(x), 0), \qquad x \in \Omega,$$

$$|\pi_x(x)| \le \sup_\Omega |\psi_x(x)|.$$

We now show that

$$0 \le u(x) \le \pi(x) \qquad \text{for} \quad x \in \Omega.$$

To this end, set $w = \min(u, \pi) \in \mathbb{K}$; then

$$\int_\Omega a_i(u_x)(w - u)_{x_i}\,dx \geq 0.$$

Also, $w - u \in H_0^1(\Omega)$ and

$$A\pi = -\frac{\partial}{\partial x_i} a_i(\pi_x) = 0 \quad \text{in} \quad \Omega,$$

so

$$\int_\Omega a_i(\pi_x)(w - u)_{x_i}\,dx = 0.$$

Subtracting, we obtain

$$\int_\Omega [a_i(w_x) - a_i(u_x)](w - u)_{x_i}\,dx = \int_\Omega [a_i(\pi_x) - a_i(u_x)](w - u)_{x_i}\,dx \leq 0.$$

Since the vector field $a(p)$ is strictly monotone, $w = u$, i.e., $u \leq \pi$. In a similar way it follows that $u \geq 0$ in Ω. With n an exterior normal direction to Ω at $x_0 \in \partial\Omega$,

$$0 \leq (1/h)[u(x_0 - hn) - u(x_0)] \leq (1/h)[\pi(x_0 - hn) - \pi(x_0)], \qquad h > 0,$$

so

$$0 \leq -(\partial u/\partial n)(x_0) \leq -(\partial \pi/\partial n)(x_0) \leq \|\psi_x\|_{L^\infty(\Omega)}.$$

Consequently, (4.4) assumes the form

$$|(\partial u/\partial \xi)(x)| \leq \|\psi_x\|_{L^\infty(\Omega)} \qquad \text{for} \quad x \in \Omega.$$

Finally, given $\varepsilon > 0$, we may choose $x_0 \in \Omega$ and $\xi \in \mathbb{R}^N$ such that

$$(\partial u/\partial \xi)(x_0) = \|u_x\|_{L^\infty(\Omega)} - \varepsilon,$$

whence it follows that

$$\|u_x\|_{L^\infty(\Omega)} \leq \|\psi_x\|_{L^\infty(\Omega)} + \varepsilon. \quad \text{Q.E.D.}$$

Our main result is

Theorem 4.3. *Let $a(p)$ be a C^1 locally coercive vector field on $\mathbb{R}^N$ and let Ω be convex with smooth boundary in $\mathbb{R}^N$. Let $\psi \in C^2(\overline{\Omega})$ satisfy* $\max_\Omega \psi > 0$ *and $\psi < 0$ on $\partial\Omega$ and set*

$$\mathbb{K} = \{v \in H_0^{1,\infty}(\Omega) : v \geq \psi \text{ in } \Omega\}.$$

Then there exists a unique

$$u \in \mathbb{K}: \quad \int_\Omega a_i(u_x)(v-u)_{x_i}\,dx \geq 0 \qquad \text{for} \quad v \in \mathbb{K}. \tag{4.5}$$

Moreover, $u \in H^{2,s}(\Omega) \cap C^{1,\lambda}(\overline{\Omega})$ *for* $1 \leq s < \infty, 0 \leq \lambda < 1$.

Proof. Let $M = \sup_\Omega |\psi_x|$ and consider a strongly coercive vector field $\tilde{a}(p)$ with the property

$$\tilde{a}(p) = \begin{cases} a(p) & \text{for} \quad |p| \leq M \\ kp & \text{for} \quad |p| \geq 3M. \end{cases}$$

Such a vector field was constructed in Lemma 4.3 of Chapter III. According to Theorem 3.6, there exists a unique solution $u \in H_0^1(\Omega)$ to the problem

$$u \in \mathbb{K}: \quad \int_\Omega \tilde{a}_i(u_x)(v-u)_{x_i}\,dx \geq 0 \qquad \text{for} \quad v \in \mathbb{K}.$$

Since by Theorem 3.6 $u \in C^1(\overline{\Omega})$, we may apply the preceding lemma to conclude

$$\|u_x\|_{L^\infty(\Omega)} \leq \|\psi_x\|_{L^\infty(\Omega)} = M.$$

Hence $\tilde{a}_i(u_x) = a_i(u_x)$ and u is a solution to (4.2). As we have already remarked, u is unique in $\mathbb{K}$. Q.E.D.

5. Another Penalization

In this section we introduce a different form of penalized equation, based on the concept of a monotone graph, and show that its solutions converge to the solution of the variational inequality. This will also provide us with a second proof of regularity for the solution. Regularity is further studied in successive sections.

We always assume that $\Omega \subset \mathbb{R}^N$ is a bounded connected domain with smooth boundary $\partial\Omega$. Choose a function $\alpha(t) \in C^\infty(\mathbb{R})$ such that

$$\alpha'(t) \geq 0 \qquad \text{and} \qquad \alpha(t) \begin{cases} = 0 & \text{for} \quad t \geq 0 \\ < 0 & \text{for} \quad t < 0 \\ \text{linear} & \text{for} \quad |t| \text{ large} \end{cases}$$

and define

$$\beta_\varepsilon(t) = \varepsilon t + (1/\varepsilon)\alpha(t), \qquad \varepsilon > 0. \tag{5.1}$$

Note that

$$\lim_{\varepsilon \to 0} \beta_\varepsilon(t) = \begin{cases} 0, & t \geq 0 \\ -\infty, & t < 0. \end{cases}$$

Let $\psi \in C^2(\bar{\Omega})$ be an obstacle, that is, $\max_\Omega \psi > 0$ and $\psi \leq 0$ on $\partial\Omega$, and let $f \in L^\infty(\Omega)$. Let $a(p) = (a_1(p), \ldots, a_N(p))$ be a strongly coercive C^1 vector field. We shall approximate the solution u to Problem 3.2 by solutions u_ε of the equation

$$u_\varepsilon \in H_0^1(\Omega) : Au_\varepsilon + \beta_\varepsilon(u_\varepsilon - \psi) = f \qquad \textit{in} \quad \Omega. \tag{5.2}$$

Note that the operator

$$v \to Av + \beta_\varepsilon(v - \psi) - f, \qquad v \in H_0^1(\Omega),$$

is strictly monotone, coercive, and continuous on finite dimensional subspaces because $\beta_\varepsilon(t)$ is increasing. Hence there exists a unique solution to (5.2). It is smooth in Ω by a well-known theorem about elliptic equations analogous to Theorem 3.3.

Lemma 5.1. *Let* u_ε, $0 < \varepsilon \leq 1$, *denote the solution to* (5.2). *Then*

$$\|Au_\varepsilon\|_{L^\infty(\Omega)} \leq c_1 \qquad \textit{and} \qquad \|u_\varepsilon\|_{H^{2,s}(\Omega)} \leq c_0 c_1, \qquad 1 \leq s < \infty,$$

where $c_1 = 2(2\|f\|_{L^\infty(\Omega)} + \|A\psi\|_{L^\infty(\Omega)} + \|\psi\|_{L^\infty(\Omega)})$ *and* $c_0 = c_0(\nu, M, s, \Omega)$ *is the constant of Theorem* 3.3.

Proof. Let $\alpha_\varepsilon(t) = (1/\varepsilon)\alpha(t) = \beta_\varepsilon(t) - \varepsilon t$. We first show that $\alpha_\varepsilon(u_\varepsilon - \psi)$ is bounded independently of ε. Now $\alpha_\varepsilon(t) = 0$ for $t \geq 0$ so $\alpha_\varepsilon(u_\varepsilon - \psi) = 0$ on $\partial\Omega$. Subtracting $A\psi$ from each side of (5.2) and multiplying the resulting equation by $[\alpha_\varepsilon(u_\varepsilon - \psi)]^{p-1}$, $p > 0$ even, we obtain

$$\begin{aligned}(Au_\varepsilon - A\psi)[\alpha_\varepsilon(u_\varepsilon - \psi)]^{p-1} &+ \beta_\varepsilon(u_\varepsilon - \psi)[\alpha_\varepsilon(u_\varepsilon - \psi)]^{p-1} \\ &= (f - A\psi)[\alpha_\varepsilon(u_\varepsilon - \psi)]^{p-1}.\end{aligned}$$

We integrate the first term by parts. There results

$$\begin{aligned}&\int_\Omega (Au_\varepsilon - A\psi)[\alpha_\varepsilon(u_\varepsilon - \psi)]^{p-1}\, dx \\ &\quad = (p-1)\int_\Omega [a_i(u_{\varepsilon x}) - a_i(\psi_x)](u_\varepsilon - \psi)_{x_i} \alpha_\varepsilon'(u_\varepsilon - \psi)[\alpha(u_\varepsilon - \psi)]^{p-2}\, dx \\ &\quad \geq 0,\end{aligned}$$

by monotonicity of $a(p)$, evenness of p, and our assumption that $\alpha_\varepsilon'(t) \geq 0$. Consequently

$$\varepsilon \int_\Omega (u_\varepsilon - \psi)[\alpha_\varepsilon(u_\varepsilon - \psi)]^{p-1}\, dx + \int_\Omega [\alpha_\varepsilon(u_\varepsilon - \psi)]^p\, dx$$
$$\leq \int_\Omega (f - A\psi)[\alpha_\varepsilon(u_\varepsilon - \psi)]^{p-1}\, dx.$$

Each term on the left-hand side is nonnegative because p is even and $t\alpha(t) \geq 0$, so

$$\int_\Omega |\alpha_\varepsilon(u_\varepsilon - \psi)|^p\, dx \leq \int_\Omega |f - A\psi|\,|\alpha_\varepsilon(u_\varepsilon - \psi)|^{p-1}\, dx.$$

Applying Hölder's inequality on the right-hand side, we see that

$$\|\alpha_\varepsilon(u_\varepsilon - \psi)\|_{L^p(\Omega)}^p \leq \|f - A\psi\|_{L^p(\Omega)} \|\alpha_\varepsilon(u_\varepsilon - \psi)\|_{L^p(\Omega)}^{p-1},$$

or

$$\|\alpha_\varepsilon(u_\varepsilon - \psi)\|_{L^p(\Omega)} \leq \|f - A\psi\|_{L^p(\Omega)}.$$

In the limit as $p \to \infty$,

$$\|\alpha_\varepsilon(u_\varepsilon - \psi)\|_{L^\infty(\Omega)} \leq \|f - A\psi\|_{L^\infty(\Omega)}. \tag{5.3}$$

We next show that u_ε itself is bounded. We write (5.2) in the form

$$Au_\varepsilon + \varepsilon u_\varepsilon = f - \alpha_\varepsilon(u_\varepsilon - \psi) + \varepsilon\psi = g,$$

where $g \in L^\infty(\Omega)$. Multiplying both sides of the above by $(u_\varepsilon)^{p-1}$, p even, and integrating, we deduce that

$$(p - 1) \int_\Omega a_i(u_{\varepsilon x}) u_{\varepsilon x_i} (u_\varepsilon)^{p-2}\, dx + \varepsilon \int_\Omega u_\varepsilon^p\, dx = \int_\Omega g u_\varepsilon^{p-1}\, dx.$$

The first term is nonnegative, so

$$0 \leq \varepsilon \int_\Omega u_\varepsilon^p\, dx \leq \int_\Omega g u_\varepsilon^{p-1}\, dx.$$

Applying Hölder's inequality as before, it follows that

$$\varepsilon \|u_\varepsilon\|_{L^p(\Omega)} \leq \|g\|_{L^p(\Omega)},$$

whence it follows that

$$\varepsilon \|u_\varepsilon\|_{L^\infty(\Omega)} \leq \|g\|_{L^\infty(\Omega)}. \tag{5.4}$$

Returning now to Eq. (5.2) and applying (5.3) and (5.4), we conclude that for $0 < \varepsilon \leq 1$

$$\begin{aligned}\|Au_\varepsilon\|_{L^\infty(\Omega)} &\leq 2(\|f\|_{L^\infty(\Omega)} + \|f - A\psi\|_{L^\infty(\Omega)} + \varepsilon\|\psi\|_{L^\infty(\Omega)}) \\ &\leq 2(2\|f\|_{L^\infty(\Omega)} + \|A\psi\|_{L^\infty(\Omega)} + \|\psi\|_{L^\infty(\Omega)}) \equiv c_1.\end{aligned}$$

The conclusion now follows from Theorem 3.3. Q.E.D.

Theorem 5.2. *Let u denote the solution to Problem* (3.2) *and $u_\varepsilon, 0 < \varepsilon \leq 1$, the solution to* (5.2). *Then*

$$u_\varepsilon \to u \qquad \textit{weakly in} \quad H^{2,s}(\Omega), \qquad 1 \leq s < \infty.$$

Proof. Let $\{u_{\varepsilon'}\}$ be a subsequence of $\{u_\varepsilon\}$ which converges weakly to a function $\tilde{u} \in H^{2,s}(\Omega)$. We appeal to a familiar argument to show that $\tilde{u}$ is the solution of the variational inequality. Since

$$v \to Av + \beta_\varepsilon(v - \psi) - f, \qquad v \in H_0^1(\Omega),$$

is monotone, Minty's lemma implies that

$$\begin{aligned}\int_\Omega a_i(v_x)(v - u_\varepsilon)_{x_i}\,dx &+ \int_\Omega \beta_\varepsilon(v - \psi)(v - u_\varepsilon)\,dx \\ &\geq \int_\Omega f(v - u_\varepsilon)\,dx \qquad \text{for} \quad v \in H_0^{1,\infty}(\Omega).\end{aligned}$$

Suppose now that $v \geq \psi + \delta$ for some $\delta > 0$. Then

$$\beta_\varepsilon(v - \psi) = \varepsilon(v - \psi) \to 0 \qquad \text{as} \quad \varepsilon \to 0,$$

so by weak convergence

$$\int_\Omega a_i(v_x)(v - \tilde{u})_{x_i}\,dx \geq \int_\Omega f(v - \tilde{u})\,dx.$$

If $\tilde{u}(x_0) - \psi(x_0) < -2\delta < 0$ for some $x_0 \in \Omega$ and $\delta > 0$, there exist a neighborhood $B_\rho(x_0)$ and an $\varepsilon_0 > 0$ such that for $\varepsilon' < \varepsilon_0$

$$u_{\varepsilon'}(x) - \psi(x) < -\delta < 0 \qquad \text{for} \quad x \in B_\rho(x_0).$$

Hence

$$\begin{aligned}\beta_{\varepsilon'}(u_{\varepsilon'}(x) - \psi(x)) &= \varepsilon'[u_{\varepsilon'}(x) - \psi(x)] + (1/\varepsilon')\alpha(u_{\varepsilon'}(x) - \psi(x)) \\ &\geq \varepsilon' C + (1/\varepsilon')\alpha(-\delta) \qquad \text{for} \quad x \in B_\rho(x_0)\end{aligned}$$

and some constant $C \geq 0$. Hence

$$\|\beta_{\varepsilon'}(u_{\varepsilon'} - \psi)\|_{L^\infty(\Omega)} \geq \varepsilon' C + (1/\varepsilon')|\alpha(-\delta)| \to \infty \qquad \text{as} \quad \varepsilon' \to 0.$$

This is a contradiction. Hence $\tilde{u} \in \mathbb{K}$.

Since any $v \in \mathbb{K}$ may be approximated by functions $v \in \mathbb{K}$ with $v > \psi$, we see that

$$\int_\Omega a_i(v_x)(v - \tilde{u})_{x_i}\, dx \geq \int_\Omega f(v - \tilde{u})\, dx \qquad \text{for all} \quad v \in \mathbb{K}.$$

Applying Minty's lemma once again, it follows that $\tilde{u} = u$, the solution to Problem 3.2. Q.E.D.

6. Limitation of Second Derivatives

We prove now that second derivatives of the solution to a variational inequality are bounded if some smoothness assumptions are satisfied by $a(p)$, ψ, and f. This result serves in the analysis of the coincidence set of the solution u. The theorem is first proved for the approximations defined by (5.2). Although we prove a theorem relative to a strongly coercive vector field, we observe that whenever a Lipschitz function is the solution to a variational inequality of the type (3.2) relative to a locally coercive vector field, the vector field may be altered by Lemma 4.3 of Chapter III so that the function is also the solution to a problem for a strongly coercive field; cf. Section 4.

As usual, $\Omega \subset \mathbb{R}^N$ is a smooth bounded connected domain with boundary $\partial\Omega$. Suppose that

$$a(p) \in C^2(\mathbb{R}^N)$$

is strongly coercive,

$$f(x) \in C^1(\bar{\Omega}), \tag{6.1}$$

$$\psi(x) \in C^2(\bar{\Omega}) \qquad \textit{satisfies} \quad \max \psi > 0 \qquad \textit{and} \qquad \psi \leq 0 \quad \textit{on} \quad \partial\Omega.$$

Let $\alpha(t) \in C^\infty(\mathbb{R})$ satisfy

$$\alpha(t)\begin{cases} < 0 & \text{for} \quad t < 0 \\ = 0 & \text{for} \quad t \geq 0, \end{cases}$$

$$\alpha'(t) \geq 0, \quad t \in \mathbb{R}, \qquad \text{and} \qquad \alpha''(t) \leq 0, \quad t \in \mathbb{R}, \tag{6.2}$$

and set $\beta_\varepsilon(t) = \varepsilon t + (1/\varepsilon)\alpha(t)$. Note that $\beta_\varepsilon'(t) > 0$ and $\beta_\varepsilon''(t) \leq 0$ for $t \in \mathbb{R}$. We denote by u_ε the solution to the problem

$$u_\varepsilon \in H_0^1(\Omega): \quad Au_\varepsilon + \beta_\varepsilon(u_\varepsilon - \psi) = f \qquad \text{in} \quad \Omega, \quad 0 < \varepsilon \leq 1. \tag{6.3}$$

We have observed in Section 5 that the solution u_ε exists and is unique. Moreover, the functions $\{u_\varepsilon : 0 < \varepsilon < 1\}$ lie in a bounded set of $H^{2,s}(\Omega)$, for each s, $1 \le s < \infty$.

Theorem 6.1. *Assume* (6.1) *and* (6.2). *Then for each compact* $K \subset \Omega$ *there is a constant* $C = C(K) > 0$ *such that*

$$\sup_K |u_{\varepsilon x_i x_j}(x)| \le C \qquad \text{for} \quad 1 \le i, j \le N \quad \text{and all} \quad \varepsilon, \quad 0 < \varepsilon \le 1.$$

The proof depends on estimating pure second order partial derivatives from below. To accomplish this we shall consider an inequality obtained by differentiating (6.3) twice. This approach does not appear promising at first glance; however, the higher order derivatives of u_ε may be expressed as divergences of functions in $L^s(\Omega)$, $s > N$. To this situation we apply Theorem B.2 of Chapter II. Instrumental in obtaining the differential inequality is the concavity of β.

Lemma 6.2. *Let* u_ε, $0 < \varepsilon \le 1$, *be the solution of* (6.2) *and* $K \subset \Omega$ *be compact. Then there exists a* $C_2 = C_2(K)$ *such that*

$$(\partial^2/\partial\xi^2)u_\varepsilon(x) \ge C_2 \qquad \text{for} \quad x \in K, \quad 0 < \varepsilon \le 1,$$

where $\partial/\partial\xi$ *is any directional derivative.*

Proof. Let $\varepsilon \in (0, 1]$ and set $u = u_\varepsilon$. In the proof, we shall depress the dependence of the various quantities on ε and denote by const any constant independent of ε, $\varepsilon \le 1$. It will be convenient for us to assume that $u \in C^4(\Omega)$, for which we temporarily impose the conditions that ψ, f, and $a(p)$ be very smooth. Since the final estimate for C_2 will depend only on $\|(\partial/\partial\xi)f\|_{L^\infty(\Omega)}$, $\|\psi\|_{H^{2,\infty}(\Omega)}$, and $\sup_{i,j,k}|[\partial^2 a_j/(\partial p_i \partial p_k)](u_x)|$, it will be clear that it is valid assuming only (6.1).

We observe that for each j, $1 \le j \le N$,

$$\begin{aligned}\frac{\partial^2}{\partial\xi^2} a_j(u_x) &= \frac{\partial}{\partial\xi}\left\{\frac{\partial a_j}{\partial p_i}(u_x)\left(\frac{\partial u}{\partial\xi}\right)_{x_i}\right\} \\ &= \frac{\partial a_j}{\partial p_i}(u_x)\left(\frac{\partial^2 u}{\partial\xi^2}\right)_{x_i} + \frac{\partial}{\partial\xi}\left(\frac{\partial a_j}{\partial p_i}(u_x)\right)\left(\frac{\partial u}{\partial\xi}\right)_{x_i}.\end{aligned}$$

Therefore

$$\frac{\partial^2}{\partial\xi^2}Au = -\frac{\partial}{\partial x_j}\left(\frac{\partial a_j}{\partial p_i}(u_x)\left(\frac{\partial^2 u}{\partial\xi^2}\right)_{x_i}\right) - \frac{\partial}{\partial x_j}\left\{\frac{\partial}{\partial\xi}\left(\frac{\partial a_j}{\partial p_i}(u_x)\right)\left(\frac{\partial u}{\partial\xi}\right)_{x_i}\right\}.$$

Since $\beta''(t) \le 0$,

$$\frac{\partial^2}{\partial \xi^2}\beta(u-\psi) = \beta'(u-\psi)(u-\psi)_{\xi\xi} + \beta''(u-\psi)(u_\xi - \psi_\xi)^2$$
$$\le \beta'(u-\psi)(u-\psi)_{\xi\xi}.$$

Combining these two relations, we obtain

$$-\frac{\partial}{\partial x_j}\left(\frac{\partial a_j}{\partial p_i}(u_x)\left(\frac{\partial^2 u}{\partial \xi^2}\right)\right)_{x_i} + \beta'(u-\psi)\frac{\partial^2}{\partial \xi^2}(u-\psi)$$
$$\ge \frac{\partial}{\partial x_j}\frac{\partial^2 u}{\partial \xi\,\partial x_i}\frac{\partial}{\partial \xi}\frac{\partial a_i}{\partial p_i}(u_x) + \frac{\partial^2}{\partial \xi^2} f. \tag{6.4}$$

Let us denote by $L = L_\varepsilon$ the linear operator

$$Lw = -\frac{\partial}{\partial x_i}\left[\frac{\partial a_j}{\partial p_i}(u_x) w_{x_i}\right]$$

and note that

$$\nu|\lambda|^2 \le \frac{\partial a_j}{\partial p_i}(u_x)\lambda_i\lambda_j \qquad \text{for} \quad \lambda \in \mathbb{R}^N$$

and (6.5)

$$\left|\frac{\partial a_j}{\partial p_i}(u_x)\right| \le M \qquad \text{for} \quad x \in \Omega, \qquad 1 \le i, j \le N,$$

for some $0 < \nu \le M < \infty$ independent of ε. This is a consequence of the strong coerciveness of $a(p)$ and the boundedness of u_x, of Lemma 5.1.

Restricting our attention to a compact $K \subset \Omega$, let $\zeta \in C_0^\infty(\Omega)$ satisfy $\zeta = 1$ on K, $\zeta \ge 0$ in Ω. One verifies that for any $w \in C^2(\Omega)$,

$$L(\zeta w) = \zeta Lw - \frac{\partial}{\partial x_i}\left(w\frac{\partial a_j}{\partial p_i}(u_x)\zeta_{x_j}\right) - \frac{\partial}{\partial x_j}\left(w\frac{\partial a_i}{\partial p_j}(u_x)\zeta_{x_i}\right)$$
$$+ \frac{\partial}{\partial x_i}\left(\frac{\partial a_j}{\partial p_i}(u_x)\zeta_{x_j}\right). \tag{6.6}$$

Multiplying (6.4) by ζ and setting $w = u_{\xi\xi}$ in (6.6) yields

$$L(\zeta u_{\xi\xi}) + \zeta\beta'(u-\psi)(u-\psi)_{\xi\xi}$$
$$\ge -\frac{\partial}{\partial x_i}\left(u_{\xi\xi}\frac{\partial a_j}{\partial p_i}(u_x)\zeta_{x_i}\right) - \frac{\partial}{\partial x_j}\left(u_{\xi\xi}\frac{\partial a_j}{\partial p_i}(u_x)\zeta_{x_i}\right) + u_{\xi\xi}\frac{\partial}{\partial x_i}\left(\frac{\partial a_j}{\partial p_i}(u_x)\zeta_{x_j}\right)$$
$$+ \zeta\frac{\partial}{\partial x_j}\left(\frac{\partial}{\partial x_i}u_\xi\frac{\partial}{\partial \xi}\frac{\partial a_j}{\partial p_i}(u_x)\right) + \zeta\frac{\partial^2}{\partial \xi^2} f.$$

We rewrite this in the form

$$\begin{aligned} &L(\zeta u_{\xi\xi}) + \zeta\beta'(u-\psi)(u-\psi)_{\xi\xi} \\ &\geq -\frac{\partial}{\partial x_i}\left(u_{\xi\xi}\frac{\partial a_j}{\partial p_i}(u_x)\zeta_{x_j}\right) - \frac{\partial}{\partial x_j}\left(u_{\xi\xi}\frac{\partial a_j}{\partial p_i}(u_x)\zeta_{x_i}\right) \\ &\quad + u_{\xi\xi}\frac{\partial}{\partial x_i}\left(\frac{\partial a_j}{\partial p_i}(u_x)\zeta_{x_j}\right) + \frac{\partial}{\partial x_j}\left(\zeta\frac{\partial}{\partial x_i}u_\xi\frac{\partial}{\partial \xi}\left(\frac{\partial a_j}{\partial p_i}(u_x)\right)\right) \\ &\quad - \zeta_{x_j}\frac{\partial}{\partial x_i}u_\xi\frac{\partial}{\partial \xi}\frac{\partial a_j}{\partial p_i}(u_x) + \frac{\partial}{\partial \xi}\left(\zeta\frac{\partial f}{\partial \xi}\right) - \frac{\partial f}{\partial f}\frac{\partial \zeta}{\partial \xi}. \end{aligned} \tag{6.7}$$

With exception of the terms involving f, any term on the right-hand side is the product of second derivatives of u multiplied by a bounded function or else the derivative of such a quantity. So we can rewrite (6.7) in this manner:

$$L(\zeta u_{\xi\xi}) + \zeta\beta'(u-\psi)(u-\psi)_{\xi\xi} \geq g_0 + \sum_{i=1}^{N}(g_i)_{x_i}, \tag{6.8}$$

where

$$\sum_0^N \|g_i\|_{L^s(\Omega)} \leq \text{const}(\|u\|^2_{H^{2,2s}(\Omega)} + \|f_\xi\|_{L^s(\Omega)}),$$

for any s, $1 \leq s < \infty$. According to Theorem 5.5 of Chapter II there is a solution $w \in H_0^1(\Omega)$ to the problem

$$\begin{aligned} Lw &= g_0 + \sum_1^N (g_i)_{x_i} \quad &&\text{in} \quad \Omega, \\ w &= 0 &&\text{on} \quad \partial\Omega, \end{aligned}$$

and by Theorem B.2 of Chapter II

$$\|w\|_{L^\infty(\Omega)} \leq \text{const} \sum_0^N \|g_i\|_{L^s(\Omega)} = \text{const}.$$

Hence by (6.8),

$$L(\zeta u_{\xi\xi} - w) + \zeta\beta'(u-\psi)(u_{\xi\xi} - \psi_{\xi\xi}) \geq 0 \qquad \text{in} \quad \Omega.$$

Finally, let $x_0 \in \overline{\Omega}$ be a point where $\zeta u_{\xi\xi} - w$ achieves its minimum. If $\zeta(x_0) = 0$, then

$$\zeta(x)u_{\xi\xi}(x) - w(x) \geq -w(x_0) \qquad \text{in} \quad \Omega;$$

whence it follows that

$$\zeta(x)u_{\xi\xi}(x) \geq -2\|w\|_{L^\infty(\Omega)}.$$

If $\zeta(x_0) > 0$, then $x_0 \in \Omega$ and $L(\zeta u_{\xi\xi} - w)|_{x_0} \leq 0$, so

$$\zeta\beta'(u - \psi)(u_{\xi\xi} - \psi_{\xi\xi})|_{x_0} \geq 0.$$

Now, $\zeta(x_0) > 0$ and $\beta'(t) > 0$ imply that $u_{\xi\xi}(x_0) \geq \psi_{\xi\xi}(x_0)$, so

$$\begin{aligned}\zeta(x)u_{\xi\xi}(x) - w(x) &\geq \zeta(x_0)u_{\xi\xi}(x_0) - w(x_0)\\ &\geq \zeta(x_0)\psi_{\xi\xi}(x_0) - w(x_0).\end{aligned}$$

Consequently,

$$\zeta(x)u_{\xi\xi}(x) \geq -\|\psi\|_{H^{2,\infty}(\Omega)} - 2\|w\|_{L^\infty(\Omega)} = c_2(K), \qquad x \in \Omega.$$

In particular,

$$u_{\xi\xi}(x) \geq C_2, \qquad x \text{ in } K.$$

The lemma is proved. Q.E.D.

Proof of Theorem 6.1. As before, we fix a $K \subset \Omega$, compact, and depress the dependence of the various quantities on ε, $0 < \varepsilon \leq 1$. At a given point $x_0 \in K$, we assume $u_{x_i x_j}(x_0) = 0$ for $i \neq j$. Let us observe that this entails no loss in generality. Let C be an orthogonal matrix and set

$$y = (y_1, \ldots, y_N) = x_0 + C(x - x_0)$$

and

$$u'(y) = u(x) = u(x_0 + {}^tC(y - x_0)),$$

where ${}^tC$ denotes the transpose of C. It follows that u' is a solution to the equation

$$\begin{aligned}-(\partial/\partial y_i)a_i'(u_y') + \beta(u' - \psi') &= f' \quad &&\text{in} \quad \Omega',\\ u' &= 0 &&\text{on} \quad \partial\Omega',\end{aligned}$$

where $\psi'(y) = \psi(x)$, $f'(y) = f(x)$, and $\Omega' = x_0 + C(\Omega - x_0)$. The vector field $a'(p)$ is given by the formula

$$a'(p) = Ca({}^tCp),$$

where the vectors a, a', and p are written as columns. From (6.1) it is clear that

$a'(p)$ is strongly coercive; indeed,

$$(a'(p) - a'(q), p - q) = (a({}^tCp) - a({}^tCq), {}^tCp - {}^tCq) \geq \nu|{}^tCp - {}^tCq|^2 = \nu|p - q|^2,$$

where ν is the constant of coerciveness of the original a. Also

$$\left|\frac{\partial^2 a_j'(p)}{\partial p_i\, \partial p_k}\right| \leq \left(\sum_{k,l,m}\left|\frac{\partial^2 a_k(p)}{\partial p_l\, \partial p_m}\right|^2\right)^{1/2}, \qquad 1 \leq i, j, k \leq N.$$

Hence the conclusions of Lemmas 5.1 and 6.2 hold for u' with constants C_i, $i = 1, 2$, increased in modulus at most by a function of the dimension N alone. We choose C so that $u'_{y_i y_j}(x_0) = 0$ for $i \neq j$. In other words, having fixed beforehand $x_0 \in K$, we may as well assume that $u_{x_i x_j}(x_0) = 0$ for $i \neq j$. At this point

$$Au(x_0) = -\sum (\partial a_i/\partial p_i)(u_x(x_0))u_{x_i x_i}(x_0)$$

with

$$(\partial a_i/\partial p_i)(u_x(x_0)) \geq \nu > 0, \qquad i = 1, 2, \ldots, N.$$

Hence

$$C_2 \leq u_{x_i x_i}(x_0) \leq \frac{1}{\nu}\left(\|Au\|_{L^\infty(\Omega)} - (N-1)\tilde{C} \sup_{k,j}\left|\frac{\partial a_k}{\partial p_j}(u_x(x_0))\right|\right),$$

$$1 \leq i \leq N, \qquad \tilde{C} = \min(0, C_2).$$

We may now choose an arbitrary coordinate system. It follows that in any coordinate system $(x_1, \ldots, x_N)$,

$$|u_{x_i x_j}(x_0)| \leq C(K), \qquad x_0 \in K, \quad 1 \leq i, j \leq N,$$

for an appropriate $C(K)$. Q.E.D.

Theorem 6.3. *Assume* (6.1) *and let u denote the solution to Problem* 3.2. *Then for each compact $K \subset \Omega$ there is a $C(K)$ such that*

$$|u_{x_i x_j}(x)| \leq C(K), \qquad x \in K.$$

The proof is evident from the weak convergence of u_ε to u in $H^{2,s}(\Omega)$ and Theorem 6.1.

Corollary 6.4. *Assume* (6.1) *and let u denote the solution to Problem* 3.2. *If, in addition, $\psi < 0$ on $\partial\Omega$, then*

$$u \in H^{2,\infty}(\Omega).$$

Proof. Since $\psi < 0$ on $\partial\Omega$ there is a neighborhood of $\partial\Omega$ in Ω where $u > \psi$ and hence where $Au = f$. It follows from the standard regularity theory that $u \in C^2$ of this neighborhood and from Theorem 6.3 that it is in $H^{2,\infty}$ of its complement. Q.E.D.

7. Bounded Variation of Au

In this section we show that Au is a function of locally bounded variation when (6.1) is satisfied and $\psi \in C^3(\Omega)$. This will then provide us with a geometric property of the coincidence set. Precisely, for $\omega \subset \Omega$ open, we say that $f \in L^1(\omega)$ is of bounded variation in ω, or $f \in BV(\omega)$, if there exists a constant $C > 0$ such that

$$\left| \int_\omega f \zeta_{x_i} \, dx \right| \leq C \|\zeta\|_{L^\infty(\Omega)} \qquad \text{for} \quad 1 \leq i \leq N \quad \text{and all} \quad \zeta \in C^\infty(\Omega).$$

If $f \in BV(\omega)$, we define its variation $V_\omega f$ to be

$$V_\omega f = \sup\left\{ \sum_1^N \int_\omega f \zeta_{ix_i} \, dx : \zeta_i \in C^\infty(\Omega), \sum_1^N \zeta_i^2 \leq 1 \right\}. \tag{7.1}$$

If $V_\omega f < \infty$ for each $\omega \subset \bar{\omega} \subset \Omega$, we say that $f \in BV_{\text{loc}}(\Omega)$.

It is easy to see that this coincides with the classical definition of bounded variation when $N = 1$. Indeed, the functions f of $BV(\omega)$ are those functions whose distributional derivatives are (signed) measures on ω. This follows from the theorem of Schwarz because the distributions $T_i \in \mathscr{D}'(\Omega)$, $1 \leq i \leq N$, given by

$$(T_i, \zeta) = V_\omega f \cdot \int_\omega \zeta \, dx - \int_\omega \zeta_{x_i} f \, dx, \qquad \zeta \in C_0^\infty(\Omega),$$

is nonnegative. Finally, if $f_n \to f$ in L^1 weakly and $f_n \in BV(\omega)$, then

$$V_\omega f \leq \liminf_{n \to \infty} V_\omega f_n.$$

Now suppose that $E \subset \Omega$ is a Borel set and that $\varphi_E \in BV(\Omega)$. Then E is said to be of finite perimeter (in Ω) and its perimeter $P(E)$ is defined to be

$$P(E) = V_\Omega(\varphi_E).$$

If $E \subset \Omega$ is a compact manifold with smooth boundary ∂E, then $P(E)$ is the surface area of ∂E in the classical sense. So the finiteness of $P(E)$ for a general E is an indication of the area of its boundary. In particular, let us suppose that

E is a set of finite perimeter. Let us denote by μ_i, $i = 1, \ldots, N$, the measures which are the distributional partial derivatives of φ_E, that is,

$$\int_\Omega \varphi_E \zeta_{x_i}\, dx = \int_\Omega \zeta\, d\mu_i, \qquad \zeta \in C^\infty(\Omega), \quad i = 1, \ldots, N. \tag{7.2}$$

If supp $\zeta \cap \partial E = \emptyset$, then, obviously,

$$\int_\Omega \varphi_E \zeta_{x_i}\, dx = 0,$$

so

$$\operatorname{supp} \mu_i \subset \partial E, \qquad i = 1, \ldots, N. \tag{7.3}$$

Formulas (7.2) and (7.3) give us a version of Green's theorem for a set of finite perimeter.

Finally observe that if E has finite perimeter, $F \subset E$, and $\operatorname{meas}(E \setminus F) = 0$, then $P(E) = P(F)$. This is evident from (7.1).

As in the previous section, we shall prove a local result. The reader who wishes to visualize the analytic ideas of the proof is encouraged to neglect the function ζ below together with all boundary integrals.

Theorem 7.1. *Assume* (6.1) *and* (6.2). *In addition suppose that* $\psi \in C^3(\Omega)$. *Let* u_ε, $0 < \varepsilon \le 1$, *denote the solution to* (6.3). *Then for each* $\omega \subset \bar{\omega} \subset \Omega$, *open, there is a constant* $C(\omega) > 0$ *independent of* ε, $0 < \varepsilon \le 1$, *such that*

$$V_\omega(Au_\varepsilon) \le C(\omega), \qquad 0 < \varepsilon \le 1.$$

Before proving this theorem, let us mention its consequences.

Theorem 7.2. *Assume* (6.1) *and assume* $\psi \in C^3(\Omega)$. *Let u denote the solution to Problem* (3.2). *Then* $Au \in BV_{\text{loc}}(\Omega)$.

Proof. This follows immediately from the lower semicontinuity of the variation. Q.E.D.

Corollary 7.3. *Assume* (6.1) *and that* $\psi \in C^3(\Omega)$. *Suppose that*

$$A\psi - f \ne 0 \qquad \text{in} \quad \Omega.$$

Let u denote the solution to Problem (3.2). *Then the coincidence set* $I = \{x \in \Omega : u(x) = \psi(x)\}$ *of u is of locally finite perimeter in* Ω.

Proof. Since $Au \in BV_{\text{loc}}(\Omega)$ and $0 \ne A\psi - f$ is $C^1(\Omega)$,

$$\varphi_I(x) = [Au(x) - f(x)]/[A\psi(x) - f(x)] \qquad \text{a.e. in} \quad \Omega$$

is of locally bounded variation in Ω. Q.E.D.

Proof of Theorem 7.1. Again we suppress the dependence of the various quantities on ε, $0 < \varepsilon \le 1$, and denote by const any constant independent of ε. We select an approximation to sgn t, that is, a sequence of smooth functions $\gamma_\delta(t)$, $\delta > 0$, satisfying

$$|\gamma_\delta(t)| \le 1, \qquad \gamma_\delta'(t) \ge 0 \quad \text{for} \quad -\infty < t < \infty,$$

$$\gamma_\delta(0) = 0 \qquad \text{and} \qquad \lim_{\delta \to 0} \gamma_\delta(t) = \operatorname{sgn} t = \begin{cases} -1, & t < 0 \\ 0, & t = 0 \\ 1, & t > 0. \end{cases}$$

Given $\omega \subset \overline{\omega} \subset \Omega$, we select $\zeta \in C_0^\infty(\Omega)$ satisfying $\zeta = 1$ on ω and $0 \le \zeta \le 1$ in Ω.

Now differentiating (6.3) with respect to the direction ξ, we obtain

$$(\partial/\partial\xi)Au + \beta'(u - \psi)(u_\xi - \psi_\xi) = (\partial/\partial\xi)f \quad \text{in} \quad \Omega, \tag{7.4}$$

and

$$\begin{aligned} &[(\partial/\partial\xi)Au]\zeta\gamma_\delta(u_\xi - \psi_\xi) + \beta'(u - \psi)(u - \psi)_\xi\zeta\gamma_\delta(u_\xi - \psi_\xi) \\ &\quad = (\partial f/\partial\xi)\zeta\gamma_\delta(u_\xi - \psi_\xi). \end{aligned} \tag{7.5}$$

For $\gamma = \gamma_\delta(u_\xi - \psi_\xi)$, we compute that

$$\begin{aligned} \int_\Omega [(\partial/\partial\xi)Au]\zeta\gamma\, dx &= \int_\Omega (\partial/\partial\xi)a_j(u_x)\gamma_{x_j}\zeta\, dx \\ &\quad + \int_\Omega (\partial/\partial\xi)a_j(u_x)\gamma\zeta_{x_j}\, dx \\ &= \int_\Omega (\partial a_j/\partial p_i)(u_x)[(\partial u_\xi/\partial x_i) - (\partial\psi_\xi/\partial x_i)]\gamma_{x_j}\zeta\, dx \\ &\quad + \int_\Omega (\partial/\partial\xi)a_j(u_x)\gamma\zeta_{x_j}\, dx \\ &\quad + \int_\Omega (\partial a_j/\partial p_i)(u_x)(\partial/\partial x_i)\psi_\xi\gamma_{x_j}\zeta\, dx. \end{aligned}$$

The first term on the right-hand side is nonnegative, that is,

$$\begin{aligned} &\int_\Omega (\partial a_j/\partial p_i)(u_x)[(\partial u_\xi/\partial x_i) - (\partial\psi_\xi/\partial x_i)]\gamma_{x_i}\zeta\, dx \\ &\quad = \int_\Omega (\partial a_j/\partial p_i)(u_x)[(\partial u_\xi/\partial x_i) - (\partial\psi_\xi/\partial x_i)][(\partial u_\xi/\partial x_j) - (\partial\psi_\xi/\partial x_j)]\gamma'\zeta\, dx \\ &\quad \ge 0, \end{aligned}$$

because $a(p)$ is monotone and $\gamma' \geq 0$. We infer the inequality

$$-\int_\Omega [(\partial/\partial\xi)Au]\gamma\zeta\, dx \leq \int_\Omega (\partial/\partial x_j)[(\partial a_j/\partial p_i)(u_x)\psi_{\xi x_i}\zeta]\gamma\, dx - \int_\Omega (\partial/\partial\xi)a_j(u_x)\zeta_{x_j}\gamma\, dx.$$

Observe that

$$\int_\Omega \frac{\partial}{\partial x_j}\left[\frac{\partial a_j}{\partial p_i}(u_x)\psi_{\xi x_i}\zeta\right]\gamma\, dx$$
$$\leq \int_{\operatorname{supp}\zeta} \sup_{i,j}\left|\frac{\partial a_j}{\partial p_i}(u_x)\right| \sup_{i,j}(|\psi_{\xi x_i x_j}| + |\zeta_{x_j}|\,|\psi_{x_i\xi}|)\, dx$$
$$+ \int_{\operatorname{supp}\zeta} \sup|\psi_{x_i\xi}|\left|\sum \frac{\partial^2 a_j}{\partial p_i\,\partial p_k}(u_x)u_{x_j x_k}\right| dx$$

to conclude that

$$-\int_\Omega [(\partial/\partial\xi)Au]\gamma\zeta\, dx \leq \operatorname{const}(\|\psi\|_{C^3(\operatorname{supp}\zeta)} + 1)\|u\|_{H^{2,s}(\Omega)} \tag{7.6}$$

for any fixed $s \leq \infty$. This estimate is independent of ε, δ so we use (7.3) in order to derive that

$$\int_\Omega \zeta\beta'(u-\psi)\gamma_\delta(u_\xi - \psi_\xi)(u_\xi - \psi_\xi)\zeta\, dx \leq \operatorname{const} + \int |f_\xi|\,|\gamma_\delta(u_\xi - \psi_\xi)|\, dx$$
$$\leq \operatorname{const}.$$

Recall now that

$$\lim_{\delta\to 0} \frac{\partial}{\partial\xi}\beta'(u-\psi)\gamma_\delta(u_\xi - \psi_\xi) = \left|\frac{\partial}{\partial\xi}\beta(u-\psi)\right|$$

pointwise in Ω, so by the bounded convergence theorem

$$\int_\Omega \zeta|(\partial/\partial\xi)\beta(u-\psi)|\, dx \leq \operatorname{const}.$$

Returning now to (7.2),

$$\int_\Omega |(\partial/\partial\xi)Au|\zeta\, dx \leq \operatorname{const} + \int_\Omega \zeta|(\partial f/\partial\xi)|\, dx \leq \operatorname{const}$$
$$\leq (1/N)C(\omega).$$

The theorem follows after an integration by parts. Q.E.D.

8. Lipschitz Obstacles

In this section we shall refine our study of obstacle problems to include Lipschitz obstacles and their important subclass of polyhedra. For simplicity we confine ourselves to the conditions of Section 4, where $\Omega \subset \mathbb{R}^N$ is a convex domain with smooth boundary $\partial\Omega$ and $a(p)$ is a locally coercive C^1 vector field. Given an obstacle $\psi \in H^{1,\infty}(\Omega)$, namely,

$$\max_{\Omega} \psi > 0 \qquad \text{and} \qquad \psi < 0 \qquad \text{on} \quad \partial\Omega,$$

we shall consider

Problem 8.1. *To find* $u \in \mathbb{K}$: $\int_\Omega a_i(u_x)(v-u)_{x_i}\, dx \geq 0$ *for* $v \in \mathbb{K}$, *where* $\mathbb{K} = \mathbb{K}_\psi = \{v \in H_0^{1,\infty}(\Omega)\colon v \geq \psi \text{ in } \Omega\}$.

Theorem 8.2. *There exists a unique solution u to Problem* (8.1). *Moreover,* $\|u_x\|_{L^\infty(\Omega)} \leq \|\psi_x\|_{L^\infty(\Omega)}$.

Proof. This theorem is really just a corollary of Theorem 4.3. Let $M = \sup_\Omega |\psi_x|$ and let $\psi_\sigma \in C^2(\Omega)$ satisfy

$$\max_{\Omega} \psi_\sigma > 0, \qquad \psi_\sigma < 0 \qquad \text{on} \quad \partial\Omega,$$

$$\|\psi_{\sigma x}\|_{L^\infty(\Omega)} \leq M + \sigma,$$

and

$$\psi_\sigma \to \psi \qquad \text{uniformly in} \quad \overline{\Omega} \quad \text{as} \quad \sigma \to 0;$$

with u_σ the solution of (8.1) for ψ_σ, we know from Theorem 4.3 that

$$\|u_{\sigma x}\|_{L^\infty(\Omega)} \leq M + \sigma \qquad \text{and} \qquad \|u_\sigma\|_{L^\infty(\Omega)} \leq C, \tag{8.1}$$

C independent of σ for σ small. Hence we may extract a subsequence, still called u_σ, such that $u_\sigma \to u$ uniformly and weakly in $H^1(\Omega)$ to some $u \in C(\overline{\Omega}) \cap H^1(\Omega)$. Moreover, $\|u_x\|_{L^\infty(\Omega)} \leq \liminf_{\sigma\to 0} \|u_{\sigma x}\|_{L^\infty(\Omega)} = M$, and $u \geq \psi$, so $u \in \mathbb{K}$.

To show that u is a solution, we avail ourselves once again of Minty's Lemma 1.5 of Chapter III. Let $v \in \mathbb{K}$ satisfy $v > \psi$ in Ω. Then $v > \psi_\sigma$ for $\sigma \leq \sigma'$, some σ', whence it follows that

$$\int_\Omega a_i(v_x)(v - u_\sigma)_{x_i}\, dx \geq 0 \qquad \text{for} \quad \sigma \leq \sigma'.$$

As $\sigma \to 0$ we obtain

$$u \in \mathbb{K}\colon \quad \int_\Omega a_i(v_x)(v-u)_{x_i}\, dx \geq 0. \tag{8.2}$$

It is easy to see that (8.2) holds for any $v \in \mathbb{K}$, so again by Minty's lemma, we conclude that u is a solution to (8.1). Uniqueness is a consequence of the strict monotonicity of $a(p)$, as usual. Q.E.D.

Theorem 8.3. *Let G denote the class of all Lipschitz supersolutions of $Av = -(a_i(v_x))_{x_i}$ which exceed ψ in Ω and are nonnegative on $\partial\Omega$. Let u denote the solution to Problem* 8.1. *Then*

$$u(x) = \inf_{g \in G} g(x) \qquad \textit{for} \quad x \in \Omega.$$

The proof of this is almost identical to that of Theorem 6.4 of Chapter II and is omitted. As a corollary, however, we state

Corollary 8.4. *Let ψ and Φ be two obstacle functions and u, U the respective solutions to* (8.1). *If $\psi \le \Phi$ in Ω, then $u \le U$ in Ω.*

Proof. The function U is a supersolution of the operator A which satisfies $\psi \le \Phi \le U$. Now apply Theorem 8.3. Q.E.D.

Theorem 8.5. *Let ψ, ψ' be two Lipschitz obstacle functions and let u and u' denote the respective solutions to Problem* 8.1. *Then*

$$\|u - u'\|_{L^\infty(\Omega)} \le \|\psi - \psi'\|_{L^\infty(\Omega)}.$$

Proof. Let $m = \|\psi - \psi'\|_{L^\infty(\Omega)}$. Then

$$\psi \le u$$

implies that

$$\psi' \le \psi + m \le u + m.$$

Now $u + m$ is a supersolution, so by Theorem 8.3

$$u' \le u + m.$$

Similarly, $u \le u' + m$. Hence $|u - u'| \le m$. Q.E.D.

Envisage now this situation. Suppose $\psi \in H^{1,\infty}(\Omega)$ is an obstacle which satisfies

$$\psi \in C^2(U) \qquad \text{for an open} \quad U \subset \Omega.$$

The sequence $\psi_\sigma \to \psi$ of Theorem 8.2 may be chosen so that $\psi_\sigma \to \psi$ uniformly together with its first and second derivatives on compact subsets of U. For a fixed $\sigma > 0$, consider $u_\sigma \in \mathbb{K}_{\psi_\sigma}$ the solution to Problem 8.1 and let u_ε denote the approximations to u_σ defined by (3.3). The fact that the vector field considered here is only locally coercive is of no importance since we may replace $a(p)$ by a strongly coercive $\tilde{a}(p)$ independent of σ (see Lemma 4.3 of Chapter III). So u_ε satisfies

$$\begin{aligned} Au_\varepsilon &= \max(A\psi_\sigma, 0)\vartheta_\varepsilon(u_\varepsilon - \psi_\sigma) \qquad &\text{in } \Omega \\ u_\varepsilon &= 0 &\text{on } \partial\Omega. \end{aligned}$$

Given open subsets $U'' \subset U' \subset U$ it is possible to localize the estimate of Corollary 3.4 by replacing u_ε by ζu_ε, ζ an appropriate cutoff function. There results an estimate

$$\|Au_\varepsilon\|_{L^\infty(U'')} \leq \text{const}(\|A\psi_\sigma\|_{L^\infty(U')} + \|u_\varepsilon\|_{L^\infty(U')}).$$

Passing to the limit as $\varepsilon \to 0$.

$$\|Au_\sigma\|_{L^\infty(U'')} \leq \text{const}(\|A\psi_\sigma\|_{L^\infty(U')} + \|u_\sigma\|_{L^\infty(U')}),$$

so

$$\|u_\sigma\|_{H^{2,s}(U'')} \leq \text{const}(\|A\psi_\sigma\|_{L^\infty(U')} + \|\psi_\sigma\|_{L^\infty(\Omega)})$$

keeping in mind that $\|u_\sigma\|_{L^\infty(\Omega)} \leq \|\psi_\sigma\|_{L^\infty(\Omega)}$. Since $u_\sigma \to u$ weakly in $H^1(\Omega)$ we conclude that

$$u_\sigma \to u \qquad \text{weakly in} \quad H^{2,s}(U''), \quad 1 \leq s < \infty,$$

and

$$\|u\|_{H^{2,s}(U'')} \leq \text{const}(\|A\psi\|_{L^\infty(U')} + \|\psi\|_{L^\infty(\Omega)}).$$

We have proven

Theorem 8.6. *Let ψ be a Lipschitz obstacle and u the solution to Problem 8.1. If $\psi \in C^2(U)$ for some open $U \subset \Omega$, then $u \in H^{2,s}(U'')$ for every $U'' \subset \overline{U}'' \subset U$.*

As a consequence we state this corollary, referred to as the principle of the face of contact.

Corollary 8.7. *Let $a(p)$ be a C^2 locally coercive vector field and ψ a Lipschitz obstacle satisfying $\psi \in C^2(U)$ for some $U \subset \Omega$ open. Suppose that $A\psi = 0$ in U. Then $Au = 0$ in U and either*

$$u(x) > \psi(x) \qquad \textit{for all} \quad x \in U,$$

or

$$u(x) = \psi(x) \qquad \textit{for all} \quad x \in \overline{U},$$

where u denotes the solution to Problem 8.1 for ψ.

Proof. According to the previous theorem, $u \in H^{2,s}(U'')$ for any $U'' \subset \overline{U''} \subset U$ open. Hence we may compute Au by Theorem 8.6. So $Au = 0$ in $U'' \cap (\Omega - I)$, obviously. In $U'' \cap I$,

$$u = \psi, \qquad u_{x_i} = \psi_{x_i}, \qquad 1 \le i \le N,$$

and so

$$u_{x_i x_j}(x) = \psi_{x_i x_j}(x) \qquad \text{a.e. in} \quad U'' \cap I.$$

Therefore,

$$Au = A\psi = 0 \qquad \text{in} \quad U''$$

(in the sense of distributions). Smoothness of $a(p)$ implies that $u, \psi \in C^{2,\lambda}(U'')$ for some $\lambda > 0$. By the mean value theorem,

$$L(u - \psi) = -(\partial/\partial x_i)[\alpha_{ij}(x)(\partial/\partial x_j)(u - \psi)] = 0 \qquad \text{in} \quad U''$$

where

$$\alpha_{ij}(x) = \int_0^1 (\partial a_i/\partial p_j)\{\psi_x(x) + t[u_x(x) - \psi_x(x)]\}\, dt.$$

So $u - \psi$ is a solution to the equation

$$Lv = 0,$$

where

$$Lv = -(\partial/\partial x_i)[\alpha_{ij}(x)(\partial v/\partial x_i)].$$

Owing to the smoothness of u, ψ, and $a_i(p)$, $\alpha_{ij} \in C^{1,\lambda}(U'')$. We may apply a classical maximum principle to infer that

$$u(x) - \psi(x) > \inf_{\partial U''}(u - \psi), \qquad x \in U'',$$

unless

$$u(x) - \psi(x) = 0, \qquad x \in U''. \quad \text{Q.E.D.}$$

For example, if ψ is the height function of a polyhedron, the coincidence set consists of a number of faces and parts of edges and may well consist only of portions of edges.

We may use the preceding theorem to study the obstacle problem for obstacles defined only on a portion of Ω. Suppose that $E \subset \Omega$ is a compact subset and $\psi \in C(E)$ admits an extension to $H^1(\Omega)$. Let $a(p)$ be a strictly coercive C^2 vector field and, as before, set

$$A\colon H_0^1(\Omega) \to H^{-1}(\Omega),$$
$$Av = (\partial/\partial x_j)a_j(v_x), \qquad v \in H_0^1(\Omega).$$

From Theorem 1.7 and Corollary 1.8 of Chapter III, we infer the existence of a solution u of the variational inequality

$$u \in \mathbb{K}_\psi: \quad \langle Au, v - u \rangle \geq 0 \quad \text{for} \quad v \in \mathbb{K}_\psi, \qquad (8.3)$$
$$\mathbb{K}_\psi = \{v \in H_0^1(\Omega) : v \geq \psi \text{ on } E\}.$$

Now let $w \in H^1(\Omega \backslash E)$ denote the solution of the problem

$$Aw = 0 \quad \text{in} \quad \Omega \backslash E,$$
$$w = \psi \quad \text{on} \quad \partial E,$$
$$w = 0 \quad \text{on} \quad \partial \Omega,$$

and define

$$\tilde{\psi}(x) = \begin{cases} \psi(x), & x \in E \\ w(x), & x \in \Omega \backslash E. \end{cases}$$

We shall impose the hypothesis

$$\tilde{\psi}(x) \quad \textit{is Lipschitz in} \quad \Omega.$$

Hence there is a Lipschitz solution $\tilde{u}$ of the problem

$$\tilde{u} \in \mathbb{K}_{\tilde{\psi}}: \quad \langle A\tilde{u}, v - \tilde{u} \rangle \geq 0 \quad \textit{for} \quad v \in \mathbb{K}_{\tilde{\psi}}, \qquad (8.4)$$
$$\mathbb{K}_{\tilde{\psi}} = \{v \in H_0^1(\Omega) : v \geq \tilde{\psi} \text{ in } \Omega\}.$$

We shall prove that $u = \tilde{u}$. In particular, (8.3) admits a Lipschitz solution.

Since $A\tilde{\psi} = 0$ in $\Omega \backslash E$, by the previous theorem, $\tilde{u} = \tilde{\psi}$ or $\tilde{u} > \tilde{\psi}$ in $\Omega \backslash E$. In either case

$$A\tilde{u} = 0 \quad \text{in} \quad \Omega \backslash E. \qquad (8.5)$$

Given $v \in \mathbb{K}_\psi$ and $\varepsilon > 0$, define

$$v_\varepsilon' = \max(\tilde{\psi} - v - \varepsilon, 0)$$

and note that $v_\varepsilon' = 0$ near $\partial\Omega \cup E$. Hence $v_\varepsilon' \in H_0^1(\Omega \backslash E)$. Set

$$v_\varepsilon'' = v + v_\varepsilon' \in \mathbb{K}_{\tilde{\psi}}.$$

Then

$$\langle A\tilde{u}, v - \tilde{u} \rangle = \langle A\tilde{u}, v_\varepsilon'' - \tilde{u} \rangle - \langle A\tilde{u}, v_\varepsilon' \rangle = \langle A\tilde{u}, v_\varepsilon'' - \tilde{u} \rangle$$

by (8.5). Now allowing $\varepsilon \to 0$,

$$v_\varepsilon'' \to v_0'' = v + v_0' \in \mathbb{K}_{\tilde{\psi}}.$$

Hence by (8.4)

$$\langle A\tilde{u}, v - \tilde{u} \rangle = \langle A\tilde{u}, v_0'' - \tilde{u} \rangle \geq 0.$$

A frequently studied class of obstacles like ψ above is the class of thin obstacles, or obstacles defined on lower dimensional submanifolds of Ω.

9. A Variational Inequality with Mixed Boundary Conditions

Here we show how the preceding theory leads to the resolution of a problem which bears a formal resemblence to a mixed boundary value problem. Namely, we consider a variational inequality for a linear elliptic second order operator on a bounded domain Ω of $\mathbb{R}^N$ whose solution lies above a given obstacle and assumes prescribed boundary values only on a portion of $\partial\Omega$.

Let Ω be a bounded open set with smooth boundary $\partial\Omega$, for sake of simplicity, and let $\partial_1\Omega$, $\partial_2\Omega$ be two smooth and disjoint open subsets of $\partial\Omega$ such that $\partial\Omega = \overline{\partial_1\Omega} \cup \overline{\partial_2\Omega}$. Set

$$V = \{v \in H^1(\Omega) : v = 0 \text{ on } \partial_1\Omega\}.$$

Let $\psi \in H^1(\Omega)$ satisfy $\psi \le 0$ on $\partial_1\Omega$ and let

$$\mathbb{K} = \{v \in V : v \ge \psi \text{ in } \Omega\}.$$

Now consider a uniformly elliptic second order operator

$$Au = -(a_{ij}u_{x_i})_{x_j}, \qquad a_{ij}(x) \in L^\infty(\Omega),$$

where the coefficients $a_{ij}(x)$ satisfy the usual condition of ellipticity. The operator A defines the bilinear form

$$a(u, v) = \int_\Omega a_{ij}(x)u_{x_i}(x)v_{x_j}(x)\,dx.$$

Let T be a distribution determined by

$$\langle T, v\rangle = \int_\Omega [f_0 v + f_i v_{x_i}]\,dx + \int_{\partial_2\Omega} gv\,d\sigma, \qquad v \in V,$$

where $f_0, f_1, \ldots, f_N, g$ belong to suitable L^p spaces.

If $\partial_1\Omega$ is sufficiently large so that the Poincaré inequality holds for the functions of V, then the form $a(u, v)$ is coercive on V. Therefore the theorem of existence, Theorem 2.1 of Chapter II, applies and we may conclude that there exists a unique solution of the variational inequality

$$u \in \mathbb{K}: \quad a(u, v - u) \ge \langle T, v - u\rangle \qquad \text{for all} \quad v \in \mathbb{K}. \tag{9.1}$$

Our discussion centers about the regularity of the solution to the variational inequality (9.1):

Theorem 9.1. *If $u \in \mathbb{K}$ is the solution of the variational inequality* (9.1), *where*

$$\psi \in L^\infty(\Omega) \cap H^1(\Omega) \quad \textit{with} \quad \psi \leq 0 \quad \text{on} \quad \partial_1\Omega,$$

$$f_0 \in L^{p/2}(\Omega), \quad f_j \in L^p(\Omega), \quad j = 1, 2, \ldots, N, \quad \textit{with} \quad p > N,$$

and

$$g \in L^q(\partial_2\Omega) \quad \textit{with} \quad q > N - 1,$$

then we have, for a suitable constant C,

$$\psi(x) \leq u(x) \leq \max\left(\max_\Omega \psi, 0\right) + C\left[\|f_0\|_{L^{p/2}(\Omega)} + \sum_{i=1}^N \|f_i\|_{L^p(\Omega)} + \|g\|_{L^q(\partial_2\Omega)}\right]$$

almost everywhere in Ω.

For any real number $k \geq k_0 = \max(\max_\Omega \psi, 0)$, let $v = \min(u, k) \in \mathbb{K}$. Substituting this v in the variational inequality (9.1) and repeating the arguments similar to that used in Theorem B.2 we get for the function

$$\mu(k) = \text{meas } A(k) + \{\text{meas}[A(k) \cap \partial_2\Omega]\}^{N/(N-1)}$$

an inequality of the type

$$u(h) \leq [c/(h-k)^{2^*}]|\mu(k)|^\beta, \quad \beta > 1 \quad \text{for all} \quad h > k > k_0.$$

Employing Lemma B.1 the theorem follows. Q.E.D.

Theorem 9.2. *If $u \in \mathbb{K}$ is the solution of variational inequality* (9.1), *where the assumptions of the theorem above hold, then*

$$u \in C^{0,\lambda}(\overline{\Omega}),$$

where λ, $0 < \lambda < 1$, depends only on Ω, N, p, q.

The proof of this theorem is similar to that of Theorem C.2 of Chapter II and we omit it.

The aim of the theorem above is to show the Hölder continuity of u. When ψ is absent, that is, in the case of an equation we are led to a regularity theorem for the mixed boundary value problem which can be treated by the same method of truncation as in Theorem C.2 of Chapter II. One could attempt to extend to this problem the regularity theorems proved in Sections 4 and 5. However, with this in mind, it is easy to see that the solution does not belong to $H^{2,p}(\Omega)$ for all $p < \infty$. Indeed, this result is known not to be true even if ψ is absent unless the data of the problem satisfy some compatibility conditions. We refer here to the result of Shamir [1] on mixed boundary value problems.

We consider now the solution of the variational inequality (9.1) with $f_i = 0$ for $i = 1, 2, \ldots, N$ and $g = 0$. It can be approximated in $C^{0,\lambda} \cap V$ by solutions of certain quasi-linear mixed boundary value problems associated with the elliptic operator A.

Assume that $A\psi$ is a measure on Ω and that the outward conormal derivative

$$\partial\psi/\partial\nu = a_{ij}\psi_{x_j}\nu_i, \qquad \nu = (\nu_1, \ldots, \nu_N) \qquad \text{the outward normal,}$$

is a measure on $\partial_1\Omega$ which fulfills the conditions

$$\max(A\psi - f, 0) \in L^p(\Omega), \quad p > \tfrac{1}{2}N,$$
$$\max(\partial\psi/\partial\nu, 0) \in L^q(\partial_2\Omega), \quad q > N - 1.$$

Let $\vartheta(t)$ be a nonincreasing Lipschitz function on $\mathbb{R}$, $0 \le \vartheta(t) \le 1$. Consider the nonlinear mixed boundary value problem:

$$\begin{aligned} Au &= \max(A\psi - f, 0)\vartheta(u - \psi) + f \qquad \text{in} \quad \Omega \\ u &= 0 \qquad \text{on} \quad \partial_1\Omega, \\ \partial u/\partial\nu &= \max(\partial\psi/\partial\nu, 0)\vartheta(u - \psi). \qquad \text{on} \quad \partial_2\Omega. \end{aligned} \tag{9.2}$$

The variational formulation of the mixed problem (9.2) can be defined by means of the quasi-linear form

$$\begin{aligned} b_\vartheta(u, v) = a(u, v) &- \int_\Omega \max(A\psi - f, 0)\vartheta(u - \psi)v\,dx \\ &- \int_{\partial_2\Omega} \max(\partial\psi/\partial\nu, 0)\vartheta(u - \psi)v\,d\sigma. \end{aligned} \tag{9.3}$$

Indeed, (9.2) is equivalent to the problem of finding a solution of

$$u \in V\colon \quad b_\vartheta(u, v) = \int_\Omega fv\,dx \qquad \text{for all} \quad v \in V. \tag{9.4}$$

Then the general theory of monotone operators, which we have treated in Chapter III, yields the existence of a solution of (9.4).

We consider two sequences of functions of the type $\vartheta(t)$ defined as follows:

$$\vartheta'_m(t) = \begin{cases} 1 & \text{for} \quad t \le -1/m \\ -mt & \text{for} \quad -1/m \le t \le 0 \\ 0 & \text{for} \quad t \ge 0 \end{cases}$$

and

$$\vartheta''_m(t) = \begin{cases} 1 & \text{for} \quad t < 0 \\ -mt + 1 & \text{for} \quad 0 \le t \le 1/m \\ 0 & \text{for} \quad t \ge 1/m. \end{cases}$$

Then $\vartheta'_m(t)$ is a nondecreasing sequence of functions each of which is Lipschitz and not increasing while $\vartheta''_m(t)$ is a nonincreasing sequence of functions with the same properties. Both sequences "converge" to the multi-valued function $\tilde{\vartheta}(t)$ defined by

$$\tilde{\vartheta}(t) = \begin{cases} 1 & \text{for} \quad t < 0 \\ [0, 1] & \text{for} \quad t = 0 \\ 0 & \text{for} \quad t > 0. \end{cases}$$

Let ψ be a function such that $\psi \leq 0$ on $\partial_1 \Omega$ and let $\tilde{\psi}$ be the unique solution of the Dirichlet problem

$$\Delta \tilde{\psi} = f \quad \text{in} \quad \Omega, \qquad \tilde{\psi} = \psi \qquad \text{on} \quad \partial\Omega,$$

and assume that $\tilde{\psi} \in H^{2,p}(\Omega)$. Consider the closed convex subsets of V:

$$\mathbb{K} = \{v \in V : v \geq \psi \text{ on } \partial\Omega\}, \qquad \tilde{\mathbb{K}} = \{v \in V : v \geq \tilde{\psi} \text{ in } \Omega\};$$

then we have the following:

Theorem 9.3. *If u is the solution of the variational inequality*

$$u \in \tilde{\mathbb{K}}: \quad a(u, v - u) \geq \int_\Omega f(v - u)\, dx \qquad \textit{for all} \quad v \in \tilde{\mathbb{K}}, \tag{9.5}$$

then u resolves the variational inequality

$$u \in \mathbb{K}: \quad a(u, v - u) \geq \int_\Omega f(v - u)\, dx \qquad \textit{for all} \quad v \in \mathbb{K}. \tag{9.6}$$

Proof. $u \in \mathbb{K}$ since $\tilde{\mathbb{K}} \subset \mathbb{K}$. It is enough to show that (9.6) holds true for all $v \in \mathbb{K}$.

If $v \in \mathbb{K}$, then either $v \geq \tilde{\psi}$ in Ω or $v < \tilde{\psi}$ in some open subset of Ω of positive capacity. In the first case (9.5) is nothing but (9.6). So we have only to consider the case in which $v \geq \tilde{\psi}$ in Ω does not hold true.

In this case, we write v as a sum by defining $v = v_1 + v_2$, where $v_1 = \max(v, \tilde{\psi})$ and $v_2 = v - v_1$.

Then $v_1 \in \tilde{\mathbb{K}}$ and $v_2 \in V$ with

$$v_2 = \begin{cases} 0 & \text{in} \quad \{x \in \Omega : v \geq \tilde{\psi}\} \\ v - \tilde{\psi} & \text{in} \quad \{x \in \Omega : v < \tilde{\psi}\}. \end{cases}$$

Since $v \in \mathbb{K}$ by assumption, so that $v \geq \psi = \tilde{\psi}$ on $\partial\Omega$, it follows that $\operatorname{supp} v_2 \subset \Omega$ and thus $v_2 \in H_0^1(\Omega) \subset V$. We can now write

$$a(u, v - u) = a(u, v_1 - u) + a(u, v_2)$$

and we shall show that

$$a(u, v_2) = \int_\Omega f v_2 \, dx. \tag{9.7}$$

In fact, u is the weak limit in $V \cap C^{0,\lambda}(\overline{\Omega})$ of the sequence (a subsequence) u'_m of solutions of the mixed boundary value problems

$$u'_m \in V: \quad a(u'_m, \eta) = \int_\Omega [\max(A\tilde{\psi} - f, 0)\vartheta'_m(u'_m - \tilde{\psi}) + f]\eta \, dx$$

$$+ \int_{\partial_2\Omega} [\max(\partial\psi/\partial v, 0)\vartheta'_m(u'_m - \tilde{\psi})]\eta \, d\sigma \quad \text{for all } \eta \in V.$$

Since $A\tilde{\psi} = f$ in Ω and since supp $v_2 \subset \Omega$ we find that

$$a(u'_m, v_2) = \int_\Omega f v_2 \, dx,$$

which, passing to the limit, proves (9.7).

Finally, since u is the solution of (9.5) and since $v_1 \in \tilde{\mathbb{K}}$, we find

$$a(u, v - u) \geq \int_\Omega f(v_1 - u) \, dx + \int_\Omega f v_2 \, dx = \int_\Omega f(v - u) \, dx.$$

Since the last inequality holds for all $v \in \mathbb{K}$, Theorem 9.3 is proved. Q.E.D.

Appendix A. Proof of Theorem 3.3

The main purpose of this appendix is to sketch the proof of the following theorem stated in Section 3.

Theorem A.1. *Let Ω be a smooth bounded open set of $\mathbb{R}^N$. Let $a(p)$ be a C^1 vector field such that*

$$(\partial a_i/\partial p_j)(p)\xi_i\xi_j \geq v|\xi|^2, \qquad \xi \in \mathbb{R}^N,$$

and

$$|\partial a_i/\partial p_j| \leq M, \qquad 0 < v \leq M < +\infty.$$

Let $g \in L^\infty(\Omega)$ and suppose that

$$u \in H_0^1(\Omega): \quad Au = -(\partial/\partial x_i)a_i(u_x) = g \qquad \text{in} \quad \Omega. \tag{A.1}$$

Then for each $s > 1$, there is a $C_0 = C_0(s, \nu, M, \Omega)$ such that

$$\|u\|_{H^{2,s}(\Omega)} \le C_0 \|g\|_{L^\infty(\Omega)}. \tag{A.2}$$

The proof of this theorem is based on several steps.

Lemma A.2. *Under the assumption of Theorem A.1, $u \in H^2(\Omega)$.*

Recall that by assumption

$$u \in H_0^1(\Omega): \quad \int_\Omega a_i(u_x)\varphi_{x_i}\, dx = \int_\Omega g\, dx \qquad \text{for all} \quad \varphi \in H_0^1(\Omega). \tag{A.3}$$

Now let V be the solution of the Dirichlet problem

$$V \in H_0^1(\Omega): \quad \int_\Omega V_{x_i}\eta_{x_i}\, dx = \int_\Omega g\eta\, dx \qquad \text{for all} \quad \eta \in H_0^1(\Omega).$$

It is known by the Calderon–Zygmund theorem (see Section 2) that for $g \in L^s(\Omega)$ it follows that $V \in H^{2,s}(\Omega)$ and thus $V_{x_i} \in H^{1,s^*}(\Omega)$, where $1/s^* = (1/s) - (1/N)$. So (A.3) can be written

$$\int_\Omega a_i(u_x)\eta_{x_i}\, dx = \int_\Omega V_{x_i}\eta_{x_i}\, dx \qquad \text{for all} \quad \eta \in H_0^1(\Omega). \tag{A.3$'$}$$

(a) First of all we shall prove that $u \in H^2_{\text{loc}}(\Omega)$. Denote by $\varphi_{s,h}$ the first difference quotient

$$\frac{\varphi(x_1, \ldots, x_s + h, \ldots, x_N) - \varphi(x_1, \ldots, x_N)}{h}$$

of the function $\varphi(x)$. Let $\zeta(x)$ be any function in $H^1(\Omega)$ with compact support in Ω. Let Ω' be an open set such that $\overline{\Omega'} \subset \Omega$; then if $x \in \Omega'$ and h is small enough we can insert the function

$$(1/h)[\zeta(x_1, \ldots, x_N) - \zeta(x_1, \ldots, x_s - h, \ldots, x_N)]$$

instead of φ in Eq. (A.3$'$). By changing variables and using the integral form of the mean value theorem, we obtain an identity of the form

$$\int_\Omega [\alpha_{ij}(x, h)(u_{x_i})_{s,h}\zeta_{x_j} - (V_{x_i})_{s,h}\zeta_{x_i}]\, dx = 0, \tag{A.4}$$

where the functions $\alpha_{ij}(x, h)$ satisfy uniformly with respect to h

$$\nu|\xi|^2 \le \alpha_{ij}(x, h)\xi_i\xi_j, \qquad |\alpha_{ij}(x, h)| \le M.$$

Let α be a C^∞ function with compact support in a ball $B_R(x) \subset \Omega'$ and equal to 1 in $B_\rho(x)$, $\rho < R$. Taking $\zeta = \alpha^2 u_{s,h}$, by a standard computation we get

$$\int_{B_\rho(x)} (u_{s,h})_x^2 \, dx \le \text{const}\left\{[1/(R-\rho)^2] \int_{B_R(x)} u_{s,h}^2 \, dx + \int_{B_R(x)} g^2 \, dx\right\}$$

with $\rho < R$, where the constant in front is independent of h.

Passing to the limit and remembering that $\overline{\Omega}'$ is compact, we obtain

$$\|u\|_{H^2(\Omega')} \le K(\|u\|_{H^1(\Omega)} + \|g\|_{L^2(\Omega)}). \tag{A.5}$$

From this estimate we can infer that $u \in H^2_{\text{loc}}(\Omega)$ and that u satisfies a.e. in Ω the equation

$$-\frac{\partial a_i}{\partial p_j}(u_x)\frac{\partial^2 u}{\partial x_i \, \partial x_j} = g. \tag{A.6}$$

(b) In order to prove that $u \in H^2(\Omega)$ we straighten a portion $\partial_1\Omega$ of $\partial\Omega$ and obtain in the new coordinates an identity of the type of the one above. In a neighborhood of $\partial_1\Omega$ we examine only these difference quotients $u_{s,h}$ which are taken in directions parallel to $\partial_1\Omega$. These difference quotients vanish on $\partial_1\Omega$.

Using the same device as above we can deduce that the $H^1(\Omega)$ norms of these difference quotients are bounded uniformly with respect to h. The derivative remaining in the normal direction may then be estimated directly from Eq. (A.6). Thus $u \in H^2(\Omega)$.

Proof of theorem A.1. Passing to the limit in (A.4) as $h \to 0$ (through a subsequence) we get

$$\int_\Omega [(\partial a_i/\partial p_j)(u_x)(u_{x_s})_{x_i}\zeta_{x_j} + V_{x_i x_s}\zeta_{x_i}] \, dx = 0. \tag{A.7}$$

Then from Theorem C.2 of Chapter II we get that $u_{x_s} \in C^{1,\lambda}(\Omega') \cap H^1(\Omega')$ for a suitable λ with $0 \le \lambda < 1$. Hence we have a.e. in Ω

$$-\left[\frac{\partial a_i}{\partial p_j}(u_x)(u_{x_s})_{x_i}\right]_{x_j} - \frac{\partial}{\partial x_i} V_{x_i x_s} = 0. \tag{A.8}$$

If we straighten a portion $\partial_1\Omega$ of $\partial\Omega$ we have in the new variables an identity like the one above and we examine only those difference quotients $u_{s,h}$ which are taken in directions parallel to $\partial_1\Omega$; we deduce that these derivatives u_{x_s} are Hölder continuous up to the boundary.

Moreover, we know from (b) above that

$$u_{x_s x_i} \in L^2(\Omega).$$

From Remark D.6 of Chapter II we may deduce that

$$\int_{\Omega(x,\rho)} z^2 \, dx \leq c\rho^{N-2+2\lambda},$$

where z is any second derivative of u except $u_{\nu\nu}$, i.e., except the second pure normal derivative. But $u_{\nu\nu}$ may be expressed by mean of the other derivatives through Eq. (A.6). So we deduce that

$$\int_{\Omega(x,\rho)} u_{x_s x_i}^2 \, dx \leq c\rho^{N-2+2\lambda}$$

for $s, i = 1, 2, \ldots, N$ and for a λ such that $0 \leq \lambda < 1$.

From a well-known lemma due to Morrey we infer that the derivatives u_{x_s} are Hölder continuous up to the boundary of Ω.

Since Eq. (A.6) can considered as a linear equation with continuous coefficients $(\partial a_i/\partial p_j)(u_x)$, using again the result of Calderon–Zygmund we deduce that (A.2) holds.

Comments and Bibliographical Notes

It is unfortunately beyond the scope of these brief notes to survey the extensive regularity theory associated to variational inequalities. The principal references for the material of this chapter are the papers Lewy and Stampacchia [1] (Section 2), [2] (Sections 3, 4, 8); Brezis and Kinderlehrer [1] (Sections 5, 6, and Appendix); Murthy and Stampacchia [1] (Section 9); and Kinderlehrer [1] (Section 9). A detailed presentation of the subject may be found in Brezis [1] (cf. also Brezis and Stampacchia [1]). A concise treatment of penalization and some of its consequences, especially in regard to interpolation theorems, appears in Lions [1].

The notion of finite perimeter and its properties, due to De Giorgi, is developed in Miranda [1]. Existence and smoothness of the solution of Problem 4.1 when $a(p) = p/(1 + p^2)^{1/2}$ was also considered by Giaquinta and Pepe [1] and Miranda [2], who studied the parametric problem.

The question of lower dimensional obstacles for the Dirichlet integral was considered in Lewy [6]. Additional results about the smoothness of the solution have been obtained in Lewy [7], Frehse [3, 4], and Caffarelli [3].

General existence theorems about the minimal surface case may be found in Giusti [1], Kinderlehrer [1], and Stampacchia and Vignoli [1].

For problems with mixed boundary data, we note also the work in Beirao da Veiga [1].

The relationship between coercivity of the vector field $a(p)$ and the smoothness of the solution was studied by Mazzone [1]. The special case of assigned mean curvature, namely, the inhomogeneous problem with $a(p) = p/(1 + p^2)^{1/2}$ was considered by Mazzone [2] and Gerhardt [1]. Gerhardt [2] also gave a different proof of Theorem 6.3.

Exercises

1. Supply the details for Theorem 2.5.

2. Let Ω denote the unit cube $\{x \in \mathbb{R}^N : 0 < x_i < 1, i = 1, \ldots, N\}$ and let $f \in L^s(\Omega)$, $g \in H^{2,s}(\Omega)$, for some s, $N < s < \infty$. Suppose that

$$\begin{aligned} w \in H^1(\Omega): \quad -\Delta w &= f \quad \text{in} \quad \Omega, \\ w &= g \quad \text{in} \quad \partial\Omega. \end{aligned}$$

Show that $w \in H^{2,s}(\Omega)$. [*Hint*: After superposition, we may take $g = 0$. Extend w to $\tilde{\Omega} = \{x \in \mathbb{R}^N : -1 < x_N < 0, 0 < x_\sigma < 1, \sigma = 1, \ldots, N-1\}$ by $w(x', x_N) = -w(x', -x_N)$, $x = (x', x_N) \in \tilde{\Omega}$. Show that the extended w is the solution to a weak equation in $\Omega \cup \tilde{\Omega}$. This may be used to show that w is smooth on the faces $x_N = x_\sigma = 0$, $1 \le \sigma \le N-1$. Now continue this procedure, performing further reflections.]

3. With Ω, f, g as in Exercise 2, set

$$\partial_2\Omega = \{x \in \mathbb{R}^N : x_N = 0, 0 < x_\sigma < 1, \sigma = 1, \ldots, N-1\}$$

and let $\partial_1\Omega$ denote the union of the remaining faces of $\partial\Omega$. Consider

$$\begin{aligned} w \in H^1(\Omega): \quad -\Delta w &= f \quad \text{in} \quad \Omega, \\ w &= g \quad \text{on} \quad \partial_2\Omega, \\ \partial w/\partial v &= 0 \quad \text{on} \quad \partial_1\Omega. \end{aligned}$$

Assume that $g = |g_x| = 0$ on the boundary of $\partial_2\Omega$. Prove that $w \in H^{2,s}(\Omega)$. [*Hint*: First extend w to $x_N < 0$ by $w(x', x_N) = w(x', -x_N)$, $-1 < x_N < 0$.]

4. Let Ω be convex with smooth boundary $\partial\Omega$ and let $\psi \in H^{1,\infty}(\Omega)$ with $\psi = 0$ on $\partial\Omega$. Let $\mathbb{K}$ denote the set of Lipschitz functions v in $\overline{\Omega}$ which satisfy $v \geq \psi$ in Ω and $v = 0$ on $\partial\Omega$. Given a locally coercive vector field $a(p)$ let $u = T(a)$ denote the solution of

$$u \in \mathbb{K}: \quad \int_\Omega a_i(u_x)(v - u)_{x_i}\, dx \geq 0 \qquad \text{for} \quad v \in \mathbb{K}.$$

Now define

$$U(x) = \sup_a T(a)(x), \qquad x \in \Omega.$$

Show that U exists and that U is the height function of the convex hull of $\max(\psi, 0)$ in Ω.

5. Consider Exercise 13 of Chapter II. Show that under suitable hypotheses, the solution $u = (u^1, u^2)$ has components $u' \in H^{2,s}(\Omega) \cap H^{2,\infty}_{\mathrm{loc}}(\Omega)$, $1 < s < \infty$. (*Hint*: formulate a penalized problem.)

CHAPTER V

Free Boundary Problems and the Coincidence Set of the Solution

1. Introduction

In the previous chapters our study emphasized the existence and smoothness of the solution of a variational inequality. The smoothness of the solution, we have seen, was limited by the constraints which defined the convex set under consideration and, indeed, this necessitated the development of a special regularity theory. For an obstacle problem, in particular, the effect of these constraints is also manifested by the presence of the coincidence set. Now we shall investigate some of its properties, with our major goal the establishment of criteria assuring the regularity of its boundary.

Such problems are of interest for several reasons. From one point of view, they illustrate how a set with certain topological properties is necessarily smooth. For example, consider Problem 4.1 of Chapter IV when $\Omega \subset \mathbb{R}^2$ is strictly convex and the obstacle ψ is strictly concave and smooth. In this case we shall be able to conclude that ∂I is a smooth Jordan curve.

In addition, the formulation of many physical problems requests an analytic free boundary as part of the solution. Examples of this are steady cavitational flow or the flow of water through a homogeneous porous medium.

Let us begin by suggesting formally what constitutes a free boundary. We consider an obstacle problem. Let $\Omega \subset \mathbb{R}^N$ be a bounded, open, connected set with smooth boundary $\partial\Omega$ and $\psi \in C^\infty(\bar{\Omega})$ an obstacle satisfying

$$\max_{\Omega} \psi > 0 \qquad \text{and} \qquad \psi < 0 \quad \text{on} \quad \partial\Omega.$$

We set $\mathbb{K} = \{v \in H_0^1(\Omega) : v \geq \psi \text{ in } \Omega\}$. Consider the solution u of

$$u \in \mathbb{K}: \quad \int_\Omega u_{x_i}(v - u)_{x_i}\, dx \geq 0 \qquad \text{for all} \quad v \in \mathbb{K} \tag{1.1}$$

and its coincidence set

$$I = I(u) = \{x \in \Omega : u(x) = \psi(x)\}. \tag{1.2}$$

As we have observed, $u - \psi \in C^1(\Omega)$ and attains its minimum at any point $x \in I$; hence

$$u = \psi \qquad \text{and} \qquad u_{x_i} = \psi_{x_i} \qquad \text{in} \quad I, \quad 1 \leq i \leq N.$$

Also, $\Delta u = 0$ in $\Omega - I$. In this fashion we may regard u as the solution of the Cauchy problem

$$\begin{aligned} \Delta u &= 0 \qquad \text{in} \quad \Omega - I, \\ u &= \psi \\ u_{x_i} &= \psi_{x_i} \qquad \text{on} \quad \partial I. \end{aligned} \tag{1.3}$$

Since the Cauchy problem for Δ is not well posed even for a smooth initial surface ∂I, the mere existence of a solution suggests restrictive conditions on ∂I. The set ∂I is what we call a "free boundary."

We want to indicate the reasoning which leads to the conclusion that ∂I is analytic for the solution of (1.1) with an analytic ψ and also some of the difficulties encountered in adapting it to other problems. First we discuss a two dimensional situation. Our attention is limited to the analytical questions. A discussion of the topological ones is delayed until Section 6.

Before we begin let us remark about how we regard functions of two variables $g(x) = g(x_1, x_2)$. Introducing the complex variable $z = x_1 + ix_2$ and its conjugate $\bar{z} = x_1 - ix_2$ we frequently write $G(z, \bar{z}) = g(x_1, x_2) = g(\frac{1}{2}(z + \bar{z}), (1/2i)(z - \bar{z}))$. We also write $G(z)$ for $G(z, \bar{z})$ without necessarily intending that G be holomorphic in z. Suppose, for instance, that $g(x)$ is real analytic near $x = c$; thus,

$$g(x) = \sum_{j,k=0}^{\infty} a_{jk}(x_1 - c_1)^j (x_2 - c_2)^k,$$

where the series is absolutely and uniformly convergent for $|x - c|$ small. Then, with $\gamma = c_1 + ic_2$,

$$G(z, \bar{z}) = \sum_{j,k=0}^{\infty} a_{jk}\left(\frac{1}{2}(z + \bar{z}) - c_1\right)^j \left(\frac{1}{2i}(z - \bar{z}) - c_2\right)^k$$

$$= \sum_{j,k=0}^{\infty} \alpha_{jk}(z - \gamma)^j(\bar{z} - \bar{\gamma})^k$$

for suitable α_{jk} where the series is absolutely and uniformly convergent for $|z - \gamma|$ and $|\bar{z} - \bar{\gamma}|$ sufficiently small. Let $\zeta = \xi_1 + i\xi_2$ denote a second complex variable. The series $\sum \alpha_{jk}(z - \gamma)^j(\zeta - \bar{\gamma})^k$ is absolutely and uniformly convergent for small $|z - \gamma|$ and $|\zeta - \bar{\gamma}|$ and is holomorphic in both z and ζ. Thus if $g(x)$ is real analytic near $x = c$ we may regard $G(z, \bar{z})$ to be the restriction to $\zeta = \bar{z}$ of the holomorphic function of two complex variables

$$G(z, \zeta) = \sum_{j,k=0}^{\infty} \alpha_{jk}(z - \gamma)^j(\zeta - \bar{\gamma})^k$$

Theorem 1.1. *Suppose that ω is a simply connected bounded domain in the $z = x_1 + ix_2$ plane with $\Gamma \subset \partial\omega$ a Jordan arc, $\Gamma = \partial\omega \cap U$ for some open $U \subset \mathbb{R}^2$. Consider functions u, ψ satisfying*

$$u \in C^1(\omega \cup \Gamma),$$

ψ real analytic in a neighborhood of $\omega \cup \Gamma$,

$$\begin{aligned} \Delta u &= 0 && \text{in} \quad \omega, \\ u &= \psi && \\ u_{x_j} &= \psi_{x_j} && \text{on} \quad \Gamma, j = 1, 2, \\ \Delta\psi &\neq 0 && \text{in} \quad \overline{\omega \cup \Gamma}. \end{aligned} \tag{1.4}$$

Then Γ admits an analytic parametrization.

Proof. We shall show that there is a function $z^*(z)$ analytic in ω such that $z^*(z) = \bar{z}$ on Γ. From the existence of such a function, we then conclude that Γ is analytic.

Suppose that $z = 0 \in \Gamma$. The complex gradient $f(z) = u_{x_1} - iu_{x_2}$ is holomorphic in ω and has continuous boundary values $F(z, \bar{z}) = \psi_{x_1} - i\psi_{x_2}$ for $z \in \Gamma$. In a neighborhood of $z = 0$, consider the equation

$$f(z) = F(z, z^*(z)), \tag{1.5}$$

keeping in mind that $z^*(z) = \bar{z}$ is a solution of (1.5) whenever $z \in \Gamma$. Recalling that

$$\frac{\partial}{\partial z^*} F(z, z^*)\bigg|_{(z, z^*)=(0,0)} = \frac{1}{2}\Delta\psi(0) \neq 0,$$

we infer the existence of a solution $z^* = \zeta(z, f)$ of (1.5) holomorphic in (z, f) near $(0, f(0))$. In particular

$$z^*(z) = \zeta(z, f(z)), \qquad z \in \omega, \quad |z| \quad \text{small},$$

is holomorphic in z. By the uniqueness statement of the implicit function theorem

$$z^*(z) = \bar{z}, \qquad z \in \Gamma, \quad |z| \quad \text{small}. \tag{1.6}$$

At this stage, we have represented $\bar{z}$ on Γ as the boundary value of a function holomorphic in ω. This implies that Γ is analytic by a reflection principle. To verify this conclusion, let

$$\varphi: G \to \omega, \qquad G = \{t = t_1 + it_2 : |t| < 1, \operatorname{Im} t > 0\},$$
$$\varphi(0) = 0,$$

be a conformal mapping of G onto ω which transforms the real segment $(-1, 1)$ onto Γ. We know that φ exists and $\varphi \in C(G \cup (-1, 1))$. Define

$$\Phi(t) = \begin{cases} \varphi(t), & \operatorname{Im} t \geq 0, \quad |t| < 1 \\ \overline{z^*(\varphi(\bar{t}))}, & \operatorname{Im} t \leq 0, \quad |t| < 1\text{'} \end{cases} \tag{1.7}$$

a holomorphic function in $G \cup \{t : |t| < 1, \operatorname{Im} t < 0\}$. When t is real, (1.6) ensures that Φ is continuous. By Morera's theorem Φ is holomorphic in a neighborhood of $t = 0$. Consequently,

$$t \to \Phi(t), \qquad t \text{ real}, \quad |t| \quad \text{small},$$

exhibits an analytic parametrization of a portion of Γ. Q.E.D.

The topological requirement that Γ be a Jordan curve will be weakened considerably. The proof fails, evidentally, when ψ is not analytic. Indeed, the definition $z^* = \zeta(z, f)$ is suspect in this situation.

From a slightly different viewpoint, (1.7) offers an extension of φ to a function Φ satisfying the homogeneous analytic equation $(\partial/\partial\bar{t})\Phi = 0$ in $|t| < 1$. In general, it is not possible to achieve an extension of this nature, so an alternative idea has been devised which might be interpreted as a combination of classical potential theory and some ideas about Sobolev spaces, especially their trace classes.

2. The Hodograph and Legendre Transformations

We introduce here another method for determining the smoothness of ∂I, the boundary of the coincidence set of (1.1). Although this method will serve us in any dimension, some initial regularity of ∂I is required for its application. It has the effect of "straightening" the free boundary at the expense of replacing the equation by a highly nonlinear one.

Suppose that $u(x)$ is a solution of (1.1) with

$$\psi \quad \text{analytic} \qquad \text{and} \qquad -\Delta\psi > 0 \qquad \text{in} \quad \Omega. \tag{2.1}$$

In addition, assume that

$$\begin{aligned} &\Gamma \subset \partial I \qquad \text{is a} \quad C^1 \text{ hypersurface and} \\ &u_{ij} \in C(\Gamma \cup (\Omega \backslash I)), \qquad 1 \leq i, \quad j \leq N, \end{aligned} \tag{2.2}$$

where $u_{ij} = u_{x_i x_j}$. We assume that $0 \in \Gamma$ and that the inward normal to $\Omega \backslash I$ at 0 is in the direction of the positive x_1 axis. Set $w = u - \psi$. Then, analogously to (1.3), suppose that

$$\Delta w = a \qquad \text{in} \quad \Omega - I, \tag{2.3}$$

$$\left.\begin{aligned} &w = 0 \\ &w_i = 0, \qquad 1 \leq i \leq N, \end{aligned}\right\} \quad \text{on} \quad \Gamma, \tag{2.4}$$

where $a = -\Delta\psi$ is analytic in a neighborhood of $x = 0$. We introduce the change of variables

$$y_1 = -w_1 \equiv -w_{x_1}, \qquad y_\alpha = x_\alpha, \qquad 2 \leq \alpha \leq N, \tag{2.5}$$

and the function

$$v(y) = x_1 y_1 + w(x). \tag{2.6}$$

We refer to (2.5) as a partial hodograph transformation and to v in (2.6) as the Legendre transform of w. Note that (2.5) and (2.6) differ from the customary definitions by a change in sign.

It is easy to see that (2.5) is 1 : 1 near $x = 0$. In fact, since $w_i = 0$ on Γ and $(1, 0, \ldots, 0)$ is normal to Γ at 0,

$$w_{i\alpha}(0) = 0, \qquad 1 \leq i \leq N, \quad 2 \leq \alpha \leq N,$$

so from (2.3)

$$w_{11}(0) = a(0) > 0.$$

Hence $dy/dx(0)$ is nonsingular. Under the mapping (2.5), a neighborhood of $x = 0$ in $\Omega \backslash I$ is mapped onto a set

$$U \subset \{y : y_1 < 0\}$$

and a neighborhood of $x = 0$ in Γ is mapped onto

$$\Sigma \subset \{y : y_1 = 0\}.$$

The property of the Legendre transform is that

$$dv = x_1\, dy_1 + y_1\, dx_1 + dw = x_1\, dy_1 + \sum_{\alpha > 1} w_\alpha\, dy_\alpha$$

or

$$v_1 = x_1, \qquad v_\alpha = w_\alpha, \qquad 2 \le \alpha \le N. \tag{2.7}$$

Here the subscripts of v refer to differentiation with respect to y while those of w refer to differentiation with respect to x. In particular, a portion Γ' of Γ admits the parametrization

$$\Gamma': \quad x_1 = \frac{\partial v}{\partial y_1}(0, x_2, \ldots, x_N), \qquad (0, x_2, \ldots, x_N) \in \Sigma. \tag{2.8}$$

The smoothness of Γ' becomes a question of that of v in $U \cup \Sigma$. We calculate the equation satisfied by v in U. Set $y' = (y_2, \ldots, y_N)$.

First observe that

$$\frac{dy}{dx} = \begin{pmatrix} -w_{11} & -w_{12} & \cdots & -w_{1N} \\ 0 & 1 & & 0 \\ & & \ddots & \\ 0 & 0 & & 1 \end{pmatrix}$$

and from (2.7),

$$\frac{dx}{dy} = \begin{pmatrix} v_{11} & v_{12} & \cdots & v_{1N} \\ 0 & 1 & & 0 \\ & & \ddots & \\ 0 & 0 & & 1 \end{pmatrix}.$$

Since

$$\frac{dy}{dx} = \left(\frac{dx}{dy}\right)^{-1},$$

we obtain that

$$w_{11} = -1/v_{11} \qquad \text{and} \qquad w_{1\alpha} = +(v_{1\alpha}/v_{11}), \qquad 2 \le \alpha \le N.$$

Hence

$$w_{\alpha\alpha} = \frac{\partial v_\alpha}{\partial x_\alpha} = v_{\alpha\alpha} - \frac{v_{1\alpha}^2}{v_{11}}, \qquad 2 \leq \alpha \leq N.$$

Now we see from (2.3) and (2.4) that

$$-\frac{1}{v_{11}} - \frac{1}{v_{11}} \sum_{\alpha>1} v_{1\alpha}^2 + \sum_{\alpha>1} v_{\alpha\alpha} - a(v_1, y') = 0 \quad \text{in} \quad U, \qquad v = 0 \quad \text{on} \quad \Sigma. \tag{2.9}$$

Note that $-v_{11}(0) = 1/w_{11}(0) > 0$. The equation of (2.9) is elliptic and analytic near $y = 0$ in U; hence, by a well-known theorem (cf. Morrey [1], Section 6.7), $v(y)$ is analytic near $y = 0$ in $U \cup \Sigma$. Therefore, in view of (2.8), a portion of Γ' is analytic near $x = 0$.

In summary, once (2.2) is assumed, the free boundary is analytic provided the obstacle is. Similarly, if $\psi \in C^{m,\alpha}(\Omega)$, then Γ is of class $C^{m-1,\alpha}$, $0 < \alpha < 1$. This idea may be implemented in many other problems. We shall discuss this in Chapter VI.

So once some regularity is assumed, we may conclude higher regularity of Γ. But the hypothesis (2.2) is not always satisfied. For example, ∂I may exhibit cusps, as we shall illustrate in Section 4. We shall summarize what has been recently developed about the first stages of regularity of Γ in Chapter VI.

3. The Free Boundary in Two Dimensions

In two dimensional problems, the smoothness of the free boundary may be determined by a method analogous to our analytic extension of Section 1. Here we shall prove a theorem useful in a variety of situations. It is more general than the discussion of Section 1 in two ways: first, the data and the equation are not required to be analytic; second, the free boundary need not be a Jordan arc. We then apply our theorem to a case of Problem 1.1.

The argument entails the extension of a conformal mapping, so let us review this subject first. Throughout we denote by ω a bounded, connected, and simply connected domain in the $z = x_1 + ix_2$ plane and we set $G = \{t = t_1 + it_2 : |t| < 1, \operatorname{Im} t > 0\}$. Suppose that

$$\varphi : G \to \omega \tag{3.1}$$

is a conformal mapping of G onto ω. Then $\varphi(t)$ is an analytic function bounded in $\overline{G}$; but moreover we may express the area A of ω by the formula

$$A = \int_G |\varphi'(t)|^2 \, dt_1 \, dt_2.$$

Since ω is bounded, $A < \infty$, so $\varphi \in H^1(G)$. By our trace theorem (Chapter II, Appendix A),

$$\varphi(t_1) \equiv \lim_{\varepsilon \to 0^+} \varphi(t_1 + i\varepsilon) \qquad \text{exists a.e.,} \qquad -1 < t_1 < 1,$$

and $\varphi(t_1) \in L^2(-1, 1)$.

Now set

$$\begin{aligned} \gamma = \{z \in \partial\omega : &\text{there exists a sequence } t_n \in G, \\ &t_n \to t \in (-1, 1), \text{ such that } z = \lim \varphi(t_n)\}. \end{aligned} \tag{3.2}$$

It is clear that γ contains the graph of $\varphi(t_1)$, $-1 < t_1 < 1$, and that γ is compact. It is elementary to check that γ is connected. Observe that if γ is a closed Jordan arc in $\partial\omega$, then there is a conformal representation φ of ω for which γ admits the representation (3.2) and $\varphi \in C(G \cup (-1, 1))$.

Theorem 3.1. *Let $g \in H^{1,\infty}(\omega)$ and $\alpha \in C^{0,\lambda}(U)$, U a neighborhood of $\overline{\omega}$ and $0 < \lambda < 1$, satisfy*

$$g_{\bar{z}} = \alpha \qquad \text{in} \quad \omega, \tag{3.3}$$

$$g(z) = 0 \qquad \text{for} \quad z \in \gamma \subset \partial\omega, \tag{3.4}$$

$$|\alpha(z)| \geq \alpha_0 > 0 \qquad \text{for} \quad z \in U.$$

Suppose that φ is a conformal mapping of G onto ω and that $\gamma \subset \partial\omega$ admits the representation (3.2). Then $\varphi(t) \in C^{1,\tau}(\overline{G} \cap B_R)$ for each $\tau < \lambda$ and each $R < 1$.

To interpret the hypotheses of the theorem, consider the solution u of (1.1) when $N = 2$. Set

$$g(z) = \frac{\partial}{\partial z}(u(z) - \psi(z)), \qquad z \in \Omega,$$

$$\frac{\partial}{\partial z} = \frac{1}{2}\left(\frac{\partial}{\partial x_1} - i\frac{\partial}{\partial x_2}\right),$$

which is Lipschitz on compact subsets of Ω by Chapter IV, Theorem 6.3, when $\psi \in C^2(\Omega)$. We compute that

$$g_{\bar{z}}(z) = \tfrac{1}{4}\Delta(u - \psi) = -\tfrac{1}{4}\Delta\psi \qquad \text{in} \quad \Omega \backslash I;$$

hence, the hypotheses about ψ and u are satisfied in $\Omega - I$ if $\psi \in C^{2,\lambda}(\Omega)$, $0 < \lambda < 1$, and has a nonvanishing Laplacian. Conditions which provide us with an appropriate choice of $\gamma \subset \partial I$ will be discussed in the sequel.

We shall begin by demonstrating that φ' has a suitable integrability property. We then extend φ to a function which is differentiable at a prescribed point of the t_1 axis.

Given $z_0 \in \gamma$, we fix a solution $g^*(z) = g^*(z, z_0)$ to the equation $\partial g^*/\partial \bar{z} = \alpha$ for $z \in U$ such that

$$g^*(z_0) = g_z^*(z_0) = 0. \tag{3.5}$$

Then $g^* \in C^{1,\lambda}(U)$ and

$$g^*(z) = \alpha(z_0)(\bar{z} - \bar{z}_0) + R(z, z_0), \qquad z \in U, \tag{3.6}$$

$$|z - z_0|^{-1}|R(z, z_0)| + |R_z(z, z_0)| + |R_{\bar{z}}(z, z_0)| \leq C|z - z_0|^\lambda \tag{3.7}$$

with a constant $C > 0$ independent of $z_0 \in \gamma$. Such a choice of g^* is easy to find. Indeed, assuming without loss in generality that $\alpha(z) \in C^{0,\lambda}(V)$ for a neighborhood V of $\bar{U}$, one may set

$$A(z) = -\frac{1}{\pi}\int_V \frac{\alpha(\zeta)}{\zeta - z}\, d\xi_1\, d\xi_2, \qquad z \in U.$$

Then $A \in C^{1,\lambda}(\bar{U})$ and for any $z_0 \in U$ admits the development

$$A(z) = A(z_0) + A_z(z_0)(z - z_0) + \alpha(z_0)(\bar{z} - \bar{z}_0) + R(z, z_0), \qquad z \in U,$$

where R satisfies (3.7) and we may take

$$g^*(z) = A(z) - A(z_0) - A_z(z_0)(z - z_0).$$

A complex valued function $f \in H^1(D)$, $D \subset \mathbb{R}^2$ is called *quasi-conformal* in D provided that $\sup_D |f_{\bar{z}}(z)/f_z(z)| = q < 1$.

Lemma 3.2. *Assume the hypotheses of Theorem* 3.1. *Then* φ *admits a quasi-conformal extension into a neighborhood of the real axis and*

$$\varphi \in H^{1,s}(G \cap B_R) \qquad \text{for all} \quad s, \quad 1 \leq s < \infty, \quad \text{and} \quad R, \quad 0 < R < 1.$$

Proof. To prove the lemma it suffices to verify the conclusion at a given point $t_0 \in (-1, 1)$. We may assume that $t_0 = 0$, and that $z_0 = \lim_{\varepsilon \to 0} \varphi(i\varepsilon)$ exists.

Choose $r > 0$ such that $g^*(z)$ defined in (3.5) satisfies

$$\left| \frac{\partial g^*}{\partial z}(z) \Big/ \frac{\partial g^*}{\partial \bar{z}}(z) \right| < \varepsilon < 1 \qquad \text{in} \quad B_r(z_0). \tag{3.8}$$

This is possible by (3.5). Let $\zeta = g^*(z)$ denote the mapping of $B_r(z_0)$ into another plane under g^*. Observe that the Jacobian of this mapping

$$|Dg^*| = \tfrac{1}{2}(|g_{\bar{z}}^*|^2 - |g_z^*|^2) \geq (\alpha_0^2/2)(1 - \varepsilon^2) > 0 \qquad \text{in} \quad B_r(z_0),$$

so there is a neighborhood $B_{r'}(z_0)$ in which g^* is $1:1$ and has a $C^{1,\lambda}$ inverse $(g^*)^{-1}$. We suppose that $B_r(z_0) \subset B_{r'}(z_0)$. It is possible to choose r independently of $z_0 \in \gamma$ for z_0 in compact subsets of γ contained in $\gamma \cap \varphi(\bar{G} \cap B_R)$, $R < 1$.

The difference

$$h(z) = g^*(z) - g(z), \qquad z \in \omega \cup \gamma,$$

is continuous in $\omega \cup \gamma$, analytic in ω, and satisfies

$$h(z) = g^*(z) \qquad \text{for} \quad z \in \gamma. \tag{3.9}$$

Moreover, $h \in H^{1,\infty}(\omega)$ since both g^* and g have this property.

We now form an extension of $\varphi(t)$. Let $\delta > 0$ be so small that $\varphi(B_\delta(0) \cap G) \subset B_r(z_0)$ and define

$$\Phi(t) = \begin{cases} \varphi(t), & \operatorname{Im} t \geq 0 \\ g^{*-1}(h(\varphi(\bar{t}))), & \operatorname{Im} t < 0. \end{cases}$$

Let us check that $\Phi(t) \in H^1(B_\delta(0))$. Since φ is a conformal mapping onto a domain of finite area, $\varphi \in H^1(B_\delta(0) \cap G)$, as we have remarked at the beginning of this section. Now $(g^*)^{-1}$ is smooth and $h(\varphi(\bar{t}))$ is the composition of a Lipschitz function with a very smooth function. Hence $g^{*-1}(h(\varphi(\bar{t})) \in H^1(B_\delta(0) \cap \{t : \operatorname{Im} t < 0\})$. We apply the H^1 matching lemma. For every $z \in \gamma$, $|z - z_0|$ small, $g^{*-1}(h(z)) = z$ by (3.9). Hence

$$\lim_{\varepsilon \to 0} g^{*-1}(h(\varphi(t + i\varepsilon))) = \varphi(t) = \lim_{\varepsilon \to 0} \varphi(t + i\varepsilon) \qquad \text{a.e.} \quad -\delta < t < \delta,$$

so the traces of Φ from above and below are equal. By Lemma A.8 of Chapter II,

$$\Phi \in H^1(B_\delta(0)).$$

We now compute that $\Phi(t)$ is a quasi-conformal function. Indeed,

$$\Phi_{\bar{t}}(t) = 0 \qquad \text{for} \quad \operatorname{Im} t > 0, \quad |t| < \delta,$$

and

$$|\Phi_{\bar{t}}(t)/\Phi_t(t)| = |g_z^{*-1}/g_{\bar{z}}^{*-1}| = |g_z^*/g_{\bar{z}}^*| < \varepsilon \qquad \text{for} \quad \operatorname{Im} t < 0, \quad |t| < \delta.$$

It follows (cf. Bers *et al.* [1], p. 276) that

$$\Phi \in H^{1,s}(B_\eta(0)), \qquad \eta < \delta,$$

with

$$s = s(\varepsilon), \qquad \lim_{\varepsilon \to 0} s(\varepsilon) = +\infty.$$

In particular, $\varphi \in H^{1,s}(G \cap B_{\delta/2}(0))$. This proves the lemma. Q.E.D.

Consider now a function, with $B = \{t : |t| < 1\}$, $w \in H^{1,s}(B)$, such that

$$|t|^{-\sigma} w_{\bar{t}} \in L^s(B) \qquad \text{for a} \quad \sigma > 0 \quad \text{and} \quad s > 2$$

satisfying (3.10)

$$\sigma - (2/s) \geq \tau > 0.$$

Lemma 3.3. *Let* w *satisfy* (3.10). *Then there exists a complex number* c *such that*

$$\left| \frac{w(z) - w(0)}{z} - c \right| \leq C_R(\|t^{-\sigma} w_{\bar{t}}\|_{L^s(B)} + \|w\|_{L^\infty(B)}) |z|^{\min(1, \tau)},$$

$$z \in B_R, \quad R < 1,$$

with

$$|c| \leq C'_R[\|t^{-\sigma} w_{\bar{t}}\|_{L^s(B)} + \|w\|_{L^\infty(B)}],$$

where C_R, C'_R *depend on* R, s, τ.

Although this lemma is straightforward to prove, condition (3.10) is not an obvious one.

Proof. Note that $w \in C^{0,\lambda}(B)$, $\lambda = 1 - (2/s)$, by Sobolev's lemma. We recall the formula of Green in complex form. Let $g = u + iv$ and $dz = dx_1 + i\, dx_2$. For any $E \subset B_1$ with ∂E smooth and $g \in H^{1,s}(E)$,

$$\int_E g_{\bar{z}}\, dx_1\, dx_2 = -\frac{i}{2} \int_{\partial E} g\, dz$$

In particular, if $\zeta(t)$ is analytic in a neighborhood of E and $w \in H^{1,s}(B)$ is given as above,

$$\int_E w_{\bar{t}} \zeta\, dt_1\, dt_2 = -\frac{i}{2} \int_{\partial E} w\zeta\, dt.$$

We choose $\zeta(t) = 1/[t(t - z)]$ for a fixed $z \in B$ and the sets $E_\varepsilon = \{t : |t| > \varepsilon, |t - z| > \varepsilon, |t| < R\}$. Performing the integration above and passing to the limit as $\varepsilon \to 0$, which exists because w is Hölder continuous, yields the well-known formula

$$\frac{w(z) - w(0)}{z} = -\frac{1}{\pi}\int_B \frac{1}{t(t-z)} w_{\bar{t}}(t)\, dt_1\, dt_2 + \frac{1}{2\pi i}\int_{\partial B} \frac{1}{t(t-z)} w(t)\, dt \qquad \text{for} \quad 0 \neq z \in B. \tag{3.11}$$

Hence, formally,

$$c = \frac{\partial w}{\partial z}(0) = \lim_{z \to 0} \frac{1}{z}[w(z) - w(0)]$$
$$= -\frac{1}{\pi}\int_B t^{-2} w_{\bar{t}}(t)\, dt_1\, dt_2 + \frac{1}{2\pi i}\int_{\partial B} t^{-2} w(t)\, dt.$$

Observe that, in fact,

$$\left|\int_B t^{-2} w_{\bar{t}}(t)\, dt_1\, dt_2\right| \leq \|t^{-\sigma} w_{\bar{t}}\|_{L^s(B)} \left\{2\pi \int_0^1 |t|^{(\sigma-2)s'+1}\, d|t|\right\}^{1/s'},$$
$$\frac{1}{s} + \frac{1}{s'} = 1.$$

Now $\sigma > 2/s$ implies that $(\sigma - 2)s' + 1 > -1$ so the right-hand side of the inequality above is finite. We set

$$c = -\frac{1}{\pi}\int_B t^{-2} w_{\bar{t}}(t)\, dt_1\, dt_2 + \frac{1}{2\pi i}\int_{\partial B} t^{-2} w(t)\, dt$$

and subtract it from (3.11). This gives

$$\frac{w(z) - w(0)}{z} - c = -\frac{z}{\pi}\int_B \frac{1}{t^2(t-z)} w_{\bar{t}}(t)\, dt_1\, dt_2 + \frac{z}{2\pi i}\int_{\partial B} \frac{1}{t^2(t-z)} w(t)\, dt.$$

Now for $z \neq 0$,

$$\left|\int_B \frac{1}{t^2(t-z)} w_{\bar{t}}(t)\, dt_1\, dt_2\right|$$
$$\leq \|t^{-\sigma} w_{\bar{t}}\|_{L^s(B)} \left\{\int_B |t|^{(\sigma-2)s'} |t - z|^{-s'}\, dt_1\, dt_2\right\}^{1/s'}.$$

The last integral on the right-hand side may be estimated in a familiar way which we indicate here. Clearly,

$$\int_B |t|^{(\sigma-2)s'}|t-z|^{s'}\,dt_1\,dt_2 \le \int_{|t|<2} |t|^{(\sigma-2)s'}|t-z|^{-s'}\,dt_1\,dt_2.$$

For $|z|<1$, let

$$A_1 = \{t : |t-z| < \tfrac{1}{2}|z|\}, \qquad A_2 = \{t : |t| < \tfrac{3}{2}|z|, |t-z| \ge \tfrac{1}{2}|z|\},$$
$$A_3 = \{t : \tfrac{3}{2}|z| \le |t| < 2\},$$

and set

$$I_j = \int_{A_j} |t|^{(\sigma-2)s'}|t-z|^{-s'}\,dt_1\,dt_2.$$

Assume now that $\sigma < 2$. Then

$$|t|^{(\sigma-2)s'} \le 2^{(2-\sigma)s'}|z|^{(\sigma-2)s'} \qquad \text{in} \quad A_1$$

so

$$I_1 \le \text{const}\,|z|^{(\sigma-2)s'} \int_0^{|z|/2} \rho^{1-s'}\,d\rho \le \text{const}\,|z|^{(\sigma-2)s'+2-s'}.$$

On the other hand,

$$|t-z|^{-s'} \le 2^{s'}|z|^{-s'} \qquad \text{for} \quad t\in A_2,$$

which leads to the estimate

$$I_2 \le \text{const}\,|z|^{-s'} \int_0^{3|z|/2} \rho^{(\sigma-2)s'+1}\,d\rho = \text{const}\,|z|^{(\sigma-2)s'+2-s'}.$$

Finally, for I_3,

$$|t/(t-z)| \le |t|/(|t|-|z|) \le 3 \qquad \text{since} \quad \tfrac{3}{2}|z| \le |t| < 2,$$

so

$$|t|^{(\sigma-2)s'}|t-z|^{-s'} \le 3^{s'}|t|^{(\sigma-2)s'-s'}.$$

Therefore

$$\begin{aligned} I_3 &\le \text{const} \int_{3|z|/2}^{2} \rho^{(\sigma-2)s'-s'+1}\,d\rho \\ &\le \text{const}\,[|z|^{(\sigma-2)s'-s'+2} - 2^{(\sigma-2)s'-s'+2}] \\ &\le \text{const}\,|z|^{(\sigma-2)s'-s'+2} \end{aligned}$$

since $(\sigma - 2)s' - s' + 2 = s'[\sigma - 3 + (2/s')] < 0$ if $\sigma < 2$. Consequently,

$$(I_1 + I_2 + I_3)^{1/s'} \le \text{const}\,|z|^{\sigma - (2/s) - 1} \qquad \text{for} \quad |z| < 1$$
$$= \text{const}\,|z|^{\tau - 1}$$

or

$$\left| \frac{z}{\pi} \int_B \frac{1}{t^2(t - z)} w_{\bar{t}}(t)\, dt_1\, dt_2 \right| \le \text{const}\,|z|^{\tau} \qquad \text{for} \quad |z| < 1.$$

The second integral is easy to control, so the lemma follows in the case $\sigma < 2$. But if $\sigma \ge 2$, then

$$\left| \int_B |t|^{(\sigma - 2)s'} |t - z|^{-s'}\, dt_1\, dt_2 \right| \le \text{const} \int_0^2 \rho^{1 - s'}\, d\rho < \infty,$$

so the conclusion holds with exponent 1. Q.E.D.

Proof of the Theorem. To prove the theorem we show that φ, known to be in $H^{1,s}(G \cap B_R)$, $0 < R < 1$, admits an extension to B_R which satisfies (3.10). By change of scale, we may assume $\varphi \in H^{1,s}(G)$. We exhibit this extension at $t = 0$, the situation being the same at other points of $(-1, 1)$.

Let us recall again the function $g^*(z)$ of (3.5) and the difference $h(z) = g^*(z) - g(z)$, which is analytic in ω and satisfies

$$h(z) = g^*(z) \qquad \text{for} \quad z \in \gamma. \tag{3.9}$$

Write

$$g^*(z) = \alpha(z_0)(\bar{z} - \bar{z}_0) + R(z, \bar{z})$$

and determine $\varphi^*(t)$ by the relation

$$h(\varphi(t)) = \alpha(z_0)[\varphi^*(t) - \bar{z}_0] + R(\varphi(t), \overline{\varphi(t)}).$$

It follows from (3.9) and (3.6) that

$$\varphi^*(t) = \overline{\varphi(t)} \qquad \text{for} \quad \operatorname{Im} t = 0, \quad |t| < 1.$$

Since $h(\varphi(t))$ is holomorphic in G, we compute that

$$\frac{\partial \varphi^*}{\partial \bar{t}}(t) = -\frac{1}{\alpha(z_0)} \frac{\partial R}{\partial \bar{z}}(\varphi(t), \overline{\varphi(t)})\, \overline{\varphi'(t)},$$

so

$$\left| \frac{\partial \varphi^*}{\partial \bar{t}}(t) \right| \le \text{const}\,|\varphi'(t)|\,|\varphi(t) - z_0|^{\lambda}, \qquad t \in G, \tag{3.12}$$

by (3.7).

According to Lemma 3.2, $\varphi' \in L^s(G)$ for every s, $1 \leq s < \infty$, so given τ, $0 < \tau < \lambda$, we may choose s so large that

$$\sigma - (2/s) \geq \tau > 0 \qquad \text{for} \quad \sigma = \lambda[1 - (2/s)].$$

Furthermore, $\varphi_t^* \in L^s(G)$, so $\varphi^* \in H^{1,s}(G)$. Now define

$$w(t) = \begin{cases} \varphi(t), & \operatorname{Im} t \geq 0, \quad |t| < 1 \\ \overline{\varphi^*(\bar{t})}, & \operatorname{Im} t \leq 0, \quad |t| < 1. \end{cases} \tag{3.13}$$

Since w is continuous in $|t| < 1$, it follows that $w \in H^{1,s}(B)$ by the H^1, or $H^{1,s}$, matching lemma, and by (3.12) it satisfies (3.10). Therefore we may apply Lemma 3.3 to infer the existence of a constant c such that

$$\left| \frac{w(t) - w(0)}{t} - c \right| \leq \text{const } |t|^\tau, \qquad |t| < R < 1.$$

When $\operatorname{Im} t \geq 0$, $w(t) = \varphi(t)$, so we have achieved the estimate

$$\left| \frac{\varphi(t) - \varphi(0)}{t} - c \right| \leq \text{const } |t|^\tau, \qquad |t| \leq R < 1.$$

Since 0 was an arbitrary point of $(-1, 1)$, what we have really shown is that given $R < 1$, for each $t_0 \in [-R, R]$ there is a complex number $c(t_0)$ such that

$$\left| \frac{\varphi(t) - \varphi(t_0)}{t - t_0} - c(t_0) \right| \leq C_R |t - t_0|^\tau, \qquad |c(t_0)| \leq C'_R, \tag{3.14}$$

where C_R, C'_R depend on R, τ, s, and some $H^{1,s}$ norms of φ. It is easy to check by Cauchy's theorem that a function φ analytic in G and satisfying (3.14) for each $R < 1$ satisfies $\varphi \in C^{1,\tau}(\bar{G} \cap B_R)$, $R < 1$. Q.E.D.

We apply the preceding theorem to the case of a concave obstacle.

Theorem 3.4. *Let Ω be a strictly convex domain in $\mathbb{R}^2$ with smooth boundary $\partial\Omega$ and let $\psi \in C^{2,\lambda}(\bar{\Omega})$ be a strictly concave obstacle, namely,*

$$\max_{\Omega} \psi > 0, \qquad \psi < 0 \quad \textit{on} \quad \partial\Omega$$

and

$$-\psi_{x_i x_j}(z)\xi_i \xi_j > 0 \qquad \textit{for} \quad \xi \in \mathbb{R}^2, \quad \xi \neq (0, 0), \quad z \in \Omega.$$

Let

$$u \in \mathbb{K}: \quad \int_\Omega u_{x_i}(v - u)_{x_i}\, dx \geq 0 \qquad \textit{for} \quad v \in \mathbb{K},$$

$$\mathbb{K} = \{v \in H_0^1(\Omega) : v \geq \psi \textit{ in } \Omega\},$$

and set

$$I = \{z \in \Omega : u(z) = \psi(z)\}.$$

Then $\Gamma = \partial I$ *is a Jordan curve which admits a* $C^{1,\tau}$ *parametrization.*

Proof. In Section 6 we shall discuss the topological nature of I. For the moment, let us assume the conclusions of Theorem 6.2, namely, that I is a connected and simply connected domain whose interior int I is also connected. Moreover, $\Omega \setminus I$ is of the topological type of an annulus. This implies the existence of a conformal mapping φ of the upper half $t = t_1 + it_2$ plane, from which an appropriate disk has been deleted, onto $\Omega \setminus I$ that maps $[-\infty, \infty]$ onto ∂I. This means that ∂I enjoys the property $\partial I = \{z \in \mathbb{R}^2 :$ there exists a sequence $t_n \in \{t : \operatorname{Im} t > 0\}$ such that $\varphi(t_n) \to z$ and $\operatorname{Im} t_n \to 0\}$. Given a particular z_0, we may assume that the points t_n with $\varphi(t_n) \to z$ satisfy $|t_n| < 1$.

We may apply Theorem 3.1 to $\omega = \varphi(G)$, the set γ defined by (3.2),

$$g(z) = \frac{\partial}{\partial z}[u(z) - \psi(z)],$$

and

$$\alpha(z) = -\tfrac{1}{4}\Delta\psi(z).$$

In a neighborhood of I, $g(z)$ is Lipschitz because the second derivatives of u are bounded there (cf. Theorem 6.3 in Chapter IV). The strict concavity of ψ implies $|\alpha(z)| \neq 0$ on compact subsets of Ω.

Hence ∂I admits the parametrization, which is of class $C^{1,\tau}$,

$$t \to \varphi(t), \qquad -\infty < t < \infty,$$

and is thereby a curve. Suppose $\varphi(t_1) = z_0 = \varphi(t_2)$ for $t_1 < t_2$. Then the arc $C_1 = \{\varphi(t) : t_1 \leq t \leq t_2\}$ separates int I, or $C_2 = \{\varphi(t) : -\infty < t \leq t_1,$ $t_2 \leq t < \infty\}$ separates int I, or one of C_1 or C_2 encloses no open set. In the first two cases we contradict that int I is connected whereas in the last, we contradict that

$$I = \overline{\operatorname{int} I},$$

for a point on the arc C_1 would not be the limit of interior points of I. Hence ∂I has no double points, and so it is a Jordan curve. Q.E.D.

The last argument, we note, did not employ the fact that the parametrization of ∂I was smooth, but only that it was continuous.

We remark that it is also possible to show that ∂I is a "regular" curve, that is, it is free of cusps, at least if $\psi \in C^{3,\lambda}(\Omega)$. This requires a deeper study of the extension $w(t)$ defined by (3.13). Indeed, w satisfies the differential inequality

$$|w_z(z)| \leq q(z) \sup_{|t|=|z|} |w(t) - w(0)|, \qquad z \in B, \tag{3.15}$$

where $q \in L^s(B)$ for some $s > 2$.

Such functions enjoy properties analogous to those of analytic functions. We state here without proof a description of these properties:

Theorem 3.5. *Let w satisfy (3.15). If*

$$\lim_{z \to 0} |z^{-n}[w(z) - w(0)]| = 0$$

for an integer $n \geq 0$, then

$$c = \lim_{z \to 0} z^{-n-1}[w(z) - w(0)]$$

exists. Moreover, if $\lim_{z \to 0} z^{-n} w(z) = 0$ for every $n \geq 0$, then $w(z) \equiv 0$ in a neighborhood of $z = 0$.

With reference to Theorem 3.4, the restriction of w to $\operatorname{Im} t > 0$ is a nonvanishing conformal mapping; hence there is a smallest interger n for which

$$\lim_{z \to 0} \frac{w(z) - w(0)}{z^n} = c \neq 0.$$

It then follows from the $1:1$ nature of the conformal mapping that $n = 1$ or $n = 2$.

It is natural to ask about higher differentiability of the curve γ. We state here a result about this; its proof is left as an exercise.

Theorem 3.6. *With the hypotheses of Theorem 3.1, assume that $\alpha \in C^{m,\lambda}(U)$, $m \geq 0$ and $0 < \lambda < 1$. Then the conformal mapping $\varphi \in C^{m+1,\tau}(\bar{G} \cap B_R)$ for each $\tau < \lambda$ and $R < 1$.*

This theorem is slightly stronger than the result we obtained with the aid of the Legendre transform because here ∂I is not required to be a regular curve.

4. A Remark about Singularities

Theorems 3.1 and 3.5 show that the only singularities displayed by the free curve γ are cusps. Such cusps may arise. We give here a simple example. Consider the mapping φ from

$$G = \{t = t_1 + it_2 : |t| < 1, \operatorname{Im} t > 0\}$$

onto a domain $\omega = \varphi(G)$ in the $z = x_1 + ix_2$ plane given by

$$z = \varphi(t) = t^2 + it^\mu, \qquad \mu \text{ odd.}$$

This maps $(-1, 1)$ onto the curve

$$\Gamma\colon \quad x_2 = \pm x_1^{\mu/2}, \qquad 0 \leq x_1 < 1.$$

We shall find, for ε sufficiently small, a function

$$u \in H^{2,\infty}(B_\varepsilon), \qquad B_\varepsilon = \{z : |z| < \varepsilon\}$$

such that

$$\begin{aligned} \Delta u = 2 \quad &\text{and} \quad u > 0 \quad &&\text{in} \quad \omega \cap B_\varepsilon \\ u = 0 \quad &\text{and} \quad u_{x_j} = 0 \quad &&\text{in} \quad B_\varepsilon \backslash \omega, \quad j = 1, 2 \end{aligned} \tag{4.1}$$

provided that $\mu = 4k + 1$, $k = 1, 2, 3, \ldots$. It follows that u is the solution to the variational inequality

$$u \in \mathbb{K}\colon \quad \int_{B_\varepsilon} u_{x_j}(v - u)_{x_j}\, dx \geq -2 \int_{B_\varepsilon} (v - u)\, dx \qquad \text{for} \quad v \in \mathbb{K} \tag{4.2}$$

where

$$\mathbb{K} = \{v \in H^1(B_\varepsilon) : v \geq 0 \text{ and } v = u \text{ on } \partial B_\varepsilon\}.$$

In view of (4.1), the set of coincidence of u,

$$I = \{z : u(z) = 0\},$$

will have free boundary $\Gamma \cap B_\varepsilon$ in B_ε.

That Γ has an analytic parametrization as the boundary values of a conformal mapping suggests the existence of a holomorphic $f(z)$, $z \in \omega$, such that

$$f(z) = \bar{z} \qquad \text{for} \quad z \in \gamma.$$

Indeed, it is given explicitly by

$$f(z) = F(t) = t^2 - it^\mu, \qquad t \in G.$$

To describe its behavior, we find z as a function of t. Clearly

$$t^2 = z - iz^{\mu/2} + \cdots, \qquad 0 \leq \arg z^{1/2} \leq \pi,$$

where $\cdots$ represents terms of degree higher than $\mu/2$, so $F(t) = t^2 - it^\mu = -z + 2t^2$ and

$$f(z) = z - 2iz^{\mu/2} + \cdots, \qquad z \in \omega, \quad |z| \quad \text{small.}$$

Our function u is found by integration:

$$u(z) = \begin{cases} \frac{1}{2}|z|^2 - \operatorname{Re} \int_0^{\bar{z}} f(\zeta)\, d\zeta, & z \in \omega \\ 0, & z \in B_\varepsilon \setminus \omega \end{cases} \tag{4.3}$$

for ε sufficiently small. Since $f(z) = \bar{z}$ on Γ,

$$u(z) = u_{x_j}(z) = 0 \qquad \text{on} \quad \Gamma, \quad j = 1, 2.$$

Now u admits the expansion

$$u(z) = x_2^2 - \frac{2}{1 + (\mu/2)} \rho^{(\mu/2)+1} \sin\left(\frac{\mu}{2} + 1\right)\theta + \cdots, \qquad z \in \omega, \quad |z| \quad \text{small,} \tag{4.4}$$

from which we see that $u(x_1) > 0$ when $x_1 < 0$ provided $\mu = 4k + 1$, $k = 1, 2, \ldots$.

To complete the proof that u satisfies (4.1), we must show that $u > 0$ in $\omega \cap B_\varepsilon$ for some $\varepsilon > 0$. Since $(\mu/2) + 1 \geq \frac{5}{2}$,

$$u_{x_2x_2}(z) = \begin{cases} 2 + O(|z|^{1/2}) & \text{if} \quad z \in \omega \cap B_\varepsilon \\ 0 & \text{if} \quad z \in \omega \setminus B_\varepsilon. \end{cases}$$

Also, $u_{x_2}(z) = -u_{x_2}(\bar{z})$, $z \in B_\varepsilon$, $u(z) = u(\bar{z})$, $z \in B_\varepsilon$.

On each vertical line $x_1 = \delta$, $|\delta| < \varepsilon$, $u(\delta + ix_2)$ is a convex $C^{1,1}$ function whose minimum is attained at a point where $u_{x_2}(\delta + ix_2) = 0$. If $\delta < 0$, the unique point which this property is $x_2 = 0$, and we have seen that $u(x_1) > 0$ for $x_1 < 0$. Suppose $\delta > 0$. Then $u(\delta + ix_2) = u_{x_2}(\delta + ix_2) = 0$ for $|x_2| \leq \delta^{\mu/2}$ whereas $u_{x_2x_2}(\delta + ix_2) > 0$ for $|x_2| > \delta^{\mu/2}$. Hence u is strictly increasing for $|x_2| > \delta^{\mu/2}$, so it is positive there.

5. The Obstacle Problem for a Minimal Surface

To commence, we review the geometry of a two dimensional nonparametric minimal surface in $\mathbb{R}^3$. A function $u \in C^2(U)$, $U \subset \mathbb{R}^2$ open, is a solution to the minimal surface equation in U if

$$-\frac{\partial}{\partial x_j}\left(\frac{u_{x_j}}{\sqrt{1 + u_x^2}}\right) = 0 \qquad \text{in} \quad U. \tag{5.1}$$

Its graph

$$S = \{x = (x_1, x_2, x_3) : x_3 = u(z),\ z = x_1 + ix_2 \in U\}$$

is called a minimal surface. More generally, the mean curvature H of the surface defined by $v \in C^2(U)$ is given by

$$H = \frac{1}{2} \frac{\partial}{\partial x_j} \left(\frac{v_{x_j}}{\sqrt{1 + v_x^2}} \right), \qquad z \in U.$$

From this, and the obvious fact that (5.1) is the Euler equation of the nonparametric area integrand

$$F(v) = \int_U \sqrt{1 + v_x^2}\, dx$$

we infer, formally speaking, that the surface which minimizes area among all those having a given boundary is a surface of vanishing mean curvature.

The metric of the surface

$$M = \{x \in \mathbb{R}^3 : x_3 = v(z), z \in U\}$$

is defined by the symmetric positive definite matrix

$$g = (g_{ij}) = \begin{pmatrix} 1 + v_{x_1}^2 & v_{x_1} v_{x_2} \\ v_{x_1} v_{x_2} & 1 + v_{x_2}^2 \end{pmatrix}. \tag{5.2}$$

To the point $x \in M$ we associate $n(x) \in S^2$, the upward normal of M at x,

$$n(x) = (-v_{x_1}/W, -v_{x_2}/W, 1/W), \qquad W = \sqrt{1 + v_x^2}.$$

This mapping induces another, possibly degenerate, metric on M via the bilinear form

$$l = (l_{ij}) = \begin{pmatrix} n_{x_1} \cdot n_{x_1} & n_{x_1} \cdot n_{x_2} \\ n_{x_1} \cdot n_{x_2} & n_{x_2} \cdot n_{x_2} \end{pmatrix}.$$

The matrix l is symmetric and nonnegative. A minimal surface S enjoys the property that the matrices g and l are proportional at each point. This is the statement that the normal mapping of S to the sphere S^2 is conformal, or anticonformal, as this function of proportionality is positive or negative. We shall verify this in the ensuing discussion.

Replacing a point $(c_1, c_2, c_3) \in S^2$ by its (negative conjugate) stereographic projection onto the equatorial plane from the south pole,

$$\zeta = -(c_1 - ic_2)/(1 + c_3),$$

and applying this to $n(x)$, we obtain what we shall refer to as the Gauss mapping of the surface S, whether it is minimal or not,

$$f(z) = \frac{u_{x_1}(z) - iu_{x_2}(z)}{1 + \sqrt{1 + u_x(z)^2}}, \qquad z \in U. \tag{5.3}$$

Since the mapping from $S^2 \to \mathbb{C}$ given by stereographic projection induces a metric on S^2 proportional to its original one, we need only check that $|d\zeta|^2$ for $\zeta = f(z)$ is proportional to the metric g.

Lemma 5.1. *Let $u \in C^2(U)$, $U \subset \mathbb{R}^2$ open, be a solution to the minimal surface equation. Then*

$$f_{\bar{z}}(z) = \mu(z) f_z(z) \qquad \text{for} \quad z \in U,$$

where

$$\mu(z) = \overline{f(z)}^2.$$

The proof of the lemma is a direct, but tedious, computation. See also Kinderlehrer [3] for a more geometric proof. The matrix g, expressed in terms of dx_1 and dx_2, namely, written as a tensor, is

$$g = g_{11}\, dx_1^2 + 2g_{12}\, dx_1\, dx_2 + g_{22}\, dx_2^2.$$

Rewriting this with $dz = dx_1 + i\, dx_2$ and $d\bar{z} = dx_1 - i\, dx_2$, we obtain

$$\begin{aligned} g = \tfrac{1}{4}[(g_{11} - g_{22} - 2ig_{12})\, dz^2 + 2(g_{11} + g_{22})\, dz\, d\bar{z} \\ + (g_{11} - g_{22} + 2ig_{12})\, d\bar{z}^2]. \end{aligned}$$

Now we compute that

$$\begin{aligned} |d\zeta|^2 &= |f_z\, dz + f_{\bar{z}}\, d\bar{z}|^2 = |f_z|^2 |dz + \mu\, d\bar{z}|^2 \\ &= |f_z|^2(\bar{\mu}\, dz^2 + (1 + |\mu|^2)\, dz\, d\bar{z} + \mu\, d\bar{z}^2). \end{aligned}$$

Observe that

$$\mu = \frac{g_{11} - g_{22} + 2ig_{12}}{(1 + W)^2}, \qquad W = \sqrt{1 + u_x^2},$$

and

$$1 + |\mu|^2 = \frac{2(g_{11} + g_{22})}{(1 + W)^2};$$

hence

$$|d\zeta|^2 = [|f_z|^2/(1 + W)^2]g.$$

We have demonstrated

Proposition 5.2. *Let $u \in C^2(U)$, $U \subset \mathbb{R}^2$ open, be a solution to the minimal surface equation. Then the Gauss mapping of*

$$S = \{x \in \mathbb{R}^3 : x_3 = u(z),\ z \in U\}$$

is conformal.

These considerations suggest that it is feasible to exploit the Gauss mapping of the surface S in place, say, of the complex gradient of u to ascertain the smoothness of the free boundary. First we recall the theorem of Korn and Lichtenstein (cf. Courant and Hilbert [1].)

Theorem 5.3. *Let $\Omega \subset \mathbb{R}^2$ be a simply connected domain with smooth boundary and set $B = \{\zeta : |\zeta| < 1\}$. Suppose that $\mu \in C^{0,\lambda}(\bar{\Omega})$ satisfies*

$$|\mu(z)| \le \mu_0 < 1 \qquad \text{in} \quad \Omega.$$

Then there exists a homeomorphism $\zeta = h(z)$ from $\bar{\Omega}$ to $\bar{B}$ such that $h \in C^{1,\lambda}(\bar{\Omega})$,

$$h_{\bar{z}} = \mu h_z \qquad \text{in} \quad \Omega,$$

$$|h_z| > 0 \qquad \text{in} \quad \Omega,$$

and $h(z_0) = 0$ for a preassigned $z_0 \in \Omega$. If $f \in C^1(U)$, $U \subset \Omega$ open, satisfies

$$f_{\bar{z}} = \mu f_z \qquad \text{in} \quad \Omega,$$

then $F(\zeta) = f(z)$ is a holomorphic function of ζ in $h(U)$.

The second statement of the theorem is elementary inasmuch as

$$f_z = F_\zeta h_z + F_{\bar{\zeta}}(\bar{h})_z \qquad \text{and} \qquad f_{\bar{z}} = F_\zeta h_{\bar{z}} + F_{\bar{\zeta}}(\bar{h})_{\bar{z}},$$

so

$$0 = \mu f_z - f_{\bar{z}} = F_\zeta(\mu h_z - h_{\bar{z}}) + F_{\bar{\zeta}}(\mu \bar{h}_{\bar{z}} - \bar{h}_z) = F_{\bar{\zeta}}(|\mu|^2 - 1)\bar{h}_z.$$

Now $|\mu| < 1$ and $h_z \ne 0$, so $F_{\bar{\zeta}} = 0$, which means that F is a holomorphic function of ζ.

Fix a domain $\Omega \subset \mathbb{R}^2$ strictly convex with smooth boundary $\partial\Omega$ and a strictly concave obstacle $\psi \in C^{2,\lambda}(\Omega)$. Consider the problem: To find

$$u \in \mathbb{K} : \int_\Omega \frac{u_{x_j}}{\sqrt{1 + u_x^2}} (v - u)_{x_j}\, dx \ge 0 \tag{5.4}$$

for $v \in \mathbb{K}$ with

$$\mathbb{K} = \{v \in H_0^{1,\infty}(\Omega) : v \ge \psi \text{ in } \Omega\},$$

and let

$$I = \{z \in \Omega : u(z) = \psi(z)\}.$$

We know u exists and moreover $u \in C^{1,\alpha}(\bar{\Omega}) \cap H^{2,\infty}(\Omega_0)$ for each $\alpha \in (0, 1)$ and $\Omega_0 \subset \bar{\Omega}_0 \subset \Omega$.

Theorem 5.4. *Let $I = I(u)$ denote the coincidence set of the solution u to the problem* (5.4). *Then ∂I is a Jordan curve which admits a $C^{1,\tau}$ parametrization for each $\tau < \lambda$.*

Proof. In the course of the proof we shall use Theorem 6.2 as we did in the proof of Theorem 3.4.

Define the Gauss mapping $f(z)$, $z \in \Omega$, by (5.3). Then $f(z)$ is Lipschitz in compact subsets of Ω and satisfies the equation

$$f_{\bar{z}} = \mu f_z \qquad \text{in} \quad \Omega - I,$$
$$\mu = \bar{f}^2$$

since

$$-\frac{\partial}{\partial x_j}\left(\frac{u_{x_j}}{\sqrt{1 + u_x^2}}\right) = 0 \qquad \text{in} \quad \Omega - I$$

according to Lemma 5.1. In addition, with

$$f^*(z) = \frac{\psi_{x_1}(z) - i\psi_{x_2}(z)}{1 + \sqrt{1 + \psi_x(z)^2}}, \qquad z \in \Omega, \tag{5.5}$$

$f^* \in C^{1,\lambda}(\Omega)$ and

$$f^*(z) = f(z) \qquad \text{in} \quad I$$

since $u_{x_j}(z) = \psi_{x_j}(z)$ in I. The next lemma derives from the concavity of the obstacle.

Lemma 5.5. *With f^* defined by* (5.5), $|f_z^*(z)| < |f_{\bar{z}}^*(z)|$ *for* $z \in \Omega$.

Proof. We may rotate coordinates so that $\psi_{x_1x_2}(z_0) = 0$ for a given z_0. This does not change the modulus of f_z^* or $f_{\bar{z}}^*$. One then computes directly that

$$f_z^* = \psi_{x_1x_1}\alpha_1 - \psi_{x_2x_2}\alpha_2 + ib$$

and

$$f_{\bar{z}}^* = \psi_{x_1x_1}\alpha_1 + \psi_{x_2x_2}\alpha_2 - ib,$$

where $\alpha_j > 0$ and b is real. Since ψ is concave, $\psi_{x_jx_j} < 0$, hence $|\operatorname{Re} f_z^*| < |\operatorname{Re} f_{\bar{z}}^*|$. Since $|\operatorname{Im} f_z^*| = |\operatorname{Im} f_{\bar{z}}^*|$, the lemma follows.

To complete the proof of the theorem, we map Ω onto B by the theorem of Korn and Lichtenstein, Theorem 5.3, with coefficient $\mu = \bar{f}^2$. More precisely, let $\zeta = h(z)$ satisfy

$$h_{\bar{z}} = \mu h_z, \qquad z \in \Omega, \quad \text{with} \quad \mu = \bar{f}^2, \tag{5.6}$$

and suppose $z_0 \in \partial I$ is given and that $h(z_0) = 0$. Define the functions

$$F^*(\zeta) = f^*(z) \qquad \text{and} \qquad F(\zeta) = f(z)$$

and set

$$g(\zeta) = F^*(\zeta) - F(\zeta), \qquad \zeta \in B,$$

a function that is Lipschitz in a neighborhood of $h(I)$. Now $F(\zeta)$ is holomorphic in $B - h(I)$, whence

$$g_{\bar\zeta}(\zeta) = F^*_{\bar\zeta}(\zeta) = f^*_z \frac{\partial h^{-1}}{\partial \bar\zeta} + f^*_{\bar z} \frac{\partial (\bar h)^{-1}}{\partial \bar\zeta} = \alpha(\zeta) \qquad \text{in} \quad B - h(I). \quad (5.7)$$

Evidently, $\alpha(\zeta)$ defined by (5.7) for all $\zeta \in B$ is in $C^{0,\lambda}(B)$, since h and its inverse h^{-1} are in $C^{1,\lambda}$ and $f^* \in C^{1,\lambda}(\Omega)$. We must check that $|\alpha(\zeta)| \neq 0$. From (5.6) one sees immediately that

$$\frac{\partial h^{-1}}{\partial \bar\zeta} = a\mu \frac{\partial h^{-1}}{\partial \zeta},$$

where $|a| = 1$. Keeping in mind that $\partial h^{-1}/\partial\zeta \neq 0$ and $|\mu| < 1$, we use the preceding lemma to calculate that

$$\begin{aligned} |\alpha(\zeta)| &= \left|\frac{\partial h^{-1}}{\partial \zeta}\right| |f^*_z \mu a + f^*_{\bar z}| \\ &\geq \left|\frac{\partial h^{-1}}{\partial \zeta}\right| (|f^*_{\bar z}| - |\mu||f^*_z|) > 0. \end{aligned}$$

At this point we wish to apply Theorem 3.1. Assuming Theorem 6.2, we argue as in the proof of Theorem 3.4. Given $\zeta_0 \in \partial I$, consider a conformal mapping $\varphi(t)$ from the upper half-plane minus a suitable disk onto $B - h(I)$, a topological image of $\Omega - I$, which maps the real axis $(-\infty, \infty)$ onto $\partial h(I)$ in the sense described by (3.1), with, say, $\varphi(0) = 0$. By Theorem 3.1, $\varphi(t) \in C^{1,\tau}(-\infty, \infty)$ for each $\tau < \lambda$ and $t \to h^{-1}(\varphi(t))$ is a $C^{1,\tau}$ parametrization of ∂I.

It follows exactly as in the proof of Theorem 5.4 that ∂I is a Jordan curve. Q.E.D.

Using this result, it is possible to give a direct proof of the analyticity of ∂I when ψ is concave and analytic. In fact, historically a significant part of the development of Section 5 was motivated by this application. The demonstration relies on the resolution of a system of differential equations in the complex domain and the use of the solution to analytically extend a conformal representation of the minimal surface $S = \{x : x_3 = u(z),\ z \in \Omega - I\}$. The idea of connecting an analytic function to its possible extension by means of a differential equation is due to Lewy [2, 4]. We state

Theorem 5.6. *Let $I = I(u)$ denote the coincidence set of the solution u to the problem (5.4). Suppose in addition that ψ is real analytic. Then ∂I is a Jordan curve which admits an analytic parametrization.*

6. The Topology of the Coincidence Set When the Obstacle is Concave

For a general obstacle ψ and domain Ω the topology of the coincidence set for a solution of the problem (1.1) is very difficult to ascertain. In the exercises we show that for any preassigned integer k there is an obstacle whose coincidence set has at least k components. However, if the obstacle is strictly concave and $\Omega \subset \mathbb{R}^2$ is strictly convex, it is possible to prove that I is a simply connected domain equal to the closure of its interior. We have already used these facts in the proofs of Theorems 3.4 and 5.4.

Throughout this section let us assume

Ω is a strictly convex domain with smooth boundary in the $z = x_1 + ix_2$ plane, (6.1)

$\psi \in C^2(\overline{\Omega})$ is a strictly concave obstacle, namely, $-\psi_{x_j x_k}(z)\xi_j\xi_k > 0$ for $(0, 0) \neq (\xi_1, \xi_2) \in \mathbb{R}^2$ and $z \in \overline{\Omega}$, (6.1′)

and

$a(p) = (a_1(p), a_2(p))$ is a locally coercive analytic vector field. (6.1″)

Consider the solution u to the variational inequality

$$u \in \mathbb{K} : \int_\Omega a_j(u_x)(v - u)_{x_j}\, dx \geq 0 \qquad \text{for} \quad v \in \mathbb{K},$$
$$\mathbb{K} = \{v \in H_0^{1,\infty}(\Omega) : v \geq \psi \text{ in } \Omega\}, \tag{6.2}$$

and its set of coincidence

$$I = \{z \in \Omega : u(z) = \psi(z)\}.$$

There is a unique solution u to (6.2) and it satisfies

$$u \in H^{2,s}(\Omega) \cap H^{2,\infty}(\Omega_0) \qquad \text{for} \quad 1 \leq s < \infty$$

and for all subdomains $\Omega_0 \subset \overline{\Omega}_0 \subset \Omega$. Since $\psi < 0$ on $\partial\Omega$, I is a compact subset of Ω (Chapter IV, Theorems 4.3 and 6.3).

As usual, we define

$$A: H_0^{1,\infty}(\Omega) \to H^{-1}(\Omega)$$

by

$$Av = -\frac{\partial}{\partial x_j} a_j(v_x), \qquad v \in H_0^{1,\infty}(\Omega).$$

The concavity of ψ implies that

$$A\psi > 0 \quad \text{in} \quad \Omega.$$

In fact,

$$\begin{aligned} A\psi &= -\frac{\partial}{\partial x_j} a_j(\psi_x(z)) = -\frac{\partial a_j}{\partial p_k}(\psi_x(z))\psi_{x_k x_j}(z) \\ &= -\text{trace } \alpha\Psi, \end{aligned}$$

where

$$\alpha = \begin{pmatrix} \dfrac{\partial a_1}{\partial p_1}(\psi_x(z)) & \dfrac{1}{2}\left(\dfrac{\partial a_1}{\partial p_2}(\psi_x(z)) + \dfrac{\partial a_2}{\partial p_1}(\psi_x(z))\right) \\ \dfrac{1}{2}\left(\dfrac{\partial a_1}{\partial p_2}(\psi_x(z)) + \dfrac{\partial a_2}{\partial p_1}(\psi_x(z))\right) & \dfrac{\partial a_2}{\partial p_2}(\psi_x(z)) \end{pmatrix}$$

and

$$\Psi = \begin{pmatrix} \psi_{x_1x_1}(z) & \psi_{x_1x_2}(z) \\ \psi_{x_1x_2}(z) & \psi_{x_2x_2}(z) \end{pmatrix}.$$

Now α is positive definite and symmetric and Ψ is negative definite symmetric. Hence $-\text{trace } \alpha\Psi > 0$.

Lemma 6.1. *The set $\Omega - I$ is connected.*

Proof. Since $u(z) = 0 > \psi(z)$ for $z \in \partial\Omega$ and $\partial\Omega$ is connected, there is only one component of $\Omega - I$ whose closure intersects $\partial\Omega$. Suppose ω' to be another component. Then $\partial\omega' \subset I$. For any $\zeta \in C_0^\infty(\omega')$, $\zeta \geq 0$,

$$0 \leq \int_{\omega'} [a_j(\psi_x) - a_j(u_x)]\zeta_{x_j}\, dx = \int_{\omega'} \alpha_{jk}(\psi - u)_{x_k}\zeta_{x_j}\, dx,$$

where

$$\alpha_{jk}(z) = \int_0^1 \frac{\partial a_j}{\partial p_k}(u_x(z) + t(\psi_x(z) - u_x(z)))\, dt$$

is a positive definite form owing to the local coercivity of $a(p)$. Hence $\psi - u$ is a supersolution to an elliptic equation, which implies

$$\psi(z) - u(z) \geq \inf_{\partial\omega'}(\psi - u), \qquad z \in \omega'.$$

But $\partial\omega' \subset I$, so $\psi - u \geq 0$ in ω', a contradiction to $\omega' \subset \Omega - I$. Q.E.D.

Consider for the moment a solution $v(z)$ to the equation

$$Av(z) = 0 \text{ in } B_r(z_0), \qquad z_0 \in \Omega, \quad r > 0. \tag{6.3}$$

By a well-known theorem, $v(z)$ is a real analytic function in $B_r(z_0)$. Writing v as a series of homogeneous polynomials,

$$v(z) = v(z_0) + \sum_{j=1}^{2} v_{x_j}(z_0)(x_j - x_{j0})$$
$$+ \sum_{m \leq n} P_n(z - z_0), \qquad |z - z_0| < r, \quad P_m(z - z_0) \neq 0,$$

unless v linear, one checks that P_m, the polynomial of lowest degree, is a solution to the elliptic equation with constant coefficients

$$\sum \frac{\partial a_j}{\partial p_k}(v_x(z_0)) \frac{\partial^2}{\partial x_j \, \partial x_k} P_m(z - z_0) = 0, \qquad |z - z_0| < r.$$

Consequently, $P_m(\zeta)$ is affinely related to the harmonic $\operatorname{Re} a\zeta^m$, for some complex number a, and $\partial P_m(\zeta)/\partial\zeta$ is affinely related to $am\zeta^{m-1}$, $m > 1$. From this we conclude that

given a solution v of (2.2) *which is not linear in a neighborhood U of z_0, $(\partial v/\partial z)(z)$ is an open mapping of U.* (6.4)

We define two mappings of $\overline{\Omega}$ onto another plane. The first is the homeomorphism of $\overline{\Omega}$

$$f^*(z) = \psi_{x_1}(z) - i\psi_{x_2}(z), \qquad z \in \overline{\Omega}.$$

The second is the continuous mapping of $\overline{\Omega}$,

$$f(z) = u_{x_1}(z) - iu_{x_2}(z), \qquad z \in \overline{\Omega},$$

which by (6.4) is open on $\Omega - I$, since otherwise u would be linear.

Theorem 6.2. *Let I denote the coincidence set of the solution u to* (6.2). *Then*

(i) *I is a connected simply connected domain which is the closure of its interior,*
(ii) $f(\Omega) = f^*(I)$,
(iii) *$\Omega - I$ is homeomorphic to an annulus, and*
(iv) *$\text{int } I$ is connected.*

Proof. Let I_1 be the set of points $z \in \Omega$ for which the tangent plane at $(z, \psi(z))$,

$$\Pi_z: \quad y_3 = v(y),$$

does not meet Ω. Since the point $\tilde{z} \in \Omega$ where ψ assumes its maximum has this property, I_1 is not empty. Since $\psi \in C^1$, I_1 contains a neighborhood of $\tilde{z}$ and is closed. Also

$$v(y) > \psi(y) \qquad \text{for} \quad y \in \Omega, \qquad y \neq z,$$

because ψ is strictly concave.

We show that $I_1 \subset I$. Indeed, $v(y)$ is a solution to $Av = 0$ in Ω and $v \geq 0$ on $\partial\Omega$. Hence by Theorem 8.3 of Chapter IV, $u \leq v$ in Ω. Therefore

$$\psi(z) \leq u(z) \leq v(z) = \psi(z) \qquad \text{at} \quad z \in I_1.$$

Now $z \in I_1$ is an interior point unless $v(y) = 0$ for some $y \in \partial\Omega$, in which case the intersection of Π_z with the plane containing Ω is tangent to $\partial\Omega$ at some point. The boundary of I_1 consists of all such points.

The converse also holds. Let $P \in \partial\Omega$ and construct a plane through the tangent to $\partial\Omega$ at P which is tangent to ψ at some $z \in \Omega$. Then $z \in I$, by the preceding argument, and the correspondence $P \to z(P)$ is a $1:1$ continuous mapping in view of the strict concavity of ψ and strict convexity of $\partial\Omega$.

With $z = z(P)$, let $w(y) = u(y) - v(y)$, which is a solution to the linear elliptic equation

$$Lw = \sum \frac{\partial a_j}{\partial p_k}(u_y(z)) w_{y_j y_k} = 0$$

in Ω near P. The maximum principle is applicable, so because $w(y) \leq 0$ near P, and attains its maximum value zero, at P,

$$\frac{\partial u}{\partial \nu}(P) > \frac{\partial v}{\partial \nu}(P),$$

where ν is outward directed normal. That $u(y) > 0$ in Ω implies that

$$0 > \frac{\partial u}{\partial \nu}(P) = -\frac{\partial u}{\partial x_1}(P)\cos\theta - \frac{\partial u}{\partial x_2}(P)\sin\theta, \tag{6.5}$$

where θ denotes the angle from the positive x_1 axis to the inward normal at P. Since v is a linear function and the tangent to ψ at $z = z(P)$,

$$0 < \frac{\partial u}{\partial x_1}(P)\cos\theta + \frac{\partial u}{\partial x_2}(P)\sin\theta < \frac{\partial \psi}{\partial x_1}(z)\cos\theta + \frac{\partial \psi}{\partial x_2}(z)\sin\theta. \tag{6.6}$$

The tangential derivative of u vanishes on $\partial\Omega$ so that

$$0 = \frac{\partial u}{\partial x_1}(P)\sin\theta - \frac{\partial u}{\partial x_2}(P)\cos\theta$$

and therefore

$$\frac{\partial u}{\partial x_1}(P) = -\frac{\partial u}{\partial \nu}(P)\cos\theta, \qquad \frac{\partial u}{\partial x_2}(P) = -\frac{\partial u}{\partial \nu}(P)\sin\theta.$$

With (6.5) this shows that

$$P \to f(P)$$

maps $\partial\Omega$ onto a star shaped curve Γ_1 with origin as reference point.

On the other hand, Π_z is tangent to $\partial\Omega$ at P; hence,

$$0 = \frac{\partial v}{\partial x_1}(P)\sin\theta - \frac{\partial v}{\partial x_2}(P)\cos\theta = \frac{\partial \psi}{\partial x_1}(z)\sin\theta - \frac{\partial \psi}{\partial x_2}(z)\cos\theta.$$

Therefore

$$P \to f^*(z(P))$$

maps $\partial\Omega$ onto a star shaped curve Γ_2. According to (6.6), Γ_1 lies inside Γ_2, and Γ_1 cannot intersect Γ_2.

Our interest is in the outer boundary β of $f(\Omega)$, which is by definition the boundary of the infinite component of the complement of $f(\Omega)$. Since $f(\Omega)$ is connected, β is connected.

Now $f(\partial\Omega) = \Gamma_1$, which lies inside $\Gamma_2 = f^*(\partial I_1) = f(\partial I_1)$; hence, $\beta \cap f(\partial\Omega) = \varnothing$. Since f is open on $\Omega - I$, $f(\Omega - I) \cap \beta = \varnothing$. Hence if $z \in \overline{\Omega}$ and $f(z) \in \beta$, then $z \in I$ and $f(z) = f^*(z)$. So $f^{*-1}(\beta) \subset I$ and is the outer boundary of $f^{*-1}(f(\Omega))$ since f^* is a homeomorphism of $\overline{\Omega}$. Consider a component V of $f(\Omega) - f(I)$. By openness of f on Ω and the fact that $f(\partial\Omega) \subset \operatorname{int} f(I)$, if $f(z) \in \partial V$, then $z \in \partial I$. Hence any component U of $f^{-1}(V)$ satisfies $U \subset \Omega - I$ and $\partial U \subset I$. By the previous lemma this is impossible; therefore $f(\overline{\Omega}) - f(I)$ is empty, so $f(\overline{\Omega}) = f(I) = f^*(I)$.

We infer that $I = f^{*-1}(f(\overline{\Omega}))$ is connected since f^* is a homeomorphism.

At this point let $z_0 \in \partial I$ and $z_n \in \Omega - I$ satisfy $z_n \to z_0$. Since f is open, $f(z_n) \in \operatorname{int} f(\Omega) = \operatorname{int} f(I)$. Moreover, f^* is a homeomorphism, so $f^{*-1}(f(z_n)) \in \operatorname{int} I$. Now

$$f^{*-1}(f(z_n)) \to f^{*-1}(f(z_0)) = z_0$$

because $f = f^*$ in I. Therefore $I = \overline{\operatorname{int} I}$.

Similarly, $\Omega - I$ is open and connected; hence $f(\Omega - I) \subset \operatorname{int} f(I) = \operatorname{int} f^*(I)$ by openness of f. Moreover, $f(\Omega - I)$ is connected, so $f^{*-1}(f(\Omega - I)) \subset I_0$, a component of int I. Let I_1 be any other component of int I and $z_1 \in \partial I_1$, $z_1 \notin \partial I_0$. Again we find a sequence $z_n \in \Omega - I$ with $z_n \to z_1$ and $f^{*-1}(f^*(z_n)) \in I_0$. But then $z_1 \in \bar{I}_0$. This contradiction implies that $I_1 = \varnothing$.

That $\Omega - I$ is topologically an annulus follows by a standard argument.

Q.E.D.

7. A Remark about the Coincidence Set in Higher Dimensions

The analysis of the coincidence set has proceeded up to this point by developing an appropriate function theory and thus has been limited to two dimensional problems. A first step toward smoothness of the coincidence set in higher dimensions was taken in Chapter IV when we provided a criterion under which I was a set of finite perimeter. Here we continue this development. As in Section 1, let $\Omega \subset \mathbb{R}^N$ be a bounded open set with smooth boundary $\partial\Omega$ and ψ an obstacle satisfying

$$\psi \in C^2(\bar{\Omega}): \quad \max_{\Omega} \psi > 0 \quad \text{and} \quad \psi < 0 \quad \text{on} \quad \partial\Omega. \tag{7.1}$$

Suppose that an inhomogeneous term f is given with

$$f \in C^1(\bar{\Omega}). \tag{7.2}$$

Our efforts are directed toward the analysis of

Problem 7.1. *Let $u \in \mathbb{K}$ satisfy*

$$\int_\Omega u_{x_i}(v - u)_{x_i}\, dx \geq \int_\Omega f(v - u)\, dx \quad \textit{for all} \quad v \in \mathbb{K}.$$

With minor technical modifications, the proofs given here extend to include Problem 4.1 of Chapter IV.

Theorem 7.2. *Let u be the solution of Problem 7.1 under the hypotheses (7.1) and (7.2) and let I be its set of coincidence. Suppose that*

$$-\Delta\psi - f \geq \eta > 0 \qquad \text{in} \quad \Omega$$

for some $\eta > 0$. Then there exist $r_0 > 0$ and λ, $0 < \lambda < 1$, such that for any $x \in \partial I$, there is a $y \in B_r(x)$ for which

$$B_{\lambda r}(y) \subset B_r(x) \cap (\Omega - I) \qquad \text{and} \qquad |x - y| = (1 - \lambda)r, \qquad r \leq r_0.$$

Moreover, for some $\alpha > 0$

$$\sup_{B_r(x)}(u - \psi) \geq \alpha r^2.$$

This density property of $\Omega - I$ leads immediately to

Corollary 7.3. *Let u be the solution of Problem 7.1 under the hypotheses (7.1) and (7.2) and let I be its set of coincidence. Suppose that*

$$-\Delta\psi - f \geq \eta > 0 \qquad \text{in} \quad \Omega$$

for some $\eta > 0$. Then

$$\text{meas } \partial I = 0.$$

Proof of Corollary 7.3. Since I is closed, $\partial I \subset I$. Now

$$\begin{aligned}\frac{\text{meas}(\partial I \cap B_r(x))}{\omega_n r^n} &\leq \frac{\text{meas}(I \cap B_r(x))}{\omega_n r^n} \\ &\leq \frac{\text{meas}(B_r(x) - B_{\lambda r}(y))}{\omega_n r^n} \\ &\leq 1 - \lambda < 1,\end{aligned}$$

with $\omega_n = \text{meas } B_1(0)$. Thus no point of ∂I can be a density point of ∂I. Hence meas $\partial I = 0$. Q.E.D.

Proof of Theorem 7.2. Let Ω_0 be a neighborhood of I in Ω, which exists because I is compact, and choose $r_0 > 0$ so that for each $x \in I$, $B_r(x) \subset \Omega_0$ for $r < 2r_0$. Note that $u \in H^{2,\infty}(\Omega_0)$ by Corollary 6.4 of Chapter IV. For $\beta > 0$ set

$$w(x) = u(x) - \psi(x) - [1/(2N)]\beta|x - x_0|^2, \qquad x \in \Omega,$$

where $x_0 \in \partial I$ is given. Then for $x \in \Omega - I$

$$\begin{aligned}-\Delta w &= -\Delta(u - \psi) + \beta = f + \Delta\psi + \beta \\ &\leq -\eta + \beta \leq -\tfrac{1}{2}\eta \qquad \text{for} \quad 0 < \beta \leq \tfrac{1}{2}\eta.\end{aligned}$$

Hence

$$-\Delta w < 0 \qquad \text{in} \quad B_r(x_0) \cap (\Omega - I), \quad r < r_0.$$

From the maximum principle, the continuous w attains its maximum on $\partial(B_r(x_0) \cap (\Omega - I)) \subset \partial B_r(x_0) \cup \partial I$. Suppose for the moment that x_0 satisfies an interior sphere condition with respect to $\Omega - I$. By this we mean that there is some ball $B_\delta(\xi) \subset \Omega - I$ with $x_0 \in \partial B_\delta(\xi)$. Now

$$w(x_0) = w_{x_j}(x_0) = 0, \qquad 1 \le j \le N,$$

so by the Hopf boundary point lemma, x_0 is not an extremum of w. Furthermore,

$$\max_{B_r(x_0) \cap (\Omega - I)} w > 0.$$

For $x \in \partial I$, $w(x) = [1/(2N)]\beta|x - x_0|^2 \le 0$, so there exists $y \in \partial B_r(x_0)$, $y \notin I$, such that

$$w(y) > 0$$

or

$$u(y) - \psi(y) \ge [1/(2N)]\beta|y - x_0|^2 = [1/(2N)]\beta r^2 > 0.$$

To extend this inequality to a ball, we use the fact that $u_{x_i x_j} \in L^\infty(\Omega_0)$. Our choice of r ensures that $B_r(y) \subset \Omega_0$. Since $(u - \psi)_x$ is Lipschitz in Ω_0, we may write

$$u(x) - \psi(x) = u(y) - \psi(y) + \sum_i [u_{x_i}(y) - \psi_{x_i}(y)](x_i - y_i) + R|x - y|^2,$$

where $R(x, y)$ is a remainder, and

$$|u_{x_i}(y) - u_{x_i}(x_0)| \le \|u_{xx}\|_{L^\infty(\Omega_0)}|y - x_0|.$$

Consequently, recalling that $u_x = \psi_x$ on I,

$$\begin{aligned} |u_{x_i}(y) - \psi_{x_i}(y)| &\le |u_{x_i}(y) - u_{x_i}(x_0)| + |\psi_{x_i}(y) - \psi_{x_i}(x_0)| \\ &\le \text{const}|y - x| \\ &= c_1 r. \end{aligned}$$

Therefore

$$\begin{aligned} u(x) - \psi(x) &\ge u(y) - \psi(y) - c_1 r|x - y| - |R||x - y|^2 \\ &\ge [1/(2N)]\beta r^2 - c_1 r|x - y| - |R||x - y|^2 \\ &\ge \{[1/(2N)]\beta - c_1\tau - |R|\tau^2\}r^2 \\ &\ge \alpha r^2 > 0 \end{aligned}$$

for $|x - y| \le \tau r$, τ sufficiently small but independent of $x_0 \in \partial I$, and $\alpha > 0$.

If $x_0 \in \partial I$ is an arbitrary point, we may find a sequence $x_k \in \partial I$, $x_k \to x_0$, with x_k possessing the interior sphere property. By continuity,

$$u(x) - \psi(x) \geq \alpha r^2 \qquad \text{for} \quad |x - y| \leq \tau r.$$

Consequently, $B_{\tau r}(y) \subset \Omega - I$. Changing the names of the variables a little, i.e., replacing r by $r + r\tau < 2r < 2r_0$ and τ by $\lambda = \tau/(1 + \tau)$, we conclude that

$$B_{\lambda r}(y) \subset B_r(x_0) \cap (\Omega - I). \quad \text{Q.E.D.}$$

A portion of ∂I may be negligible in the sense that it is not part of the boundary of int I. Indeed, consider the open set in Ω

$$\begin{aligned} \omega &= \Omega - \overline{\text{int } I} \\ &\supset \Omega - I \end{aligned}$$

and note that $\Omega - I$ is also open in Ω. Now $\omega - (\Omega - I) \subset \partial I$, so since $-\Delta u = f$ in $\Omega - I$, we may say that $-\Delta u = f$ a.e. in ω. However, ω is open, so the regularity theory for elliptic equations guarantees that

$$-\Delta u = f \qquad \text{in} \quad \omega.$$

In other words ω is, in some sense, the largest open set in which $-\Delta u = f$. In this way we regard $\partial I - \partial$ int I inessential or negligible.

Comments and Bibliographical Notes

The discussion of free boundary problems connected with variational inequalities began in Lewy and Stampacchia [1]. For the case of the Dirichlet integral, $N = 2$, the boundary of the coincidence set was shown to be an analytic Jordan curve under the hypotheses of Theorem 3.4 with ψ analytic. The reader may prove this theorem directly from Theorems 1.1 and 6.2. Section 2 is a preview of Chapter VI, where the argument is developed and references are provided.

To discuss more general problems where, for example, the Schwarz reflection principle was not directly applicable, it was useful to obtain some preliminary regularity of the free boundary prior to demonstrating, say, its analyticity. This motivated the development of Section 3 (cf. Kinderlehrer [3]). We have adapted the proof of Kinderlehrer [4]. Lemma 3.3 and Theorem 3.5 are related to a result of Hartman and Wintner [1].

So to analyze the behavior of the free boundary in the minimal surface case, ∂I was first shown to be continuously differentiable. This we have explained in Sections 3 and 5. The analyticity of ∂I when ψ is also analytic was achieved by the resolution of a system of differential equations which connects a conformal representation of the minimal surface $\{(x_1, x_2, x_3): x_3 = u(x), (x_1, x_2) \in \Omega - I\}$ with its harmonic, or analytic, extension. This idea is due to Hans Lewy, who used it to study minimal surfaces with prescribed or partly free boundaries (Lewy [2, 4]). Examples of the method are given in the exercises.

Theorem 3.6 is due to Caffarelli and Riviere [2].

Schaeffer [1] has studied the existence of singularities of ∂I. An asymptotic description was devised by Caffarelli and Riviere [3]. The content of Section 4 was adapted from Kinderlehrer and Nirenberg [1].

The results of Section 7 are due to Caffarelli and Riviere [1]. They serve as the starting point of the investigation of the coincidence set in higher dimensions.

Perturbation of the free boundary has been considered by Schaeffer [1] and Lewy [9].

Exercises

1. Let $\varphi(t)$ be analytic in $G = \{t = t_1 + it_2 : |t| < 1, \operatorname{Im} t > 0\}$ and continuous in $G \cup (-1, 1)$. Suppose that for each $t_0 \in [-R, R]$, $R < 1$, there is a complex number $c = c(t_0)$ such that

$$\left| \frac{\varphi(t) - \varphi(t_0)}{t - t_0} - c \right| \le C_R |t - t_0|^\mu, \qquad t \in G \cup (-1, 1),$$

$$|c| \le C_R$$

for some fixed μ, $0 < \mu \le 1$. Use Cauchy's representation to show that $\varphi \in C^{1,\mu}(\bar{G} \cap B_R)$.

2. Give a proof of Theorem 3.6.

3. Under the hypotheses of Theorem 3.4 show that ∂I is a "regular" curve, that is, it has a continuous tangent. [*Hint*: The only possible singularities of ∂I are cusps, in view of the remarks in the text. Given $z_0 \in \partial I$, consider the mapping from $B_\varepsilon(z_0) \cap (\Omega - I)$ into $B_\varepsilon(z_0) \cap \operatorname{int} I$ given by $\Phi(z) = f^{*-1}(f(z))$. Note that $\Phi(z) = z$ for $z \in \partial I$.]

4. Let $\Omega = \{z \in \mathbb{R}^2 : |z| < 2\}$, $\varphi_n(z) = 1 - (|z - z_n|^2/r_n^2)$, where

$$z_n = \frac{1}{n}, \qquad r_n = \frac{1}{2}\left(\frac{1}{n} + \frac{1}{n+1}\right),$$

and define

$$\psi_k(z) = \max_{n \le k} \varphi_n(z), \qquad z \in \Omega.$$

Let u_k be the solution of the problem (1.1) for the obstacle ψ_k and let I_k be its coincidence set. Show that I_k has at least k components. Show that ψ_k may be modified to a C^∞ function $\tilde{\psi}_k$ with the same solution u_k.

5. Let $G = \{z = x + iy : |z| < 1,\ y > 0\}$ and $\sigma = (-1, 1)$. Suppose that $u \in C^1(\overline{G})$ satisfies

$$\Delta u = 0 \quad \text{in} \quad G,$$

$$u_y = A(x, u, u_x) \quad \text{on} \quad \sigma,$$

where $A(x, u, p)$ is an analytic function of the variables x, u, p near $x = 0$, $u = u(0)$, $p = u_x(0)$. Show that u may be extended harmonically into a neighborhood of $z = 0$ and hence u has analytic boundary values near $z = 0$ on σ (Lewy [3]). [*Hint*: Let v be the harmonic conjugate of u in G, $v(0) = 0$, and set $f(z) = u + iv$, $z = x + iy \in G$. An analytic extension of f may be found by solving a suitable differential equation. Note that

$$\overline{f'(z)} - f(z) - 2iA(z, \tfrac{1}{2}[f(z) + \overline{f(z)}], \tfrac{1}{2}[f'(z) + \overline{f'(z)}]) = 0 \quad \text{for} \quad z \in \sigma.$$

Show that there is a unique holomorphic solution $\varphi(z)$ to

$$\varphi'(z) - f(z) - 2iA(z, \tfrac{1}{2}[\varphi(z) + f(z)], \tfrac{1}{2}[\varphi'(z) + f'(z)]) = 0, \qquad z \in G,$$

$|z|$ small, and that $\varphi(z) = \overline{f(z)}$ for $z \in \sigma$, $|z|$ small.]

6. Let Ω be a domain in the plane $z = x_1 + ix_2$ whose boundary contains a C^1 arc Γ. Suppose $z = 0 \in \Gamma$ and $u \in C^1(\Omega \cup \Gamma)$ satisfies

$$\Delta u = 0 \quad \text{in} \quad \Omega,$$

$$u = 0 \quad \text{on} \quad \Gamma,$$

$$\frac{\partial u}{\partial \nu} = A(x_1, x_2) \quad \text{on} \quad \Gamma,$$

where A is an analytic function of x_1, x_2 near $z = 0$, $A(0, 0) \neq 0$. Show that Γ is analytic near $z = 0$.

7. (Kellogg's Theorem) Let Ω be a simply connected bounded domain in the $z = x_1 + ix_2$ plane whose boundary contains the arc

$$\Gamma : x_2 = f(x_1), \qquad |x_1| < a, \quad \text{with} \quad f(0) = 0.$$

Assume that $f \in C^{1,\lambda}([-a, a])$, $0 < \lambda < 1$, and let φ be a conformal mapping of $G = \{|t| < 1 : \operatorname{Im} t > 0\}$ onto Ω which maps $(-1, 1)$ onto Γ with $\varphi(0) = 0$. Prove that $\varphi \in C^{1,\tau}(B_r \cap \overline{G})$ for each $\tau < \lambda$ and $r < 1$. [*Hint*: Consider $z^*(z) = z - 2if((z + \bar{z})/2)$ for $|z|$ small and apply Theorem 3.1.] Extend this conclusion slightly by proving that $\varphi \in C^{1,\lambda}(B_r \cap \overline{G})$, $r < 1$.

CHAPTER VI

Free Boundary Problems Governed by Elliptic Equations and Systems

1. Introduction

In this chapter we shall illustrate the use of hodograph methods and their generalizations to ascertain the smoothness of free boundaries. A hint of this technique was described in Section 2 of Chapter V, where new independent and dependent variables were defined in terms of the solution w of the variational inequality. Here we shall extend that idea, proposing, for example, in the case of a single equation, the selection of a combination of derivatives of w or even w itself as a new independent variable. The object, in all situations, is to straighten the free boundary at the expense of introducing a very nonlinear equation. To the nonlinear problem which results, we may apply a known regularity theorem to deduce smoothness.

Several complications will arise. At the purely formal level, the free boundary problem may suggest a system instead of a single equation. Or the original dependent variable or variables may be defined on both sides of the free boundary as in the variational inequality of two membranes (Chapter II, Exercise 13). To surmount the first difficulty we shall study nonlinear elliptic systems. A brief review of this topic is included in the text, Section 3, as much to establish our conventions as to inform the reader unfamiliar with the theory of its basic elements. To confront problems where the functions are defined on both sides of the free boundary, we introduce a reflection mapping defined by means of the hodograph transformation in use. This will always lead to a system of equations.

Another, perhaps more delicate, question is to decide whether the hypotheses of the elliptic regularity theory are satisfied once the problem has been rewritten with hodograph transforms, reflections, and the like. This question, at times extremely difficult and technical, with rare exceptions is beyond the scope of this book when $N > 2$. We discuss this briefly and without proof in Section 2.

Our considerations will be local. Again, our theme will be that once some smoothness of the free boundary is assumed, it must be as smooth as the data permit. In this way we wish to explore the conditions at a free boundary which ensure its regularity. This chapter is independent of Chapter V.

2. Hodograph and Legendre Transforms: The Theory of a Single Equation

Let $u \in C^2(\Omega)$, where $\Omega \subset \mathbb{R}^N$ is a bounded domain. Throughout this chapter we set $u_i = u_{x_i}$, $u_{ij} = u_{x_i x_j}$, etc. The transformation of Ω defined by

$$\begin{aligned} y_\sigma &= x_\sigma, \qquad 1 \le \sigma \le N-1, \\ y_N &= -u_N \end{aligned} \tag{2.1}$$

is called a first order (partial) hodograph transformation and the function $v(y)$ defined by

$$v(y) = x_N y_N + u(x), \qquad x \in \Omega, \tag{2.2}$$

is called the first order Legendre transform of u. Let us assume for the moment that (2.1) defines a $1:1$ mapping of Ω onto a domain U. The property of the Legendre transform is that

$$dv = x_N\, dy_N + y_N\, dx_N + du = x_N\, dy_N + \sum_{\sigma < N} u_\sigma\, dy_\sigma,$$

which means that

$$\begin{aligned} v_\sigma &= u_\sigma, \qquad 1 \le \sigma \le N-1, \\ v_N &= x_N. \end{aligned} \tag{2.3}$$

Here and in what follows we shall understand subscripts of u to denote differentiation with respect to x and those of v to indicate differentiation with respect to y. The inverse mapping to (2.1) is given by

$$\begin{aligned} x_\sigma &= y_\sigma, \qquad 1 \le \sigma \le N-1, \\ x_N &= v_N. \end{aligned} \tag{2.4}$$

Moreover, if u satisfies a second order equation, then v does also. To check this, merely note that

$$\frac{dy}{dx} = \begin{pmatrix} 1 & & 0 & 0 \\ & \ddots & & \vdots \\ 0 & & 1 & 0 \\ -u_{1N} & & \cdots & -u_{NN} \end{pmatrix}$$

and

$$\frac{dy}{dx} = \left(\frac{dx}{dy}\right)^{-1} = \begin{pmatrix} 1 & & 0 & 0 \\ & \ddots & & \vdots \\ 0 & & 1 & 0 \\ -\dfrac{v_{1N}}{v_{NN}} & \cdots & -\dfrac{v_{N-1N}}{v_{NN}} & \dfrac{1}{v_{NN}} \end{pmatrix}$$

imply that

$$\frac{\partial y_N}{\partial x_N} = -u_{NN} = \frac{1}{v_{NN}}, \qquad \frac{\partial y_N}{\partial x_\sigma} = -u_{N\sigma} = -\frac{v_{N\sigma}}{v_{NN}}. \tag{2.5}$$

In other words, with $\partial_k = \partial/\partial y_k$,

$$\frac{\partial}{\partial x_\sigma} = \partial_\sigma - \frac{v_{N\sigma}}{v_{NN}} \partial_N, \qquad \frac{\partial}{\partial x_N} = \frac{1}{v_{NN}} \partial_N. \tag{2.6}$$

In particular,

$$u_{\sigma\sigma} = v_{\sigma\sigma} - (v_{\sigma N}^2/v_{NN}), \qquad 1 \le \sigma \le N-1,$$

so that, for example,

$$\Delta u = -(1/v_{NN}) - (1/v_{NN}) \sum v_{\sigma N}^2 + \sum v_{\sigma\sigma}. \tag{2.7}$$

More generally, if u satisfies an elliptic equation, then v does also (cf. Exercise 1).

Another change of variables is given by

$$\begin{aligned} y_\sigma &= x_\sigma, \quad 1 \le \sigma \le N-1, \\ y_N &= u(x), \end{aligned} \tag{2.8}$$

which we refer to as a zeroth order hodograph transformation. To it we associate the new dependent variable

$$\psi(y) = x_N. \tag{2.9}$$

Again let us suppose that (2.8) defines a 1 : 1 mapping of Ω onto a domain U. The property of the zeroth order transformation is that

$$dy_N = du = \sum_\sigma u_\sigma \, dx_\sigma + u_N \, dx_N = \sum_\sigma u_\sigma \, dy_\sigma + u_N \, d\psi$$

or

$$\psi_\sigma = -u_\sigma / u_N, \qquad \psi_N = 1/u_N. \tag{2.10}$$

Also, note that the inverse of (2.8) is

$$\begin{aligned} x_\sigma &= y_\sigma, \qquad 1 \le \sigma \le N-1, \\ x_N &= \psi. \end{aligned} \tag{2.11}$$

From (2.10) it is obvious that

$$\begin{aligned} \frac{\partial}{\partial x_\sigma} &= \partial_\sigma - \frac{\psi_\sigma}{\psi_N} \partial_N, \qquad 1 \le \sigma \le N-1, \\ \frac{\partial}{\partial x_N} &= \frac{1}{\psi_N} \partial_N. \end{aligned} \tag{2.12}$$

We also note that

$$\begin{aligned} u_{\sigma\sigma} &= -\frac{\psi_{\sigma\sigma}}{\psi_N} + 2\frac{\psi_\sigma}{\psi_N^2}\psi_{\sigma N} - \frac{\psi_\sigma^2}{\psi_N^3}\psi_{NN}, \\ u_{NN} &= -\frac{1}{\psi_N^3}\psi_{NN}, \end{aligned}$$

so that, in particular,

$$\Delta u = -\frac{1}{\psi_N}\sum \psi_{\sigma\sigma} + \frac{2}{\psi_N^2}\sum \psi_\sigma \psi_{\sigma N} - \frac{1}{\psi_N^3}\left(1 + \sum \psi_\sigma^2\right)\psi_{NN}. \tag{2.13}$$

Again, if u satisfies an elliptic equation, then ψ does also (Exercise 2).

We now apply these transformations to the study of a simple free boundary problem. As before, let $\Omega \subset \mathbb{R}^N$ be a domain. Let $0 \in \partial\Omega$ and suppose that near 0 $\partial\Omega$ is a C^1 hypersurface Γ with inner normal at 0 in the direction of the positive x_N axis. Let $u(x) \in C^2(\Omega \cup \Gamma)$ satisfy

$$\begin{aligned} \Delta u(x) &= a(x), \qquad x \in \Omega, \\ u(x) &= 0, \\ \sum_1^N c_k u_k(x) &= b, \end{aligned} \qquad x \in \Gamma, \tag{2.14}$$

where $a(x)$ is smooth in a full neighborhood of $x = 0$ and, for simplicity, $c_1, \ldots, c_N, b \in \mathbb{R}$. The nondegeneracy condition $c_N \neq 0$ will always be required. More generally, it is necessary to know that the direction $c = (c_1, \ldots, c_N)$ is not tangential to Γ.

We shall consider two cases of (2.14). First assume that $b = 0$, $c_N \neq 0$, and $a(0) < 0$. Here we seek to apply (2.1). Indeed, since $c_N \neq 0$, $u_k(x) = 0$, $1 \leq k \leq N$, in a neighborhood of 0 in Γ, which we may take to be Γ, so (2.14) may be restated as

$$\begin{aligned} \Delta u &= a \quad \text{in} \quad \Omega, \\ u = u_k &= 0 \quad \text{on} \quad \Gamma, \quad 1 \leq k \leq N. \end{aligned} \tag{2.15}$$

To see that (2.1) is 1:1 near 0 in Ω observe that since $u_k = 0$ on Γ and $(0, \ldots, 0, 1)$ is normal to Γ at $x = 0$,

$$u_{k\sigma}(0) = 0, \qquad 1 \leq k \leq N, \quad 1 \leq \sigma \leq N - 1,$$

so from (2.15)

$$u_{NN}(0) = a(0) < 0.$$

It follows that (2.1) maps a neighborhood of 0 in Ω, say Ω, onto a domain

$$U \subset \{y : y_N > 0\}$$

and Γ onto a portion S of the hyperplane $y_N = 0$. Consequently, $v(y)$ defined by (2.2) satisfies

$$\begin{aligned} -(1/v_{NN}) - (1/v_{NN}) \sum (v_{N\sigma})^2 + \sum v_{\sigma\sigma} - a(y', v_N) &= 0 \quad \text{in} \quad U \\ v &= 0 \quad \text{on} \quad S, \\ y' &= (y_1, \ldots, y_{N-1}), \end{aligned} \tag{2.16}$$

and Γ admits the representation

$$\Gamma: \quad x_N = v_N(x', 0), \qquad (x', 0) \in S. \tag{2.17}$$

Theorem 2.1. *Let $u(x) \in C^2(\Omega \cup \Gamma)$ satisfy (2.15) with $a(0) \neq 0$. If $a \in C^{m,\alpha}(B_r(0))$, $m \geq 0$, $0 < \alpha < 1$, then there is a neighborhood $B_\rho(0)$ such that*

$$u \in C^{m+2,\alpha}(\overline{\Omega} \cap B_\rho(0))$$

and $B_\rho(0) \cap \Gamma$ is of class $C^{m+1,\alpha}$. If a is analytic in $B_r(0)$, then $\Gamma \cap B_\rho(0)$ is also analytic.

Proof. The function v is a solution of the (Dirichlet) boundary value problem (2.16). According to Agmon *et al.* [1, Theorem 11.1],

$$v(y) \in C^{m+2,\alpha}(U \cup S).$$

Hence $v_j \in C^{m+1,\alpha}(U \cup S)$ and, in particular, (2.17) is a $C^{m+1,\alpha}$ parametrization of Γ. To complete the proof note that since the mapping $x = (y', v_N(y))$ is of class $C^{m+1,\alpha}$, so is its inverse $y = y(x)$. Thus

$$u_N(x) = -y_N(x) \in C^{m+1,\alpha}(\overline{\Omega} \cap B_\rho(0))$$

and

$$u_\sigma(x) = v_\sigma(y(x)) \in C^{m+1,\alpha}(\overline{\Omega} \cap B_\rho)), \qquad 1 \leq \sigma \leq N-1.$$

If $a(x)$ is analytic, it follows that (2.17) gives an analytic parametrization of Γ (Morrey [1] or Friedman [1]). Q.E.D.

We encounter a different situation when $b \neq 0$. So, suppose now that $b \neq 0$ and $c_N \neq 0$ and introduce the new dependent variables (2.8). Again, since $u = 0$ on Γ and $(0, \ldots, 0, 1)$ is normal to Γ at $x = 0$,

$$u_\sigma(0) = 0, \qquad 1 \leq \sigma \leq N-1,$$

whence

$$u_N(0) = b/c_N \neq 0.$$

Assuming that $b/c_N > 0$, (2.8) is a $1:1$ mapping of a portion of Ω near $x = 0$, which we take to be all of Ω, onto a domain $U \subset \{y : y_N > 0\}$. Also Γ is mapped onto a portion S of $\{y : y_N = 0\}$. We now have from (2.10), (2.13), and (2.14) that

$$\begin{aligned} -\frac{1}{\psi_N}\sum \psi_{\sigma\sigma} + \frac{2}{\psi_N^2}\sum \psi_\sigma\psi_{\sigma N} - \frac{1}{\psi_N^3}(1 + \sum \psi_\sigma^2)\psi_{NN} - a(y', \psi) = 0 \quad &\text{in } U, \\ \sum c_\sigma \psi_\sigma + b\psi_N = c_N \quad &\text{on } \Gamma. \end{aligned} \tag{2.18}$$

Also we may represent Γ as

$$\Gamma: \quad x_N = \psi(x', 0), \qquad (x', 0) \in S. \tag{2.19}$$

We state this next conclusion only for the analytic case.

Theorem 2.2. *Let $u \in C^2(\Omega \cup \Gamma)$ satisfy* (2.14), *where a is analytic in a full neighborhood of the origin, and suppose that $b \neq 0$, and $c_N \neq 0$. Then Γ is analytic near $x = 0$.*

Proof. Again we appeal to Theorem 11.1 of Agmon *et al.* [1]. In this case the boundary condition is not a Dirichlet condition, but a Neumann one. The regularity theorem applies nonetheless because the boundary condition is of coercive type. We shall discuss this property in more detail in the next section. Q.E.D.

When can we verify the hypothesis of the theorems? With respect to Theorem 2.2, once u is assumed merely in $C^1(\Omega \cup \Gamma)$, Γ is also C^1 and the conclusion applies. For Theorem 2.1 we state this criterion:

Theorem 2.3. *Let $\Omega \subset \mathbb{R}^N$ be a bounded open set with $0 \in \partial\Omega$ and let $u \in H^{2,\infty}(\Omega)$ satisfy (2.15) near $x = 0$ in Ω and $\partial\Omega$. Suppose that $a(x) \in C^{0,\alpha}(\overline{\Omega})$ for some α, $0 < \alpha < 1$, and $a(0) \neq 0$. If*

$$\lim_{\rho \to 0} \rho^{-N} \operatorname{meas}(B_\rho(0) \cap \Omega) > 0, \tag{2.20}$$

then

(i) *$\partial\Omega$ is a C^1 hypersurface Γ near $x = 0$, and*
(ii) *$u \in C^2(\Omega \cup \Gamma)$.*

3. Elliptic Systems

This section is a brief introduction to the theory of elliptic systems. For the greater part, our treatment recounts well-known definitions and properties, but the reader unfamiliar with "weights" and their use will find here a self-contained explanation.

Let $\Omega \subset \mathbb{R}^N$ be a domain, which in our applications will usually be bounded or $\mathbb{R}^N_+ = \{y \in \mathbb{R}^N : y_N > 0\}$ and set

$$D = (D_1, \ldots, D_N), \qquad D_j = \frac{1}{i}\frac{\partial}{\partial y_j}, \qquad 1 \leq j \leq N.$$

Let $L_{kj}(y, D)$, $1 \leq j, k \leq n$, be linear differential operators with continuous complex valued coefficients and consider the system of equations in the dependent variables $u^1, \ldots, u^n$:

$$\sum_{j=1}^{n} L_{kj}(y, D)u^j(y) = f_k(y) \qquad \text{in} \quad \Omega, \quad 1 \leq k \leq n, \tag{3.1}$$

where the f_k are given, say smooth, functions. To each equation we assign an integer weight $s_k \leq 0$ and to each dependent variable an integer weight $t_k \geq 0$ such that

$$\begin{aligned} &\text{order } L_{kj}(y, D) \leq s_k + t_j \qquad \text{in} \quad \Omega, \quad 1 \leq k, j \leq n, \\ &\max_k s_k = 0. \end{aligned} \tag{3.2}$$

Such an assignment of weights is called *consistent*. By convention, $L_{kj}(y, D) \equiv 0$ if $s_k + t_j < 0$. Let $L'_{kj}(y, D)$ denote the part of $L_{kj}(y, D)$ of precisely order $s_k + t_j$; it is either zero or homogeneous in D of degree $s_k + t_j$.

Definition 3.1. The system (3.1) is *elliptic* (with *weights* s_k and t_j) provided that

$$\operatorname{rank}(L'_{kj}(y, \xi)) = n \qquad \text{for each} \quad 0 \neq \xi \in \mathbb{R}^N \quad \text{and} \quad y \in \Omega \tag{3.3}$$

and, in addition, for each pair of independent vectors $\xi, \eta \in \mathbb{R}^N$ and $y \in \Omega$ the polynomial in z, $p(z) = \det(L'_{kj}(y, \xi + z\eta))$, has exactly $\mu = \frac{1}{2} \deg p$ roots with positive imaginary part and $\mu = \frac{1}{2} \deg p$ roots with negative imaginary part.

The condition about the roots is automatic if $N \geq 3$, but when $N = 2$ an equation may satisfy (3.3) but not the root condition. As an example consider the Cauchy–Riemann equation

$$\frac{\partial}{\partial \bar{z}} u = f$$

or any of its powers.

The matrix $L'_{kj}(y, \xi)$ is called the *principal symbol* matrix. Suppose, for fixed $y_0 \in \Omega$, we search for solutions to the homogeneous equations with constant coefficients

$$\sum_{j=1}^{n} L'_{kj}(y_0, D)u^j(y) = 0, \qquad y \in \mathbb{R}^N, \quad 1 \leq k \leq n, \tag{3.4}$$

which have the form $u^j(y) = c^j e^{iy\xi}$, $c^j \in \mathbb{C}$, $0 \neq \xi \in \mathbb{R}^N$. Then (3.3) holds if and only if $c^1 = \cdots = c^n = 0$ for every $0 \neq \xi \in \mathbb{R}^N$, $y_0 \in \Omega$. Such solutions are called *exponential*. Thus, equivalent to (3.3) is the condition that the principal part of (3.1), which is (3.4), admits no nontrivial exponential solutions.

A general system of equations

$$F_k(y, Du^1, \ldots, Du^n) = 0, \qquad y \in \Omega, \quad 1 \leq k \leq n, \tag{3.5}$$

where here D stands for *all* derivatives and not merely first order ones, is called elliptic along the solution $u = (u^1, \ldots, u^n)$ provided that the variational equations

$$\sum_{j=1}^{n} L_{kj}(y, D)\bar{u}^j = \frac{d}{dt} F_k(y, D(u_1 + t\bar{u}^1), \ldots, D(u^n + t\bar{u}^n))\Big|_{t=0} = 0 \quad (3.6)$$

constitute an elliptic system in the sense of Definition 3.1.

We now discuss boundary conditions which we need only consider on portions S of $\partial\Omega$ contained in the hyperplane $y_N = 0$. Let $B_{hj}(y, D)$, $1 \le h \le \mu$, $1 \le j \le n$, be linear differential operators with continuous coefficients.

Definition 3.2. The set of boundary conditions

$$\sum_{j=1}^{n} B_{hj}(y, D)u^j(y) = g_h(y) \qquad \text{on} \quad S \subset \partial\Omega \cap \{y_N = 0\}, \quad 1 \le h \le \mu, \quad (3.7)$$

is *coercive for the system* (3.1) provided that

(i) the system (3.1) is elliptic and $2\mu = \sum_1^n (s_j + t_j) \ge 0$ is even,

(ii) there exist integers r_h, $1 \le h \le \mu$, such that order $B_{hj}(y, D) \le r_h + t_j$ on S, and

(iii) with B'_{hj} the part of B_{hj} of exactly order $r_h + t_j$, for each $y_0 \in S$ the homogeneous boundary value problem

$$\begin{aligned} &\sum_j L'_{kj}(y_0, D)u^j = 0 \qquad \text{in} \quad \mathbb{R}^N_+, \qquad 1 \le k \le n, \\ &\sum B'_{hj}(y_0, D)u^j = 0 \qquad \text{on} \quad y_N = 0, \qquad 1 \le h \le \mu, \end{aligned} \quad (3.8)$$

admits no nontrivial bounded exponential solutions of the form

$$u^j(y) = e^{i\xi' y'}\varphi^j(y_N), \qquad 1 \le j \le n, \quad 0 \ne \xi' \in \mathbb{R}^{N-1}.$$

[Here, as usual, $y' = (y_1, \ldots, y_{N-1})$ and $\xi' = (\xi_1, \ldots, \xi_{N-1})$.]

Similarly, we may extend the notion of coercivity to nonlinear systems. A set of boundary conditions

$$\Phi_h(y, Du^1, \ldots, Du^n) = 0 \qquad \text{on} \quad S, \quad 1 \le h \le \mu, \quad (3.9)$$

is *coercive for the system* (3.5) *along* $u = (u^1, \ldots, u^n)$ provided there exist weights $r_1, \ldots, r_\mu$ such that the set of linearized boundary conditions

$$\sum_j B_{hj}(y, D)\bar{u}^j = \frac{d}{dt} \Phi_h(y, D(u^1 + t\bar{u}^1), \ldots, D(u^n + t\bar{u}^n))\Big|_{t=0} = 0 \qquad \text{in} \quad S \quad (3.10)$$

is coercive for (3.6) on S.

In the sections which follow, our concern will be the use of hodograph and Legendre transforms to rephrase free boundary problems as nonlinear elliptic systems with coercive boundary conditions. Although such systems display many interesting properties of existence and uniqueness, our interest is in the regularity of their solutions. We state here the theorem to which we shall appeal. Let U be a neighborhood of 0 in $\mathbb{R}_+^N$ and $S = \partial U \cap \{y_N = 0\}$.

Theorem 3.3. *Suppose that $u = (u^1, \ldots, u^n)$ is a solution of*

$$\begin{aligned} F_k(y, Du^1, \ldots, Du^n) &= 0 \qquad \text{in} \quad U, \quad 1 \le k \le n, \\ \Phi_h(y, Du^1, \ldots, Du^n) &= 0 \qquad \text{on} \quad S, \quad 1 \le h \le \mu, \end{aligned} \tag{3.11}$$

where (3.11) is elliptic and coercive along u with weights s_k, t_j, r_h, $1 \le j, k \le n$, $1 \le h \le \mu$. Suppose also that F_k and Φ_h are analytic functions of y and the derivatives of u. Let $r_0 = \max_h(0, 1 + r_h)$. If $u^j \in C^{t_j + r_0, \alpha}(U \cup S)$ for some $\alpha > 0$, then the u^j are analytic in $U \cup S$, $1 \le j \le n$.

As a first example of an elliptic system, consider

$$\begin{aligned} \Delta u^j &= f_j \qquad \text{in} \quad U, \quad 1 \le j \le \mu, \\ \sum_{j=1}^{\mu} a_{hj} u^j &= g_h \qquad \text{on} \quad S, \quad 1 \le h \le \mu, \end{aligned} \tag{3.12}$$

where (a_{kj}) is a $\mu \times \mu$ constant matrix and, as usual, U is a neighborhood of 0 in $\mathbb{R}_+^N$ and $S = \partial U \cap \{y_N = 0\}$. Assign to each equation the weight $s_k = 0$, to each u^j the weight $t_j = 2$, and to each boundary condition the weight $r_h = -2$. The system corresponding to (3.8) is

$$\begin{aligned} \Delta \bar{u}^j &= 0 \qquad \text{in} \quad \mathbb{R}_+^N, \quad 1 \le j \le \mu, \\ \sum a_{hj} \bar{u}^j &= 0 \qquad \text{on} \quad \mathbb{R}^{N-1}, \quad 1 \le h \le \mu. \end{aligned}$$

A bounded exponential solution of the differential equations is given by

$$\begin{aligned} \bar{u}^j(y', t) &= e^{i\xi' y'} \varphi^j(t), \qquad t > 0, \quad y' \in \mathbb{R}^{N-1}, \\ \varphi^j(t) &= c^j e^{-|\xi'|t}, \qquad c^j \in \mathbb{C} \end{aligned}$$

for each $0 \ne \xi' \in \mathbb{R}^{N-1}$. Hence $\varphi^j(t) \equiv 0$, $1 \le j \le \mu$, if and only if $c^1 = \cdots = c^\mu = 0$ is the only solution of the system

$$\sum_1^{\mu} a_{hj} c^j = 0, \qquad 1 \le h \le \mu,$$

namely, (3.12) is coercive if and only if $\det(a_{hj}) \ne 0$. In particular, the Dirichlet problem for the Poisson equation is coercive and, indeed, the Dirichlet problem for any elliptic equation is coercive (Exercise 3).

Choosing the weights for a system can be a delicate matter, as we shall soon observe. For the moment let us examine the system

$$\begin{aligned} \Delta u^1 &= 0 \\ \Delta u^2 + \lambda u^2 &= 0 \end{aligned} \quad \text{in} \quad U, \qquad \begin{aligned} u^1 - u^2 &= 0 \\ D_N u^1 - D_N u^2 &= 0 \end{aligned} \quad \text{on} \quad S, \tag{3.13}$$

where $0 \neq \lambda \in \mathbb{R}$. Our question is what may be decided about the Cauchy data which is shared by two different equations on the hyperplane $y_N = 0$? A preliminary assignment of "obvious weights" $s_1 = s_2 = 0$, $t_1 = t_2 = 2$, $r_1 = -2$, $r_2 = -1$ leads to the "linearized system"

$$\begin{aligned} \Delta \bar{u}^1 &= 0 \\ \Delta \bar{u}^2 &= 0 \end{aligned} \quad \text{in} \quad \mathbb{R}^N,$$

$$\begin{aligned} \bar{u}^1 - \bar{u}^2 &= 0 \\ D_N \bar{u}^1 - D_N \bar{u}^2 &= 0 \end{aligned} \quad \text{on} \quad \mathbb{R}^{N-1},$$

which is not coercive. Now introduce $w = u^2 - u^1$ so that (3.13) becomes

$$\begin{aligned} \Delta u^1 &= 0 \\ \Delta w + \lambda(w + u^1) &= 0 \end{aligned} \quad \text{in} \quad U, \qquad \begin{aligned} w &= 0 \\ D_N w &= 0 \end{aligned} \quad \text{on} \quad S. \tag{3.14}$$

Set $s_1 = 0$, $s_2 = -2$, $t_1 = 2$, $t_2 = 4$, and $r_1 = -4$, $r_2 = -3$. The principal symbol of (3.14) is

$$\begin{aligned} \Delta \bar{u}^1 &= 0 \\ \Delta \bar{w} + \lambda \bar{u}^1 &= 0 \end{aligned} \quad \text{in} \quad \mathbb{R}^N_+, \qquad \begin{aligned} \bar{w} &= 0 \\ D_N \bar{w} &= 0 \end{aligned} \quad \text{on} \quad \mathbb{R}^{N-1}. \tag{3.15}$$

With $\bar{u}^1(y', t) = e^{iy'\xi'}\varphi(t)$ and $\bar{w}(y', t) = e^{iy'\xi'}\zeta(t)$ we find that $\varphi(t) = ce^{-|\xi'|t}$, so ζ is a solution of

$$\zeta''(t) - |\xi'|^2\zeta(t) = -\lambda c e^{-|\xi'|t},$$

$$\zeta(0) = 0, \tag{3.16}$$

$$\zeta'(0) = 0, \tag{3.17}$$

with the additional stipulation that $\sup|\zeta| < \infty$. Consequently, $\zeta(t) = C_0 e^{-|\xi'|t} + C_1 t e^{-|\xi'|t}$ and, by (3.16) and (3.17),

$$C_0 = 0,$$
$$-|\xi'|C_0 + C_1 = 0.$$

So $C_0 = C_1 = c = 0$; hence (3.15) admits no bounded exponential solutions. We may now apply Theorem 3.3 to deduce

Theorem 3.4. *If Δ and $\Delta + \lambda$, $\lambda \neq 0$, share Cauchy data on $y_N = 0$ for solutions defined in a neighborhood U of $y = 0$ in $y_N > 0$, then the data is analytic.*

For a second proof of the theorem, apply Δ to the second equation of (3.14). Then w is a solution of the fourth order problem

$$\begin{aligned} (\Delta^2 + \lambda\Delta)w &= 0 \quad \text{in} \quad U, \\ w &= 0 \\ D_N w &= 0 \end{aligned} \Big\} \text{ on } S. \tag{3.18}$$

This problem is coercive with the obvious weights $s = 0$, $t = 4$, $r_1 = -4$, $r_2 = -3$. Note that $t = t_2$. Hence w is analytic in $U \cup S$. Since $\lambda \neq 0$,

$$u^1 = -(1/\lambda)\,\Delta w - w,$$

is also analytic. This technique suggests that altering what appear to be the "obvious weights" of a system so it becomes coercive is analogous to differentiating the equation.

It is interesting to compare a simple transmission problem with the one we have just considered. Set $U^+ = \{y_N > 0\} \cap B_1(0)$ and $U^- = \{y_N < 0\} \cap B_1(0)$ and suppose that $u^1 \in C^1(\overline{U^+})$, $v \in C^1(\overline{U^-})$ satisfy

$$\begin{aligned} \Delta u^1 + \lambda u^1 &= 0 \quad \text{in} \quad U^+, \\ \Delta v &= 0 \quad \text{in} \quad U^-, \\ u^1 - v &= 0 \\ D_N u^1 - D_N v &= 0, \end{aligned} \Big\} \text{ on } y_N = 0, \tag{3.19}$$

where $\lambda \in \mathbb{R}$.

Defining $u^2(y) \in C^1(\overline{U^+})$ by

$$u^2(y) = v(y', -y_N),$$

we find that

$$\begin{aligned} \Delta u^1 + \lambda u^1 &= 0 \\ \Delta u^2 &= 0 \end{aligned} \quad \text{in} \quad U^+,$$
$$\begin{aligned} u^1 - u^2 &= 0 \\ D_N u^1 + D_N u^2 &= 0 \end{aligned} \quad \text{on} \quad y_N = 0. \tag{3.20}$$

It is an elementary matter to check that this system is coercive with the "obvious weights" $s_1 = s_2 = 0$, $t_1 = t_2 = 2$, $r_1 = -2$, $r_2 = -1$, *whether or not λ is zero.* Hence u and v are analytic in a full neighborhood of 0.

The equations and boundary conditions which occur in subsequent sections will be nonlinear. To test the ellipticity and coerciveness of such a system it will be convenient to have a criterion which may be applied at a single point. The remainder of this section is devoted to proving that a system with a consistent choice of weights which is elliptic and coercive "at a single point" is indeed elliptic and coercive in the sense of Definitions 3.1 and 3.2 in neighborhood $\Omega \subset \mathbb{R}_+^N$ and $S = \tilde{\Omega} \cap \{y_N = 0\}$.

Recall the system of equations

$$F_k(y, Du^1, \ldots, Du^n) = 0, \qquad y \in \Omega, \quad 1 \le k \le n, \tag{3.5}$$

its associated system of variational equations

$$\sum_1^n L_{kj}(y, D)\bar{u}^j$$
$$= \frac{d}{dt} F(y, D(u^1 + t\bar{u}^1), \ldots, D(u^n + t\bar{u}^n))\Big|_{t=0} = 0, \qquad 1 \le k \le n, \tag{3.6}$$

the boundary conditions

$$\Phi_h(y, Du^1, \ldots, Du^n) = 0, \qquad y \in S, \quad 1 \le h \le \mu, \tag{3.9}$$

and its associated system of variational equations

$$\sum_1^n B_{hj}(y, D)\bar{u}^j$$
$$= \frac{d}{dt} \Phi_h(y, D(u^1 + t\bar{u}^1), \ldots, D(u^n + t\bar{u}^n))\Big|_{t=0} = 0, \qquad 1 \le h \le \mu. \tag{3.10}$$

Suppose that s_j, t_k, r_h, $1 \le j, k \le n$, $1 \le h \le \mu$, is a consistent set of weights for this system. We assume that 0 is an interior point of S.

Theorem 3.5. *Suppose that* $u^1, \ldots, u^n$ *is a solution of* (3.5), (3.9) *in* $\Omega \cup S$ *and that* F_k, $1 \leq k \leq n$, Φ_h, $1 \leq h \leq \mu$, *are analytic functions of* $y, u^1, \ldots, u^n$, *and the derivatives of the* u^j *for* $u \in \Omega \cup S$. *Assume that* $u^j \in C^{t_j + r_0}(\Omega \cup S)$, $r_0 = \max_h(0, 1 + r_h)$. *If the variational equations* (3.6), (3.10) *are elliptic and coercive at* $y = 0$, *then* (3.5), (3.9) *are elliptic and coercive in a neighborhood* $(\Omega \cup S) \cap B_\varepsilon(0)$ *for some* $\varepsilon > 0$.

We begin our proof with

Lemma 3.6. *Assume the hypotheses of the Theorem* 3.5. *Then* (3.5) *is elliptic in a neighborhood* $(\Omega \cup S) \cap B_\varepsilon(0)$.

Proof. The coefficients of $L_{kj}(y, D)$ are continuous in $\Omega \cup S$. Hence if

$$\operatorname{rank}(L'_{kj}(y, \xi)) = n \qquad \text{for} \quad 0 \neq \xi \in \mathbb{R}^n \tag{3.21}$$

and $y = 0$, it also holds for $y \in \Omega \cup S$, $|y|$ small.

Now suppose that

$$P(y, \xi + z\eta) = \det(L'_{kj}(y, \xi + z\eta)) \tag{3.22}$$

has exactly $\mu = \frac{1}{2} \deg P$ roots with positive imaginary part and $\mu = \frac{1}{2} \deg P$ roots with negative imaginary part for each pair of independent vectors $\xi, \eta \in \mathbb{R}^N$ and $y = 0$. Since the polynomial $P(y, \xi)$ is homogeneous in ξ of degree $m = 2\mu = \sum (s_j + t_j)$, $P(y, \xi + z\eta) = 0$ if and only if

$$P\left(y, \frac{\xi}{|\xi|} + z\frac{|\eta|}{|\xi|}\frac{\eta}{|\eta|}\right) = 0.$$

Consequently the condition about the roots of $P(y, \xi + z\eta)$ is fulfulled for arbitrary independent vectors ξ, η if and only if it is fulfilled for independent unit vectors ξ, η. From this point, a simple compactness argument shows that if (3.22) holds for $y = 0$, it holds for $y \in \Omega \cup S$, $|y|$ small. Q.E.D.

To show that the coerciveness property is also open, we present an argument based on the inverse mapping theorem for Hilbert spaces. This proof is technically simple and has the advantage of exposing some of the underlying ideas in the study of elliptic boundary problems.

Lemma 3.7. *Suppose that* (3.6) *is elliptic at* $y = y_0$. *Then for each* k *there is at least one* j *such that the degree in* ξ_N *of* $L'_{kj}(y_0, \xi)$ *equals* $s_k + t_j$.

Proof. This is an elementary lemma. Obviously the degree in ξ_N of $L'_{kj}(y_0, \xi)$, $\deg_N L'_{kj}(y_0, \xi)$, does not exceed $s_k + t_j$. Suppose that for some l, $\deg_N L'_{lj}(y_0, \xi) < s_l + t_j$, $1 \leq j \leq n$. We may calculate $P(y_0, \xi)$ by expanding in cofactors across the lth row. Now the cofactor of $L'_{lj}(y_0, \xi)$, $L^{lj}(y_0, \xi)$, is a homogeneous polynomial of degree $\sum_{k \neq l} s_k + \sum_{h \neq j} t_h$, so since

$$P(y_0, \xi) = \sum_j L'_{lj}(y_0, \xi) L^{lj}(y_0, \xi),$$

$$\begin{aligned} \deg_N P(y_0, \xi) &\leq \max_j \left(\deg_N L'_{lj}(y_0, \xi) + \sum_{k \neq l} s_k + \sum_{h \neq j} t_h \right) \\ &< s_l + t_j + \sum_{k \neq l} s_k + \sum_{h \neq j} t_h = m. \end{aligned}$$

This is inconsistent with the coerciveness hypothesis as $P(y_0, (\xi', 0) + z(0,1))$ would not have m roots. Q.E.D.

It is convenient for us to introduce the notion of a coercive system of ordinary differential equations, just as an umbrella term for a long list of conditions.

Definition 3.8. The system of ordinary differential equations with constant coefficients

$$\sum_1^n p_{kj}(D_t) v^j(t) = f_k(t), \qquad 0 < t < \infty, \quad 1 \leq k \leq n, \tag{3.23}$$

$$\sum_1^n q_{hj}(D_t) v^j(0) = a_h, \qquad 1 \leq h \leq \mu, \tag{3.24}$$

$D_t = -i\, d/dt$, and weights s_k, t_j, r_h is *coercive* provided

(i) $\deg p_{kj}(\tau) \leq s_k + t_j$, $\deg q_{hk}(\tau) \leq r_h + t_j$, $1 \leq k, j \leq n$, $1 \leq h \leq \mu$,
(ii) for each k there is at least one j such that $\deg p_{kj}(\tau) = s_k + t_j$,
(iii) $p(\tau) = \det(p_{kj}(\tau))$ is of degree $m = \sum (s_j + t_j) = 2\mu$ and has exactly μ roots with positive imaginary part and μ roots with negative imaginary part, and finally,
(iv) the only solution $(v^1, \ldots, v^n)$ of the homogeneous system (3.23), (3.24) bounded for $t > 0$ is $v^1 = \cdots = v^n = 0$.

As usual, $p_{kj} = 0$ if $s_k + t_j < 0$, etc.
It is evident that if (3.5), (3.9) is coercive at y_0, then for each $0 \neq \xi' \in \mathbb{R}^{N-1}$

$$\begin{aligned} &\sum L'_{kj}(y_0, \xi', D_t) v^j(t) = f_k(t), \qquad 0 < t < \infty, \quad 1 \leq k \leq n, \\ &\sum B'_{hj}(y_0, \xi', D_t) v^j(0) = a_h, \qquad 1 \leq h \leq \mu, \end{aligned}$$

is a coercive system of ordinary differential equations.

Let us consider for a moment a solution of an arbitrary system of equations with constant coefficients

$$\sum_1^n P_{kj}(D_t)v^j(t) = 0, \qquad 0 < t < \infty, \quad 1 \le k \le n. \tag{3.25}$$

Observe that $\det(P_{kj}(D_t))v^l(t) = 0$ for each l, $1 \le l \le n$, and hence the dimension of the kernel of (3.25) is finite. To check the assertion, simply multiply the matrix $(P_{kj}(D_t))$ on the left by its classical adjoint.

By $H^\sigma(\mathbb{R}_+)$ we understand the completion of the space of complex valued $C^\infty(\mathbb{R}_+)$ functions in the H^σ norm. We take the inner product in $H^\sigma(\mathbb{R}_+)$ to be

$$(u, v)_{H^\sigma(\mathbb{R}_+)} = \int_0^\infty [u(t)\bar{v}(t) + D_t^\sigma u(t)\overline{D_t^\sigma v(t)}]\, dt, \qquad u, v \in H^\sigma(\mathbb{R}_+).$$

Given a coercive system (3.23), (3.24), for any $\sigma \ge 0$

$$p_{kj}(D_t)\colon H^{t_j+\sigma}(\mathbb{R}_+) \to H^{-s_k+\sigma}(\mathbb{R}_+), \qquad 1 \le k, j \le n,$$

continuously and because of (ii) we may assert that

$$H^{t_1+\sigma}(\mathbb{R}_+) \times \cdots \times H^{t_n+\sigma}(\mathbb{R}_+) \to H^{-s_k+\sigma}(\mathbb{R}_+),$$
$$(v^1, \ldots, v^n) \to \sum_j p_{kj}(D_t)v^j$$

is continuous. Recall that the "trace mapping"

$$H^\sigma(\mathbb{R}_+) \to \mathbb{C},$$
$$v \to D^{\sigma-1}v(0)$$

is continuous. Hence

$$q_{kj}(D_t)\colon H^{t_j+\sigma}(\mathbb{R}_+) \to \mathbb{C}, \qquad 1 \le h \le \mu, \quad 1 \le j \le n,$$
$$v \to q_{hj}(D_t)v(0)$$

is continuous for all h, j provided that $\sigma \ge r_0 = \max_h(r_h + 1, 0)$. Let us arrange our formalism. Set

$$V = H^{t_1+r_0}(\mathbb{R}_+) \times \cdots \times H^{t_n+r_0}(\mathbb{R}_+),$$
$$W = H^{r_0-s_1}(\mathbb{R}_+) \times \cdots \times H^{r_0-s_k}(\mathbb{R}_+) \times \mathbb{C}^\mu,$$

and define the continuous linear mapping

$$T\colon V \to W \tag{3.26}$$

by $Tv = f$, for $v = (v^1, \ldots, v^n) \in V, f = (f_1, \ldots, f_n, a_1, \ldots, a_\mu) \in W$, if

$$\sum_1^n p_{kj}(D_t)v^j(t) = f_k(t), \qquad 1 \le k \le n,$$

and

$$\sum_1^n q_{hj}(D_t)v^j(0) = a_h, \qquad 1 \le h \le \mu.$$

Lemma 3.9. *With* $T: V \to W$ *defined by* (3.26), $\ker T = \{0\}$ *and the dimension of the cokernel of* T *is finite.*

Proof. The first assertion is part (iv) of the definition of a coercive system. Indeed, any solution ζ of the homogeneous equation of (3.23) has components $\zeta^j(t)$ which are solutions of

$$p(D_t)\zeta^j(t) = 0, \qquad 0 < t < \infty, \quad 1 \le j \le n,$$

where $p(\tau) = \det(p_{kj}(\tau))$ [cf. (3.25)], and hence $\zeta^j(t)$, $t > 0$, is bounded if and only if it belongs to each $H^\sigma(\mathbb{R}_+)$ inasmuch as $\tau = 0$ is not a root of $p(\tau)$.

Now let $f = (f_1, \ldots, f_n, a_1, \ldots, a_\mu) \in \operatorname{coker} T$. This means that for all $v = (v^1, \ldots, v^n) \in V$

$$(Tv, f) = \sum_{k=1}^n \int_0^\infty \left[\sum_{j=1}^n p_{kj}(D_t)v^j\bar{f}_k + D_t^{r_0 - s_k} \sum_{j=1}^n p_{kj}(D_t)v^j\overline{D^{r_0 - s_k}f_k}\right]dt$$
$$+ \sum_{h=1}^\mu \sum_{j=1}^n q_{hj}(D_t)v^j(0)\bar{a}_h = 0.$$

In particular, choose $v^j \in C_0^\infty(\mathbb{R}_+)$. Then

$$\sum_{k=1}^n \int_0^\infty \left[\sum_1^n p_{kj}(D_t)v^j\bar{f}_k + D_t^{r_0 - s_k} \sum_{j=1}^n p_{kj}(D_t)v^j\overline{D^{r_0 - s_k}f_k}\right] dt = 0. \qquad (3.27)$$

Note that it is no loss in generality to assume that $f_k \in C^\infty(\mathbb{R}_+)$. Indeed, each $f_k(t)$ is the limit in $H^{-s_k + r_0}(\mathbb{R}_+)$ of a sequence $\varphi_\varepsilon * f_k(t)$ as $\varepsilon \to 0$, where φ_ε is a sequence of symmetric mollifiers with compact support. Replacing f_k by $\varphi_\varepsilon * f_k$ and rearranging the integrals in (3.27) gives (3.27) for the original f_k and new test functions $\varphi_\varepsilon * v^j \in C_0^\infty(\mathbb{R}_+)$. The validity of this depends, of course, on the fact that the $p_{kj}(D_t)$ have constant coefficients.

Integrating by parts in (3.27), placing all the derivatives on the f_k gives

$$\sum_j \int_0^\infty v_j \sum_{k=1}^n \overline{[(1 + D_t^{2r_0 - 2s_k})p_{kj}(D_t)f_k]}\, dt = 0, \qquad v^j \in C_0^\infty(\mathbb{R}^+);$$

hence $f_1, \ldots, f_n$ are solutions of the system

$$\sum_{k=1}^n (1 + D_t^{2r_0 - 2s_k})p_{kj}(D_t)f_k(t) = 0, \qquad 0 < t < \infty, \quad 1 \le j \le n.$$

As we have remarked, the dimension of this space of solutions is finite, say M.

Hence

$$\dim \operatorname{coker} T \leq M + \dim \mathbb{C}^{\mu} = M + \mu. \quad \text{Q.E.D.}$$

Theorem 3.10. *Suppose that* (3.23), (3.24) *is a coercive system. Then whenever* $v^j \in H^{r_0+t_j}(\mathbb{R}_+)$, $1 \leq j \leq n$, *and*

$$\sum p_{kj}(D_t)v^j = f_k, \qquad 0 < t < \infty, \quad 1 \leq k \leq n,$$
$$\sum q_{hj}(D_t)v^j(0) = a_h, \qquad 1 \leq h \leq \mu,$$

we have the estimate

$$\sum_1^n \|v^j\|_{H^{t_j+r_0}(\mathbb{R}_+)} \leq C\left[\sum_1^n \|f_k\|_{H^{r_0-s_k}(\mathbb{R}_+)} + \sum_1^{\mu} |a_j|\right]$$

for a constant $C > 0$, *independent of* v^j.

Proof. The mapping T defined by (3.26) is $1:1$ and has a finite dimensional cokernel E according to the last lemma. Hence T induces a continuous isomorphism $\tilde{T}: V \to W/E$. By the inverse mapping theorem, $\tilde{T}^{-1}$ is continuous. Hence if $Tv = f$ and $[f]$ denotes the equivalence class of f in W/E,

$$\begin{aligned}\|v\|_V \leq C\|[f]\|_{W/E} &\leq C \inf_{g \in E} \|f + g\|_W \\ &\leq C\|f\|_W\end{aligned}$$

for some $C > 0$, since $g = 0 \in E$. This proves the theorem. Q.E.D.

Completion of the Proof of Theorem 3.5. Since $t_j + s_k \leq r_0 + t_j$, $1 \leq j$, $k \leq n$, and $r_h + t_j \leq r_0 + t_j$, $1 \leq h \leq \mu$, $1 \leq j \leq n$, the linearized equations (3.6), (3.10) have continuous coefficients. By Lemma 3.6, the system is elliptic near $y = 0$. Consider the system of ordinary differential equations, for $0 \neq \xi' \in \mathbb{R}^{N-1}$ fixed,

$$\sum_1 L'_{kj}(y', \xi', D_t)v^j(t) = f_k(t), \qquad 0 < t < \infty, \quad 1 \leq k \leq n, \tag{3.28}$$

$$\sum_1 B'_{hj}(y', \xi', D_t)v^j(0) = a_h, \qquad 1 \leq h \leq \mu. \tag{3.29}$$

It remains to verify part (iv) of Definition 3.8 for this system when $|y'|$ is small. Given $\varepsilon > 0$, the continuity of the coefficients of the L'_{kj}, B'_{hk} insures that for some $\delta > 0$, $|y'| < \delta$ implies the coefficients of τ^l of the $L'_{kj}(y, \xi', \tau)$ and $p_{kj}(\tau)$ differ by less than $\varepsilon > 0$ for all l and the same for the coefficients of τ^l in $B'_{hj}(y, \xi', \tau)$ and $q_{hj}(\tau)$ where $p_{kj}(\tau) = L'_{kj}(0, \xi', \tau)$, $q_{hj}(\tau) = B'_{hj}(0, \xi', \tau)$.

Hence if $v \in V$, the previous theorem implies that the solution of (3.28), (3.29) satisfies

$$\sum_j \|v^j\|_{H^{t_j+r_0}(\mathbb{R}_+)} \leq C\left[\sum_1^n \|f_k\|_{H^{r_0-s_k}(\mathbb{R}_+)} + C'\varepsilon \sum_1^n \|v^j\|_{H^{t_j+r_0}(\mathbb{R}_+)} + \sum_1^\mu |a_h|\right].$$

Hence for ε sufficiently small

$$\sum_1^n \|v^j\|_{H^{t_j+r_0}(\mathbb{R}_+)} \leq C''\left[\sum_1^n \|f_k\|_{H^{r_0-s_k}(\mathbb{R}_+)} + \sum_1^\mu |a_h|\right]$$

for functions v^j, f_k satisfying (3.28) and (3.29). In particular, any bounded exponential solution of the homogeneous equations of (3.28), (3.29) vanishes. Q.E.D.

4. A Reflection Problem

Our applications to free boundary problems begin with a system which arises in the study of confined plasmas in physics. It resembles the transmission problem (3.19).

Suppose that Γ is a C^1 hypersurface in $B_1(0) \subset \mathbb{R}^N$ passing through $x = 0$ which separates $B_1(0)$ into two components, Ω^+ and Ω^-. We assume that the normal to Γ pointing into Ω^+ at 0 is $(0, \ldots, 0, 1)$. Suppose now that

$$u \in C^1(\Omega^+ \cup \Gamma \cup \Omega^-)$$

satisfies

$$\begin{aligned} \Delta u + \lambda u &= 0 \quad \text{in} \quad \Omega^+, \\ \Delta u &= 0 \quad \text{in} \quad \Omega^-, \\ &\left\{\begin{aligned} u &= 0 \\ \partial u/\partial \nu &\neq 0 \end{aligned}\right. \quad \text{on} \quad \Gamma, \end{aligned} \tag{4.1}$$

where $\lambda(x)$ is an analytic function of $x \in \overline{B_1(0)}$.

Theorem 4.1. *Let u be a solution of* (4.1). *Then Γ is an analytic hypersurface.*

Proof. In this situation we apply our zeroth order hodograph transform (2.8). We assume that $u > 0$ in Ω^+ and $u < 0$ in Ω^-. Then the mapping

$$\begin{aligned} y_\sigma &= x_\sigma, \qquad \sigma < N, \\ y_N &= u(x), \end{aligned} \qquad x \in B_1(0), \tag{4.2}$$

is a $1:1$ mapping of $B_\varepsilon(0)$ onto a neighborhood U of $y = 0$. We may assume that (4.2) is $1:1$ in a neighborhood U of $B_1(0)$. Then (4.2) maps Ω^+ onto $U^+ = \{y \in U : y_N > 0\}$, Ω^- onto $U^- = \{y \in U : y_N < 0\}$, and Γ onto $S = \{y \in U : y_N = 0\}$. As before, we define

$$\psi(y) = x_N, \qquad y \in U, \quad x \in B_1(0).$$

Consequently, from (2.13)

$$-\frac{1}{\psi_N}\sum \psi_{\sigma\sigma} + \frac{2}{\psi_N^2}\sum \psi_\sigma \psi_{\sigma N} - \frac{1}{\psi_N^3}(1 + \sum \psi_\sigma^2)\psi_{NN} + \lambda(y', \psi)y_N = 0 \text{ in } U^+,$$

$$-\frac{1}{\psi_N}\sum \psi_{\sigma\sigma} + \frac{2}{\psi_N^2}\sum \psi_\sigma \psi_{\sigma N} - \frac{1}{\psi_N^3}(1 + \sum \psi_\sigma^2)\psi_{NN} = 0 \text{ in } U^-,$$

$$\psi \in C^1(U^+ \cup S \cup U^-). \tag{4.3}$$

As usual, the sums extended on σ are $1 \le \sigma \le N - 1$. The system (4.3) resembles a transmission problem and we treat it in the same fashion. Introduce

$$\varphi(y) = \psi(y', -y_N), \qquad y \in U^+.$$

This leads to the system

$$-\frac{1}{\psi_N}\sum \psi_{\sigma\sigma} + \frac{2}{\psi_N^2}\sum \psi_\sigma \psi_{\sigma N} - \frac{1}{\psi_N^3}(1 + \sum \psi_\sigma^2)\psi_{NN} + \lambda(y', \psi)y_N = 0$$

$$\text{in } U^+,$$

$$-\frac{1}{\varphi_N}\sum \varphi_{\sigma\sigma} + \frac{2}{\varphi_N^2}\sum \varphi_\sigma \varphi_{\sigma N} - \frac{1}{\varphi_N^3}(1 + \sum \varphi_\sigma^2)\varphi_{NN} = 0 \tag{4.4}$$

with the boundary conditions

$$\left.\begin{aligned} \varphi - \psi &= 0 \\ \varphi_N + \psi_N &= 0 \end{aligned}\right\} \quad \text{on } S. \tag{4.5}$$

To prove the theorem it suffices to check that this system is coercive. In view of our local criterion Theorem 3.5, we need only check this at $y = 0$. By our choice of coordinates

$$\psi_N(0) = \frac{1}{u_N(0)} > 0, \qquad \psi_\sigma(0) = -\frac{u_\sigma(0)}{u_N(0)} = 0, \qquad 1 \le \sigma \le N - 1,$$

hence the linearized equations for (4.4) with the obvious weights $s_1 = s_2 = 0$, $t_1 = t_2 = 2$, are, for variations $\bar{\psi}$, $\bar{\varphi}$,

$$\left.\begin{aligned}\sum_\sigma \bar{\psi}_{\sigma\sigma} + a\bar{\psi}_{NN} = 0\\ \sum_\sigma \bar{\varphi}_{\sigma\sigma} + a\bar{\varphi}_{NN} = 0\end{aligned}\right\} \quad \text{in } \mathbb{R}^N_+, \tag{4.6}$$

where $a = u_N(0) > 0$. The linearized boundary conditions are just

$$\left.\begin{aligned}\bar{\varphi} - \bar{\psi} = 0\\ \bar{\varphi}_N + \bar{\psi}_N = 0\end{aligned}\right\} \quad \text{on} \quad \mathbb{R}^{N-1}. \tag{4.7}$$

It is easy to see that (4.6), (4.7) admits no bounded exponential solutions; hence, the system (4.3), (4.4) is coercive. Q.E.D.

5. Elliptic Equations Sharing Cauchy Data

Suppose we are given solutions to two different elliptic equations defined in the same domain which share Cauchy data on a portion of its boundary. What can be said of this boundary portion? This question, a free boundary analog of (3.13) and Theorem 3.4, will be discussed here in its simplest form. The problem certainly has some interest in its own right and it will also prepare us for the study of the free boundaries which occur in the double membrane variational inequality and in variational inequalities of higher order. In order to exhibit the overdetermined nature of our free boundary with a minimum amount of technical detail we shall assume that the various quantities possess a good deal of initial smoothness.

Let $\Omega \subset \mathbb{R}^N$ be a domain whose boundary near $0 \in \partial\Omega$ is a smooth hypersurface Γ passing through $x = 0$ with inward pointing normal $(0, \ldots, 0, 1)$ at $x = 0$. Suppose that

$$\left.\begin{aligned}\Delta u^1 + \lambda u^1 = f_1\\ \Delta u^2 = f_2\end{aligned}\right\} \quad \text{in} \quad \Omega, \tag{5.1}$$

$$\left.\begin{aligned}u^1 = u^2 = 0\\ \frac{\partial u^1}{\partial \nu} = \frac{\partial u^2}{\partial \nu}\end{aligned}\right\} \quad \text{on } \Gamma, \tag{5.2}$$

where f_1, f_2 are analytic functions in a neighborhood of $\Omega \cup \Gamma$ and $\lambda \in \mathbb{R}$. It is not sufficient to know merely that $u^1 = u^2$.

We shall consider two cases of this problem, each suggesting a new choice of variables defined in terms of $w(x) = u^1(x) - u^2(x)$. In the first case, we suppose that $f_1(0) - f_2(0) \neq 0$. Here a first order hodograph transform will be appropriate, leading to a coercive elliptic system. In the second case we suppose that $f_1 \equiv f_2$. This will suggest a fourth order problem for w alone, which, after suitable transformation, also leads to an elliptic system, although one rather different from the first case.

Theorem 5.1. *Let u^1, u^2 be a solution of* (5.1), (5.2) *and suppose that $f_1(0) \neq f_2(0)$. Then Γ is analytic near $x = 0$.*

Proof. Setting $w = u^1 - u^2$ permits us to express (5.1), (5.2) in the form

$$\left.\begin{aligned}\Delta w + \lambda(w + u^2) &= f_1 - f_2\\ \Delta u^2 &= f_2\end{aligned}\right\} \quad \text{in} \quad \Omega, \tag{5.3}$$

$$\left.\begin{aligned}u^2 = w &= 0\\ \frac{\partial w}{\partial v} &= 0\end{aligned}\right\} \quad \text{on} \quad \Gamma. \tag{5.4}$$

The boundary condition for w means that $w_i = 0$ on Γ, $1 \leq i \leq N$. Let us introduce the hodograph transform

$$y = (x, -w_N), \qquad x \in \Omega \cup \Gamma, \tag{5.5}$$

and the new dependent variables

$$\left.\begin{aligned}v^1(y) &= x_N y_N + w(x),\\ v^2(y) &= u^2(x),\end{aligned}\right\} \quad x \in \Omega \cup \Gamma. \tag{5.6}$$

Since the normal to Γ at $x = 0$ is in the x_N direction,

$$w_{i\sigma}(0) = 0, \qquad 1 \leq i \leq N, \quad 1 \leq \sigma \leq N - 1, \tag{5.7}$$

so

$$w_{NN}(0) = f_1(0) - f_2(0) \neq 0.$$

Assuming that $w_{NN}(0) < 0$, we see that (5.5) maps a neighborhood of $x = 0$ in Ω, which we may take to be all of Ω, onto a domain $U \subset \{y : y_N > 0\}$ and maps Γ onto a portion S of $y_N = 0$. According to (2.7), v^1 satisfies a second order nonlinear equation which we shall write down later ((5.8)).

Let us turn our attention to the equation satisfied by $v^2(y)$. According to (2.6),

$$\frac{\partial^2}{\partial x_N^2} u^2 = \frac{1}{v_{NN}^1} \partial_N\left(\frac{1}{v_{NN}^1} \partial_N v^2\right) = \left(\frac{1}{v_{NN}^1}\right)^2 \partial_N^2 v^2 + \frac{\partial_N v^2}{v_{NN}^1} \partial_N\left(\frac{1}{v_{NN}^1}\right)$$

and

$$\frac{\partial^2}{\partial x_\sigma^2} u^2 = \left(\partial_\sigma - \frac{v_{N\sigma}^1}{v_{NN}^1}\partial_N\right)^2 v^2$$

$$= \partial_\sigma^2 v^2 - 2\frac{v_{N\sigma}^1}{v_{NN}^1}\partial_\sigma\partial_N v^2 + \left(\frac{v_{N\sigma}^1}{v_{NN}^1}\right)^2 \partial_N^2 v^2$$

$$+ \partial_N v^2\left[\frac{v_{N\sigma}^1}{v_{NN}^1}\partial_N\left(\frac{v_{N\sigma}^1}{v_{NN}^1}\right) - \partial_\sigma\left(\frac{v_{N\sigma}^1}{v_{NN}^1}\right)\right].$$

Thus it is apparent that our equation, though of second order in v^2, is of third order in v^1. Some care in choosing the weights will be necessary! Our new system for v^1, v^2 is [cf. (2.7)]

$$-\frac{1}{v_{NN}^1} - \frac{1}{v_{NN}^1}\sum (v_{\sigma N}^1)^2 + \sum v_{\sigma\sigma}^1 + \lambda v^2$$

$$= f_1(v_N^1, y') - f_2(v_N^1, y') + \lambda(y_N v_N^1 - v^1) \qquad \text{in} \quad U, \tag{5.8}$$

$$\partial_N v^2\left\{\sum\left[\frac{v_{N\sigma}^1}{v_{NN}^1}\partial_N\left(\frac{v_{N\sigma}^1}{v_{NN}^1}\right) - \partial_\sigma\left(\frac{v_{\sigma N}^1}{v_{NN}^1}\right)\right] + \frac{1}{v_{NN}^1}\partial_N\left(\frac{1}{v_{NN}^1}\right)\right\}$$

$$+ \sum\left[\partial_\sigma^2 v^2 - 2\frac{v_{N\sigma}^1}{v_{NN}^1}\partial_\sigma\partial_N v^2 + \left(\frac{v_{\sigma N}^1}{v_{NN}^1}\right)^2\partial_N^2 v^2\right] + \left(\frac{1}{v_{NN}^1}\right)^2\partial_N^2 v^2$$

$$= f_2(v_N^1, y') - \lambda v^2 \qquad \text{in} \quad U, \tag{5.9}$$

with

$$\left.\begin{aligned} v^1 &= 0 \\ v^2 &= 0 \end{aligned}\right. \qquad \text{on} \quad S. \tag{5.10}$$

The summations above are on σ, $1 \le \sigma \le N-1$.

We choose weights $s_1 = -2$, $s_2 = 0$, $t_1 = 4$, $t_2 = 2$, $r_1 = -4$, and $r_2 = -2$. In this way, $t_1 + s_2 = 4$, so the principal symbol will not contain any third derivatives of v^1.

To complete the proof, it remains to show that (5.8), (5.9), (5.10) is elliptic and coercive with respect to our chosen weights. By our localization principle, it suffices to check that at $y = 0$, where we may assume $v_{NN}^1(0) = -1/w_{NN}(0) = 1$. Also, from (5.7),

$$v_{i\sigma}^1(0) = 0, \qquad 1 \le i \le N, \quad 1 \le \sigma \le N-1.$$

For variations $\bar{v}^1$ and $\bar{v}^2$ we find the system

$$\left.\begin{aligned} \Delta\bar{v}^1 + \lambda\bar{v}^2 &= 0 \\ \Delta\bar{v}^2 &= 0 \end{aligned}\right. \qquad \text{in} \quad \mathbb{R}_+^N, \tag{5.11}$$

$$\bar{v}^1 = \bar{v}^2 = 0 \qquad \text{in} \quad \mathbb{R}^{N-1}. \tag{5.12}$$

The principal symbol of (5.11) is

$$\begin{pmatrix} -\xi^2 & \lambda \\ 0 & -\xi^2 \end{pmatrix},$$

which is of rank 2; hence the system is elliptic. It is easy to check that (5.11), (5.12) has no bounded exponential solutions. Hence Theorem 3.3 applies, so v^1 and v^2 are analytic in $U \cup S$. As a consequence Γ is also analytic inasmuch as it permits the representation [cf. (2.17)]

$$\Gamma: \quad x_N = v_N^1(x', 0), \qquad (x', 0) \in S. \quad \text{Q.E.D.}$$

We turn now to the discussion of (5.1), (5.2) when $f_1(x) = f_2(x)$. Here (5.3) assumes the more concise form

$$\begin{aligned} \Delta w + \lambda(w + u^2) &= 0 \\ \Delta u^2 &= f_2 \end{aligned} \qquad \text{in} \quad \Omega. \tag{5.13}$$

Applying Δ to the first equation and substituting the second in it gives

$$\Delta^2 w + \lambda \, \Delta w = -\lambda f_2 \qquad \text{in} \quad \Omega. \tag{5.14}$$

To assess the behavior of w on Γ we state this elementary lemma.

Lemma 5.2. *Let w, u^2 be a solution of* (5.13) *with the boundary conditions*

$$u^2 = w = w_i = 0 \qquad on \quad \Gamma, \quad 1 \le i \le N.$$

Then $w_{ij}(x) = 0$ on Γ for $1 \le i, j \le N$.

Note that the condition on Γ is simply (5.4).

Proof. At a given point $x^0 \in \Gamma$, choose coordinates $\xi_1, \ldots, \xi_N$ with the ξ_N axis normal to Γ at x^0. Then since $w_{x_i} = 0$ on Γ for all i, $w_{\xi_i} = 0$ on Γ for all i, and hence

$$\frac{\partial^2 w}{\partial \xi_i \, \partial \xi_\sigma}(x^0) = 0 \qquad \text{for} \quad 1 \le i \le N, \quad 1 \le \sigma \le N - 1.$$

Now from (5.13)

$$\frac{\partial^2 w}{\partial \xi_N^2}(x^0) = -\lambda[w(x^0) + u^2(x^0)] - \sum_\sigma \frac{\partial^2 w}{\partial \xi_\sigma^2}(x^0) = 0.$$

The lemma follows. Q.E.D.

Our new transform is based on the choice of $-w_{NN}$ as a new independent variable. For this to be valid, we shall assume that $\lambda \neq 0$ and $\partial u^2/\partial v(0) = \partial u^1/\partial v(0) \neq 0$. We shall first state our theorem in terms of the system (5.1), (5.2) and then in terms of the single function w.

Theorem 5.3. *Let u^1, u^2 be a solution of the system*

$$\begin{aligned} \Delta u^1 + \lambda u^1 &= f \\ \Delta u^2 &= f \end{aligned} \qquad in \quad \Omega, \tag{5.15}$$

$$\begin{aligned} u^1 = u^2 &= 0 \\ \frac{\partial u^1}{\partial v} &= \frac{\partial u^2}{\partial v} \end{aligned} \qquad on \quad \Gamma, \tag{5.16}$$

where f is analytic in a neighborhood of $x = 0$ and $\lambda \in \mathbb{R}$. Suppose that $\lambda \neq 0$ and $\partial u^1/\partial v(0) \neq 0$. Then Γ is analytic in a neighborhood of $x = 0$.

Proof. We reduce the proof of this theorem to the one below. With $w = u^1 - u^2$, as usual, we have that

$$\Delta^2 w + \lambda\, \Delta w = -\lambda f \qquad \text{in} \quad \Omega$$

and

$$D^\alpha w = 0 \qquad \text{on} \quad \Gamma$$

for any multi-index α, $|\alpha| \leq 2$, in view of Lemma 5.2. Now note that

$$\frac{\partial}{\partial x_N} \frac{\partial^2}{\partial x_i\, \partial x_j} w(0) = \frac{\partial}{\partial x_i} \frac{\partial^2}{\partial x_N\, \partial x_j} w(0) = 0$$

whenever $i + j < 2N$ because $(\partial^2/\partial x_N\, \partial x_j)w = 0$ on Γ. Hence from (5.13),

$$\begin{aligned} \frac{\partial^3}{\partial x_N^3} w(0) &= -\lambda \left[\frac{\partial w}{\partial x_N}(0) - \frac{\partial u^2}{\partial x_N}(0) \right] - \sum_\sigma \frac{\partial^3 w}{\partial x_N\, \partial x_\sigma^2}(0) \\ &= \lambda \frac{\partial u^2}{\partial x_N}(0) \neq 0. \end{aligned}$$

Theorem 5.4. *Let $w \in C^4(\Omega \cup \Gamma)$ satisfy*

$$\Delta^2 w + \lambda\, \Delta w = g \qquad in \quad \Omega \tag{5.17}$$

$$\begin{aligned} &D^\alpha w = 0 \\ &\frac{\partial^3}{\partial x_N^3} w(0) \neq 0 \end{aligned} \qquad on \quad \Gamma, \quad for \quad |\alpha| \leq 2, \tag{5.18}$$

where λ and g are analytic in a neighborhood of $x = 0$. Then Γ is analytic near $x = 0$.

Proof. Recall that the x_N axis is normal to Γ at $x = 0$. Hence (5.18) insures that the mapping

$$y = (x', -w_{NN}(x)), \qquad x \in \Omega \tag{5.19}$$

is 1 : 1 in a neighborhood of 0 in Ω, which as usual we take to be all of Ω, and maps it onto a region U which we may assume to be in $y_N > 0$. The hypersurface Γ is mapped in this way onto a portion S of $y_N = 0$. We set

$$\begin{aligned} \varphi^1(y) &= x_N y_N + w_N(x), \\ \varphi^2(y) &= \sum_{\sigma \leq N-1} w_{\sigma\sigma}(x), \end{aligned} \qquad x \in \Omega. \tag{5.20}$$

We determine equations for φ^1 and φ^2 from (5.17). Now φ^1 is merely the Legendre transform of w_N with respect to (5.19) so

$$\begin{aligned} \varphi^1_\sigma &= w_{\sigma N}, \qquad 1 \leq \sigma \leq N-1, \\ \varphi^1_N &= x_N, \end{aligned}$$

where, as usual, subscripts of φ^i refer to differentiation with respect to y and those of w to differentiation with respect to x. In addition,

$$\begin{aligned} &w_{NNN} = -1/\varphi^1_{NN}, \qquad w_{NNNN} = \varphi^1_{NNN}/(\varphi^1_{NN})^3, \\ &w_{NN\sigma\sigma} = (\varphi^1_{N\sigma}/\varphi^1_{NN})_\sigma - (\varphi^1_{N\sigma}/\varphi^1_{NN})(\varphi^1_{N\sigma}/\varphi^1_{NN})_N, \qquad 1 \leq \sigma \leq N-1. \end{aligned}$$

Also

$$\begin{aligned} \sum_{\sigma<N} w_{\sigma\sigma\tau} &= \varphi^2_\tau - (\varphi^1_{N\tau}/\varphi^1_{NN})\varphi^2_N, \qquad 1 \leq \tau \leq N-1, \\ \sum_{\sigma<N} w_{\sigma\sigma\tau\tau} &= \varphi^2_{\tau\tau} - 2(\varphi^1_{N\tau}/\varphi^1_{NN})\varphi^2_{N\tau} + (\varphi^1_{N\tau}/\varphi^1_{NN})^2\varphi^2_{NN} \\ &\quad - (\varphi^1_{N\tau}/\varphi^1_{NN})_\tau \varphi^2_N + (\varphi^1_{N\tau}/\varphi^1_{NN})_N(\varphi^1_{N\tau}/\varphi^1_{NN})\varphi^2_N. \end{aligned}$$

Now we write (5.17) as

$$w_{NNNN} + 2\sum_{\sigma<N} w_{\sigma\sigma NN} + \sum_{\sigma,\tau<N} w_{\sigma\sigma\tau\tau} + \lambda\left(w_{NN} + \sum_\sigma w_{\sigma\sigma}\right) = g,$$

so that

$$F_1(\varphi^1, \varphi^2, y) = 0 \qquad \text{in} \quad U, \tag{5.21}$$

where

$$\begin{aligned} F_1(\varphi^1, \varphi^2, y) &= \frac{\varphi^1_{NNN}}{(\varphi^1_{NN})^3} + 2\sum_{\sigma<N}\left[\left(\frac{\varphi^1_{N\sigma}}{\varphi^1_{NN}}\right)_\sigma - \frac{\varphi^1_{N\sigma}}{\varphi^1_{NN}}\left(\frac{\varphi^1_{N\sigma}}{\varphi^1_{NN}}\right)_N\right] \\ &\quad + \sum_{\sigma,\tau<N}\left[\varphi^2_{\tau\tau} - 2\frac{\varphi^1_{N\tau}\varphi^2_{N\tau}}{\varphi^1_{NN}} + \left(\frac{\varphi^1_{N\tau}}{\varphi^1_{NN}}\right)^2\varphi^2_{NN} - \left(\frac{\varphi^1_{N\tau}}{\varphi^1_{NN}}\right)_\tau \varphi^2_N \right. \\ &\quad \left. + \left(\frac{\varphi^1_{N\tau}}{\varphi^1_{NN}}\right)_N\left(\frac{\varphi^1_{N\tau}}{\varphi^1_{NN}}\right)\varphi^2_N\right] + \lambda(\varphi^2 - y_N) - g(y', \varphi^1_N) \end{aligned} \tag{5.22}$$

in U. The second equation is obtained by differentiating φ^2 in two ways:

$$\sum_{\sigma<N} w_{\sigma\sigma N} = \frac{\varphi^2_N}{\varphi^1_N}$$

and

$$\sum_{\sigma<N} w_{\sigma N\sigma} = \frac{\partial}{\partial x_\sigma}\sum_{\sigma<N}\varphi^1_\sigma = \sum_{\sigma<N}\left[\varphi^1_{\sigma\sigma} - \frac{(\varphi^1_{\sigma N})^2}{\varphi^1_{NN}}\right].$$

So

$$F_2(\varphi^1, \varphi^2) = \sum(\varphi^1_{\sigma\sigma}\varphi^1_{NN} - (\varphi^1_{N\sigma})^2) - \varphi^2_N = 0 \qquad \text{in} \quad U. \tag{5.23}$$

We have deduced that φ^1, φ^2 is a solution to (5.21), (5.23) subject to the boundary conditions

$$\varphi^1 = 0, \qquad \varphi^2 = 0 \qquad \text{on} \quad S. \tag{5.24}$$

Choose weights $s_1 = 0, s_2 = -1, t_1 = 3, t_2 = 2, r_1 = -3$, and $r_2 = -2$. To linearize the equations at $y = 0$, notice that

$$\varphi^j(0) = \varphi^j_\sigma(0) = \varphi^j_{\sigma\tau}(0) = \varphi^1_{N\sigma}(0) = \varphi^2_N(0) = 0$$

for $1 \le \sigma, \tau \le N-1, j = 1, 2$. For variations $\bar{\varphi}^1$ and $\bar{\varphi}^2$ we find from (5.22) and (5.23) that

$$\begin{aligned} L_{11}\bar{\varphi}^1 + L_{12}\bar{\varphi}^2 &= \frac{\bar{\varphi}^1_{NNN}}{\varphi^1_{NN}(0)^3} + \frac{2}{\varphi^1_{NN}(0)}\sum_{\sigma<N}\bar{\varphi}^1_{N\sigma\sigma} + \sum_{\sigma<N}\bar{\varphi}^2_{\sigma\sigma} = 0, \\ L_{21}\bar{\varphi}^1 + L_{22}\bar{\varphi}^2 &= \varphi^1_{NN}(0)\sum_{\sigma<N}\bar{\varphi}^1_{\sigma\sigma} - \bar{\varphi}^2_N = 0 \end{aligned} \tag{5.25}$$

in $\mathbb{R}^N_+$, with

$$\bar{\varphi}^1 = \bar{\varphi}^2 = 0 \qquad \text{on} \quad \mathbb{R}^{N-1}. \tag{5.26}$$

The symbol of (5.25) is

$$\begin{pmatrix} \dfrac{-i\xi_N^3}{a^3} - \dfrac{2i}{a}\sum \xi_\sigma^2 \xi_N & -\sum\limits_\sigma \xi_\sigma^2 \\ -a\sum\limits_\sigma \xi_\sigma^2 & -i\xi_N \end{pmatrix}$$

where $a = \varphi_{NN}^1(0) \neq 0$. It has determinant

$$\begin{aligned} L(\xi) &= -[(\xi_N^4/a^3) + (2/a)\sum \xi_N^2\xi_\sigma^2 + a(\sum \xi_\sigma^2)^2] \\ &= -a[(\xi_N^2/a^2) + \sum \xi_\sigma^2]^2 = -a|(\xi', \xi_N/a)|^4 \qquad \textit{for} \quad 0 \neq \xi \in \mathbb{R}^N. \end{aligned}$$

Hence the system is elliptic near $y = 0$.

Finally, we check the coerciveness of this system. Writing

$$\bar{\varphi}^1(y', t) = \zeta_1(t)e^{i\xi' y'}, \qquad \bar{\varphi}^2(y', t) = \zeta_2(t)e^{i\xi' y'}, \qquad 0 \neq \xi' \in \mathbb{R}^{N-1},$$

we obtain for ζ_1 and ζ_2 the system of ordinary differential equations

$$\begin{aligned} (1/a^3)\zeta_1''' - (2/a)|\xi'|^2\zeta_1' - |\xi'|^2\zeta_2 &= 0, \\ a|\xi'|^2\zeta_1 + \zeta_2' &= 0, \\ \zeta_1(0) = \zeta_2(0) &= 0 \end{aligned}$$

for $t > 0$. Differentiating the first equation and substituting the second into it gives, after multiplication by a^3,

$$\zeta_1^{\text{iv}} - 2a^2|\xi'|^2\zeta_1'' + a^4|\xi'|^4\zeta_1 = 0.$$

Hence, assuming $a > 0$,

$$\zeta_1(t) = c_1 t e^{-a|\xi'|t} + c_2 e^{-a|\xi'|t}.$$

From the boundary condition, $c_2 = 0$. Now

$$0 = |\xi'|^2\zeta_2(0) = (1/a)[(1/a^2)\zeta_1'''(0) - 2|\xi'|^2\zeta_1'(0)] = (c_1/a)|\xi'|^2,$$

so $c_1 = 0$. Hence $\zeta_1 = \zeta_2 = 0$. The theorem is proved. Q.E.D.

We offer some comments about the hypotheses of Theorems 5.1 and 5.3. The transformation (5.5) is defined and the resulting equations have continuous coefficients if we only assume that $u^1, u^2 \in C^{2,\alpha}(\Omega \cup \Gamma)$, say. To invoke Theorem 3.3, however, we must know that $v^1 \in C^{4,\alpha}(U \cup S)$. Because of our choice of weights, this may be achieved by first applying Theorem 2.1 to the equation for w alone in (5.3). Since

$$f_1 - f_2 - \lambda(w + u^2) \in C^{2,\alpha}(\Omega \cup \Gamma),$$

it follows that $w \in C^{4,\alpha}(\Omega \cup \Gamma)$, $\Gamma \in C^{3,\alpha}$, and the transformation $y = (x', w_N)$ and its inverse are $C^{3,\alpha}$. Since $v_\sigma^1(y) = w_\sigma(x)$, $1 \le \sigma \le N-1$, and $v_N^1 = x_N$, all derivatives of v^1 are in $C^{3,\alpha}$, that is, $v \in C^{4,\alpha}(U \cup S)$.

In a similar way we may weaken the hypotheses of Theorem 5.3. Differentiating the equation for w in (5.13) with respect to x_N, we obtain the problem

$$\Delta w_N + \lambda(w_N + u_N^2) = 0 \quad in \quad \Omega,$$
$$w_{Ni} = 0 \quad on \quad \Gamma, \quad 1 \le i \le N,$$

subject to the restriction $w_{NNN}(0) \ne 0$ provided the hypotheses of Theorem 5.3 are in force. Now $w_N, u_N^2 \in C^{1,\alpha}(\Omega \cup \Gamma)$, so Theorem 2.1 may be applied. After some manipulations (Exercise 13) we conclude that $w \in C^{4,\alpha}(\Omega \cup \Gamma)$ and φ^1 and φ^2 defined by (5.20) are in $C^{3,\alpha}(U \cup S)$ and $C^{2,\alpha}(U \cup S)$, respectively. Indeed, in this fashion it is possible to "bootstrap" to a C^∞ result, in distinct contrast to the problem of section 6, which we are about to consider.

6. A Problem of Two Membranes

In the previous section we learned how to assess the behavior of a free boundary when several functions were defined on the same side of it. The question here consists in the determination of the smoothness of a hypersurface when two functions are defined on one side of it and a third defined on the other side. This suggests incorporating into our argument a reflection mapping, as in Section 4.

First we state our problem, then we illustrate how it arises from a variational inequality. Let Γ be a smooth hypersurface in $\mathbb{R}^N$ passing through $x = 0$ whose normal at $x = 0$ is in the x_N direction. Suppose that Γ separates a neighborhood of the origin into domains Ω^+ and Ω^- with, say, $(0, \ldots, 0, 1)$ the inward pointing normal to Ω^+ at $x = 0$. Let $u^1, u^2 \in C^2(\Omega^+ \cup \Gamma) \cap C^2(\Omega^- \cup \Gamma) \cap C^1(\Omega^+ \cup \Gamma \cup \Omega^-)$ satisfy

$$\begin{aligned} \Delta u^1 + \lambda u^1 &= f_1 \\ \Delta u^2 &= f_2 \end{aligned} \quad \text{in} \quad \Omega^+ \tag{6.1}$$

and

$$\begin{aligned} u^1 &= u^2 \\ u_i^1 &= u_i^2 \\ \Delta u^1 + (\lambda/2)u^1 &= \tfrac{1}{2}(f_1 + f_2) \end{aligned} \quad \text{in} \quad \Omega^-, \quad 1 \le i \le N, \tag{6.2}$$

where f_1, f_2 are given smooth functions in $\Omega^+ \cup \Gamma \cup \Omega^-$ and $\lambda \in \mathbb{R}$. The "boundary" conditions satisfied by u^j implied by (6.1) and (6.2) are that the u^j are C^1 across Γ.

Now let $\Omega \subset \mathbb{R}^N$ be a bounded domain with smooth boundary $\partial\Omega$ and suppose given $h^1, h^2 \in C^2(\partial\Omega)$ with $h^1 > h^2$. Denote by $\mathbb{K}$ the convex set of pairs of functions

$$\mathbb{K} = \{(v^1, v^2) : v^2 \leq v^1 \text{ in } \Omega,\ v^j = h^j \text{ on } \partial\Omega, \text{ and } v^j \in H^1(\Omega), j = 1, 2\}.$$

For given, say smooth, functions f_1, f_2 in $\overline{\Omega}$ let (u^1, u^2) be the solution of the variational inequality

$$\int_\Omega [u^1_i(v^1 - u^1)_i + u^2_i(v^2 - u^2)_i - \lambda u^1(v^1 - u^1)]\, dx$$
$$\geq -\int_\Omega [f_1(v^1 - u^1) + f_2(v^2 - u^2)]\, dx \qquad \text{for} \quad (v^1, v^2) \in \mathbb{K}. \tag{6.3}$$

The existence and smoothness properties of the u^j are the subject of Exercise 13, Chapter II, and Exercise 5, Chapter IV. We suppose that $u^j \in H^{2,s}(\Omega)$, $1 \leq s < \infty$, and define

$$I = \{x \in \Omega : u^1(x) = u^2(x)\},$$

the coincidence set of u^1 and u^2. One easily checks that

$$\left.\begin{aligned} \Delta u^1 + \lambda u^1 &= f_1 \\ \Delta u^2 &= f_2 \end{aligned}\right\} \quad \text{in} \quad \Omega \backslash I. \tag{6.4}$$

Let $\zeta \in C_0^\infty(\Omega)$ and set $v^1 = u^1 + \zeta$, $v^2 = u^2 + \zeta$. Then $v^1 \geq v^2$ in Ω, so it follows that $(v^1, v^2) \in \mathbb{K}$. Substituting this (v^1, v^2) into (6.3) leads to the equation

$$\Delta u^1 + \Delta u^2 + \lambda u^1 = f_1 + f_2 \qquad \text{a.e. in} \quad \Omega.$$

Hence, since $u^1_i = u^2_i$ in I, $1 \leq i \leq N$, whence $u^1_{ij} = u^2_{ij}$ a.e. in I, we obtain

$$\Delta u^1 + \frac{\lambda}{2} u^1 = \frac{1}{2}(f_1 + f_2) \qquad \text{a.e. in} \quad I. \tag{6.5}$$

From these considerations we infer that (6.4), (6.5) may be recast as (6.1), (6.2) near suitably smooth and small portions Γ of ∂I.

However, it is far from obvious that (6.1), (6.2) is overdetermined. In fact, neglecting the equation in Ω^- temporarily gives the problem

$$\left.\begin{aligned} \Delta u^1 + \lambda u^1 &= f_1 \\ \Delta u^2 &= f_2 \end{aligned}\right\} \quad \text{in} \quad \Omega^+$$

and

$$u^1 = u^2$$

$$\frac{\partial u^1}{\partial \nu} = \frac{\partial u^2}{\partial \nu} \quad \text{on} \quad \Gamma,$$

where ν is the normal direction to Γ, which is generally coercive, and hence solvable when Γ has only a limited degree of smoothness. Evidently, we must utilize the information given by (6.2) in Ω^-.

To assist us in the resolution of this difficulty, let us introduce $w = u^1 - u^2$ and $w' = u^1 + u^2$. Then

$$\begin{aligned} [\Delta + (\lambda/2)]w + (\lambda/2)w' &= f_1 - f_2 && \text{in} \quad \Omega^+, \\ (\lambda/2)w + [\Delta + (\lambda/2)]w' &= f_1 + f_2 && \text{in} \quad \Omega^+ \cup \Gamma \cup \Omega^-, \\ w = w_i &= 0 && \text{in} \quad \Gamma \cup \Omega^-, \quad 1 \le i \le N. \end{aligned} \tag{6.6}$$

From the second equation

$$\Delta w' = f_1 + f_2 - (\lambda/2)(w' + w) \in H^{2,s}(\Omega^+ \cup \Gamma \cup \Omega^-), \qquad s < \infty.$$

The elliptic regularity theory applied here informs us that

$$w' \in H^{4,s}(\Omega^+ \cup \Gamma \cup \Omega^-) \subset C^{3,\alpha}(\Omega^+ \cup \Gamma \cup \Omega^-), \quad 1 \le s < \infty, \quad 0 \le \alpha < 1.$$

Now we impose the nondegeneracy criterion

$$f_1(0) - f_2(0) - \lambda u^1(0) \ne 0. \tag{6.7}$$

Then w is a solution of

$$\begin{aligned} \Delta w(x) = a(x, w) &= f_1(x) - f_2(x) - (\lambda/2)[w'(x) + w(x)] \quad \text{in} \quad \Omega^+, \\ w = w_i &= 0 \quad \text{on} \quad \Gamma, \quad 1 \le i \le N, \end{aligned} \tag{6.8}$$

with $a(x, w)$ analytic in w and $C^{3,\alpha}$ in x, $a(0, 0) \ne 0$. Thus we are led to

Lemma 6.1. *Let*

$$u^1, u^2 \in C^2(\Omega^+ \cup \Gamma) \cap C^2(\Omega^- \cup \Gamma) \cap C^1(\Omega^+ \cup \Gamma \cup \Omega^-)$$

satisfy (6.1), (6.2), *where* Γ *is of class* C^1. *Assume that* (6.7) *holds. Then there is a neighborhood* $B_\rho(0)$ *such that* $w = u^1 - u^2 \in C^{5,\alpha}((\Omega^+ \cup \Gamma) \cap B_\rho(0))$ *and* Γ *is of class* $C^{4,\alpha}$ *near* $x = 0$.

Proof. This is direct application of Theorem 2.1 to w which satisfies (6.8). Q.E.D.

Unfortunately this technique cannot be iterated to prove that $\Gamma \in C^\infty$.

Let us proceed to introduce the first order hodograph mapping

$$y = (x', -w_N(x)), \qquad x \in \Omega^+, \tag{6.9}$$

which, we may assume, maps Ω^+ in a $1:1$ manner onto a region U in $y_N > 0$ and Γ $1:1$ onto a portion S of $y_N = 0$. Set

$$v(y) = x_N y_N + w(x), \qquad x \in \Omega^+, \quad y \in U. \tag{6.10}$$

The inverse mapping to (6.9) is given by

$$g^+(y) = (y', v_N(y)), \qquad y \in U \cup S, \tag{6.11}$$

which carries U into Ω^+, of course. Now let C be any constant larger than $\|Dv_N\|_{L^\infty(U)}$ and define

$$g^-(y) = (y', v_N(y) - Cy_N), \qquad y \in U \cup S. \tag{6.12}$$

When $y = (y', 0) \in S$, $g^-(y) = (y', v_N(y', 0)) = g^+(y) \in \Gamma$. But due to our choice of constant C, g^- maps a neighborhood of 0 in U, say all of U, into Ω^-. Using this new reflection mapping we define new dependent variables

$$\begin{aligned} \varphi^+(y) &= w'(g^+(y)), \\ \varphi^-(y) &= w'(g^-(y)), \end{aligned} \qquad y \in U \cup S. \tag{6.13}$$

We shall prove that the three functions v, φ^+, and φ^- satisfy a coercive elliptic system.

Theorem 6.2. *Suppose that*

$$u^1, u^2 \in C^2(\Omega^+ \cup \Gamma) \cap C^2(\Omega^- \cup \Gamma) \cap C^1(\Omega^+ \cup \Gamma \cup \Omega^-)$$

is a solution of (6.1), (6.2). *Suppose that f_1 and f_2 are analytic in a neighborhood of $x = 0$ and that*

$$f_1(0) - f_2(0) - \lambda u^1(0) \neq 0. \tag{6.7}$$

Then Γ is analytic in a neighborhood of $x = 0$.

Proof. We begin by calculating the system of equations satisfied by v, φ^+, and φ^-. Our point of departure is the system (6.6). With $\partial_k = \partial/\partial y_k$,

$$\left.\begin{aligned} \frac{\partial}{\partial x_\sigma} &= \partial_\sigma - \frac{v_{N\sigma}}{v_{NN}}\partial_N, \qquad 1 \le \sigma \le N-1, \\ \frac{\partial}{\partial x_N} &= \frac{1}{v_{NN}}\partial_N \end{aligned}\right\} \quad \text{for} \quad x \in \Omega^+$$

and

$$\left.\begin{aligned}\frac{\partial}{\partial x_\sigma} &= \partial_\sigma - \frac{v_{N\sigma}}{v_{NN} - C}\partial_N, \qquad 1 \le \sigma \le N-1,\\ \frac{\partial}{\partial x_N} &= \frac{1}{v_{NN} - C}\partial_N\end{aligned}\right\} \quad \text{for} \quad x \in \Omega^-.$$

Reserving for the moment our calculation of the various equations, let us determine the boundary conditions obeyed by v, φ^+, φ^- on $y_N = 0$. As usual for a Legendre transform, $v = 0$ on $y_N = 0$. Since w' is continuous across Γ, $\varphi^+ = \varphi^-$ on $y_N = 0$. Finally, $(\partial/\partial x_N)w'$ is continuous across Γ; hence

$$\frac{1}{v_{NN}}\partial_N \varphi^+ = \frac{\partial}{\partial x_N} w' = \frac{1}{v_{NN} - C}\partial_N \varphi' \qquad \text{on} \quad y_N = 0.$$

Summarizing,

$$\left.\begin{aligned} v &= 0\\ \varphi^+ - \varphi^- &= 0 \qquad \text{on } S\\ \frac{1}{v_{NN}}\partial_N\varphi^+ - \frac{1}{v_{NN} - C}\partial_N\varphi^- &= 0\end{aligned}\right. \tag{6.14}$$

The first equation of (6.6) transforms in a familiar manner. From (2.7) we have that

$$-\frac{1}{v_{NN}} - \frac{1}{v_{NN}}\sum_{\sigma<N}(v_{\sigma N})^2 + \sum v_{\sigma\sigma} = F_1(v_N, v, \varphi^+, y) \qquad \text{in} \quad U, \tag{6.15}$$

where

$$F_1(v_N, v, \varphi^+, y) = f_1(y', v_N) - f_2(y', v_N) - (\lambda/2)(-y_N v_N + v + \varphi^+)$$

is an analytic function of its arguments.

We continue in a manner analogous to the proof of Theorem 5.1. Considering the second equation of (6.6) and restricting our attention to $x \in \Omega^+$, we obtain, as in (5.9),

$$\begin{aligned}&\partial_N\varphi^+\left\{\sum_{\sigma<N}\left[\frac{v_{N\sigma}}{v_{NN}}\partial_N\left(\frac{v_{N\sigma}}{v_{NN}}\right) - \partial_\sigma\left(\frac{v_{\sigma N}}{v_{NN}}\right)\right] + \frac{1}{v_{NN}}\partial_N\left(\frac{1}{v_{NN}}\right)\right\}\\ &\quad + \sum_{\sigma<N}\left[\partial_\sigma^2\varphi^+ - 2\frac{v_{N\sigma}}{v_{NN}}\partial_N\partial_\sigma\varphi^+ + \left(\frac{v_{N\sigma}}{v_{NN}}\right)^2\partial_N^2\varphi^+\right] + \left(\frac{1}{v_{NN}}\right)^2\partial_N^2\varphi^+\\ &\quad = F_2(v_N, v, \varphi^+, y) \qquad \text{in} \quad U,\end{aligned} \tag{6.16}$$

where

$$F_2(v_N, v, \varphi^+, y) = f_1(y', v_N) + f_2(y', v_N) - (\lambda/2)(-y_N v_N + v + \varphi^+).$$

Evidently, F_2 is an analytic function of its arguments. For $x \in \Omega^-$ note that $w(x) = 0$. Consequently,

$$\begin{aligned}
\partial_N \varphi^- &\left\{ \sum_{\sigma<N} \left[\frac{v_{\sigma N}}{v_{NN} - C} \partial_N \left(\frac{v_{\sigma N}}{v_{NN} - C} \right) - \partial_\sigma \left(\frac{v_{\sigma N}}{v_{NN} - C} \right) \right] \right. \\
&\left. + \frac{1}{v_{NN} - C} \partial_N \left(\frac{1}{v_{NN} - C} \right) \right\} \\
&+ \sum_{\sigma<N} \left[\partial_\sigma^2 \varphi^- - 2 \frac{v_{N\sigma}}{v_{NN} - C} \partial_N \partial_\sigma \varphi^- + \left(\frac{v_{N\sigma}}{v_{NN} - C} \right)^2 \partial_N^2 \varphi^- \right] \\
&+ \left(\frac{1}{v_{NN} - C} \right)^2 \partial_N^2 \varphi^- = F_3(v_N, \varphi^-, y) \qquad \text{in } U,
\end{aligned} \tag{6.17}$$

where

$$F_3(v_N, \varphi^-, y) = f_1(y', v_N) + f_2(y', v_N) - (\lambda/2)\varphi^-.$$

We assign the weights $s_1 = -2$ to (6.15) and $s_2 = s_3 = 0$ to (6.16) and (6.17), $t_v = 4$, $t_+ = 2$, $t_- = 2$ to v, φ^+, and φ^-, and $r_1 = -4$, $r_2 = -2$, $r_3 = -1$, respectively, to the boundary conditions of (6.14). To linearize these equations at $y = 0$, we assume $v_{NN}(0) = a > 0$. Note that $v_{N\sigma}(0) = 0$, $1 \le \sigma \le N - 1$. The terms of order three in v do not occur in the linearized equations, nor do the terms of order two in v in the last boundary condition occur. For variations $\bar{v}$, $\bar{\varphi}^+$, $\bar{\varphi}^-$ we find the system

$$\begin{aligned}
\sum_{\sigma<N} \bar{v}_{\sigma\sigma} + \frac{1}{a^2} \bar{v}_{NN} + \frac{\lambda}{2} \bar{\varphi}^+ &= 0 \\
\sum_{\sigma<N} \bar{\varphi}^+_{\sigma\sigma} + \frac{1}{a^2} \bar{\varphi}^+_{NN} &= 0 \qquad \text{in } \mathbb{R}^N_+, \\
\sum_{\sigma<N} \bar{\varphi}^-_{\sigma\sigma} + \frac{1}{(C - a)^2} \bar{\varphi}^-_{NN} &= 0
\end{aligned} \tag{6.18}$$

$$\begin{aligned}
\bar{v} &= 0 \\
\bar{\varphi}^+ - \bar{\varphi}^- &= 0 \qquad \text{on } \mathbb{R}^{N-1}. \\
\frac{1}{a} \bar{\varphi}^+_N + \frac{1}{C - a} \bar{\varphi}^-_N &= 0
\end{aligned} \tag{6.19}$$

It is easy to see that this system is elliptic and has no nonzero bounded exponential solutions, keeping in mind that $C - a > 0$.

The coerciveness of our problem, (6.15), (6.16), (6.17) with boundary conditions (6.14), is established. By Lemma 6.1, v, φ^+, and φ^- are sufficiently differentiable to apply Theorem 3.3. Hence v, φ^+, φ^- are analytic near 0 in $U \cup S$. This proves the theorem. Q.E.D.

Comments and Bibliographical Notes

The use of hodograph and Legendre transforms is well established in topics as diverse as fluid dynamics and convex bodies. For example, they occur in Friedrichs' study of cavitational flow (Friedrichs [1]) and Lewy's study of the Monge–Ampere equation (Lewy [1]). Our treatment here is an adaption of (Kinderlehrer and Nirenberg [1]). The regularity criterion Theorem 2.3 is due to Caffarelli [2]. Boundary value problems for elliptic equations as well as elliptic systems have been extensively studied. We refer to Agmon, Douglis, and Nirenberg [1, 2], Lions and Magenes [1], and Morrey [1] for example.

The topics in Sections 4–6 are drawn from Kinderlehrer, Nirenberg, and Spruck [1]. For more details about plasma type problems, we refer to Temam [1, 2] and Kinderlehrer and Spruck [1]. Lewy's elegant original proof of Theorem 3.4 consists in extending analytically $u^1_{x_1} - iu^1_{x_2}$ by employing the Riemann function of the equation $\Delta + \lambda$, when $N = 2$, in the spirit of Chapter V (Lewy [5]). This method may also be used to prove Theorem 5.3 when $N = 2$. Other generalizations and techniques for the study of higher order free boundary problems are given in (Kinderlehrer, Nirenberg, and Spruck [2]).

The existence and regularity theory for the problem of two membranes was developed by G. Vergara-Caffarelli [2, 3, 4]. The latter papers deal with nonlinear membranes.

Exercises

1. Let v be the Legendre transform of u given by (2.2) with respect to (2.1). Show that if $F(D^2u, Du, u, x)$ is elliptic along u and $G(D^2v, Dv, v, y) = F(D^2u, Du, u, x)$, then G is elliptic along v.

2. Let ψ be the zeroth order Legendre transform (2.9) of u given with respect to (2.8). Show that if F is elliptic along u and $H(D^2\psi, D\psi, \psi, y) = F(D^2u, Du, u, x)$, then H is elliptic along ψ.

3. Let $L(D)$ be a linear elliptic operator with smooth coefficients of order 2μ, $\mu \geq 1$, defined in a neighborhood U of $x = 0$. Show that the problem

$$\begin{aligned} L(D)u &= f \qquad \text{in} \quad U \cap \{y_n > 0\}, \\ D_N^h u &= g_h \qquad \text{on} \quad S = U \cap \{y_N < 0\}, \quad 1 \leq h \leq \mu, \end{aligned}$$

is coercive.

Study this question for the problem

$$\begin{aligned} L(D)u &= f \qquad \text{in} \quad U \cap \{y_N > 0\}, \\ D_N^{h_i} u &= g_i \qquad \text{on} \quad S = \{y_N = 0\} \cap U, \end{aligned}$$

where $0 \leq h_1 < h_2 < \cdots < h_\mu \leq 2m - 1$.

In Exercises 4–7 determine whether or not or under what conditions the systems below are elliptic.

4.

$$\left.\begin{aligned} \Delta u^1 + cu^2 &= f_1 \\ \left(\frac{\partial}{\partial x_k}\right)^p u^1 + \Delta u^2 &= f_2 \end{aligned}\right\} \quad \text{in} \quad \mathbb{R}^N.$$

5.

$$\left.\begin{aligned} \frac{\partial^3}{\partial x_N^3} u^1 + \sum_{\sigma < N} \frac{\partial^2}{\partial x_\sigma^2} u^2 &= f_1 \\ \sum_{\sigma < N} \frac{\partial^2}{\partial x_\sigma^2} u^1 + c \frac{\partial^2}{\partial x_N^2} u^2 &= f_2 \end{aligned}\right\} \quad \text{in} \quad \mathbb{R}^N.$$

6.

$$\left.\begin{aligned} \frac{\partial}{\partial x_1} u^1 + \frac{\partial}{\partial x_2} u^2 &= 0 \\ \frac{\partial}{\partial x_1} u^0 - u^1 &= 0 \\ \frac{\partial}{\partial x_2} u^0 - u^2 &= 0 \end{aligned}\right\} \quad \text{in} \quad \mathbb{R}^2.$$

7.

$$\left.\begin{aligned} \Delta^m u^1 + c_{12} u^2 &= f_1 \\ \frac{\partial^p}{\partial x_k^p} u^1 + c_{22}\, \Delta^n u^2 &= f_2 \end{aligned}\right\} \quad \text{in} \quad \mathbb{R}^N,$$

under various assumptions about m, p, n.

Check the coercivity of problems 8–10 below.

8.

$$\Delta u = 0 \quad \text{in} \quad x_N > 0,$$

$$au + b\frac{\partial}{\partial x_N}u = 0 \quad \text{on} \quad x_N = 0.$$

9.

$$\left.\begin{aligned} \Delta u^1 &= 0 \\ \Delta u^2 + \lambda u^2 &= 0 \end{aligned}\right\} \quad \text{in} \quad x_N > 0,$$

$$\left.\begin{aligned} u^1 - u^2 &= 0 \\ \frac{\partial}{\partial x_N}u^1 + \frac{\partial}{\partial x_N}u^2 &= 0 \end{aligned}\right\} \quad \text{on} \quad x_N = 0.$$

10.

$$\left.\begin{aligned} \Delta u^1 &= f_1 \\ \Delta u^2 + \lambda u^1 &= f_2 \end{aligned}\right\} \quad \text{in} \quad x_N > 0$$

$$u^1 = \frac{\partial}{\partial x_N}u^1 = 0 \quad \text{on} \quad x_N = 0$$

11. The boundary value problem

$$\frac{\partial}{\partial \bar{z}}u = f \quad \text{in} \quad \Omega = \{z \in \mathbb{C} : |z| < 1\},$$

$$u = g \quad \text{on} \quad \partial\Omega = \{z \in \mathbb{C} : |z| = 1\}$$

fails to meet the definition of coerciveness because the operator $\partial/\partial\bar{z} = \frac{1}{2}[\partial/\partial x_1) + i(\partial/\partial x_2)]$ does not satisfy the condition about roots. Nevertheless, the mapping

$$T\colon \ H^s(\Omega) \to H^{s-1}(\Omega) \times H^{s-(1/2)}(\partial\Omega)$$

satisfies $\ker T = \{0\}$. Show, however, that $\dim \operatorname{coker} T = \infty$. [*Hint*: $(f, g) \in \operatorname{coker} T$ if and only if

$$\int_\Omega \frac{\partial u}{\partial \bar{z}}\bar{f}\,dx + \int_0^{2\pi} u\bar{g}\,d\theta = 0 \qquad \text{for all} \quad u \in C^\infty(\overline{\Omega}).$$

12. Suppose that Γ is a C^1 hypersurface which divides a neighborhood of the origin into Ω^+ and Ω^- as in Section 6. Assume

$$\Delta u = 0 \quad \text{in} \quad \Omega^+,$$

$$\Delta u = 0 \quad \text{in} \quad \Omega^-,$$

$$\left\{\begin{aligned} u &= 0 \\ u_\nu^+ + u_\nu^- - \varphi(x) &= 0 \end{aligned}\right. \quad \text{on} \quad \Gamma,$$

where u_ν^+, u_ν^- are the normal derivatives of u evaluated on approach to Ω^+ and Ω^-, respectively, and ν is taken pointing into Ω^+. Assume that $u_\nu^+ \neq 0$ on Γ. What can be said of the regularity of Γ?

13. Prove that the hypotheses of Theorem 5.3 imply that

$$w \in C^{4,\alpha}(\Omega \cup \Gamma), \qquad \varphi^1 \in C^{3,\alpha}(U \cup S), \qquad \text{and} \qquad \varphi^2 \in C^{2,\alpha}(U \cup S).$$

14. Let $p(z)$ be a polynomial of degree 2μ with roots $a_1, \ldots, a_\mu$ in the upper half-plane and $a_{\mu+1}, \ldots, a_{2\mu}$ in the lower half-plane. Show that a general solution of

$$p(D_t)u(t) = 0, \qquad -\infty < t < \infty, \quad D_t = -i\frac{d}{dt},$$

is given by

$$u(t) = \int_\Gamma e^{izt}\frac{q(z)}{p(z)}\,dz, \qquad -\infty < t < \infty,$$

where Γ is any contour enclosing the roots $a_1, \ldots, a_{2\mu}$ of p and $q(z)$ is a polynomial with $\deg q \leq 2\mu - 1$. If $u(t) \to 0$ as $t \to +\infty$, then

$$u(t) = \int_\Gamma e^{izt}\frac{q(z)}{p^+(z)}\,dz, \qquad -\infty < t < \infty,$$

where $p^+(z) = (z - a_1)\cdots(z - a_\mu)$, $\deg q \leq \mu - 1$, and Γ is a contour enclosing $a_1, \ldots, a_\mu$. Show also that

$$D^k u(0) = 0 \qquad \text{if} \quad k + \deg q \leq \mu - 1.$$

15. Consider this variation of the problem of the confined plasma: Given $\lambda > \lambda_0$, λ_0 = smallest eigenvalue of $+\Delta$ for the Dirichlet problem in Ω, find $u \in H^1(\Omega)$ such that

$$-\Delta u + \lambda \min(u, 0) = 0 \quad \text{in} \quad \Omega,$$

$$u = 1 \quad \text{on} \quad \partial\Omega.$$

Reformulate this question in terms of "Bellman's problem," Chapter III, Exercises 8–10.

CHAPTER VII

Applications of Variational Inequalities

1. Introduction

One of the attractions of the theory of variational inequalities is its application to many questions of physical interest. Several of them are described here; specifically, a problem of lubrication (Section 2), the steady filtration of a liquid through a porous membrane in both two and three dimensions (Sections 3–6), the motion of a fluid past a given profile (Sections 7 and 8), and the small deflections of an elastic beam (Section 9). The formulations of these problems are all variational inequalities involving elliptic operators and convex sets of admissible functions of obstacle type. In the next chapter we study the Stefan problem, which is connected to the heat operator.

In several cases the problem as first given requires finding the solution of a free boundary problem which itself is not the solution of a variational inequality. A new dependent variable which is the solution of a variational inequality is then introduced by means of an auxiliary transformation. We show that this new unknown may be adopted as the solution of a Cauchy problem involving the original one.

2. A Problem in the Theory of Lubrication

In this section we present a simple mechanical problem, the lubrication of a complete journal bearing, whose resolution is facilitated by the use of variational inequalities. One seeks the pressure distribution of a thin film of lubricant, a fluid of given viscosity η, contained in the narrow clearance G between two surfaces, the shaft Σ_s and the bearing Σ_b, which are in relative motion. In the region G, p must satisfy Reynolds' equation, which will be written below in a form appropriate for us, together with certain boundary conditions.

Let us direct our attention to a case of special interest in engineering (see Fig. 1). Consider a full cylindrical bearing where Σ_s and Σ_b are portions of circular cylinders with parallel, though distinct, axes of height $2b$. We assume that Σ_b is inside Σ_s, Σ_s is fixed, and Σ_b rotates with constant angular velocity ω. Under normal conditions of operation the pressure reaches a certain minimum value p_c related to the vapor pressure of the lubricant below which the film separates from Σ_s or Σ_b forming a region of cavitation. This region of cavitation is occupied by a vapor at constant pressure p_c. We take $p_c = 0$, an approximation to atmospheric pressure. Our naive formulation of the problem must be extended to account for this phenomenon. We seek a

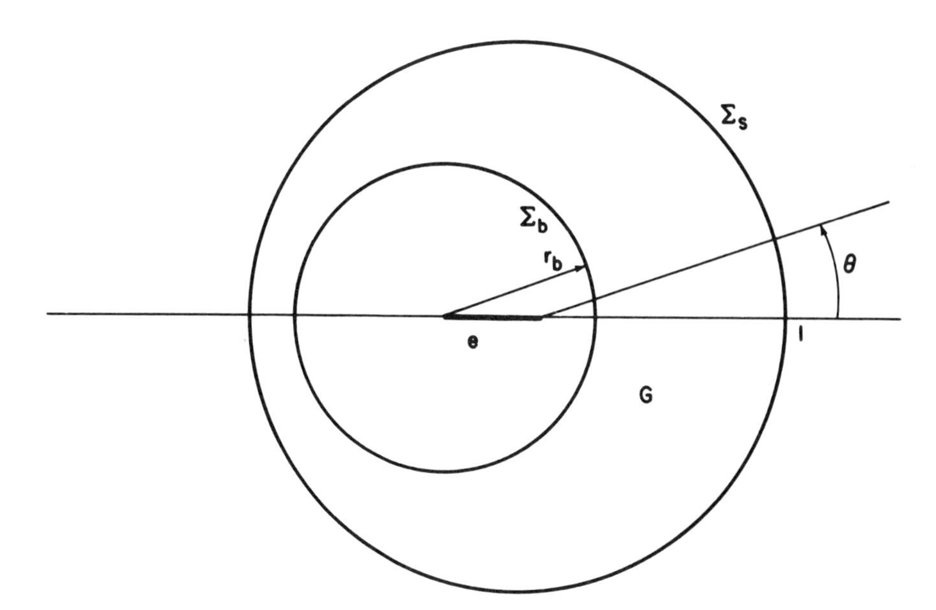

Figure 1. The cylindrical journal bearing.

function $p \geq 0$ in G satisfying Reynolds' equation where $p > 0$. Since no mass is transferred across the boundary of the set I of cavitation, $\partial p/\partial \nu = 0$ on $\partial I \cap G$, ν the normal to ∂I.

Introduce cylindrical coordinates r, θ, z with origin at the center of Σ_s and θ, $0 \leq \theta \leq 2\pi$, measured from the line of maximum clearance of the centers (Fig. 1). Since the gap between the bearing and the shaft is assumed very small, we take p to be a a function of θ and z alone, which permits a formulation in the θ, z plane. To write Reynold's equation, we introduce some parameters. Let $r_s = 1$ be the radius of Σ_s, r_b the radius of Σ_b, e the distance between the axes of Σ_s and Σ_b, and $\varepsilon = e/(1 - r_b)$, $0 \leq \varepsilon < 1$, the eccentricity ratio of the bearing. Set $\Omega = \{(\theta, z) : 0 < \theta < 2\pi, |z| < b\}$. The question we are discussing takes the form

Problem 2.1. *Find a function $p(\theta,z) \in C^1(\bar{\Omega})$ and a region $I \subset \Omega$ such*

$$A(p) = -\frac{\partial}{\partial \theta}\left(\alpha \frac{\partial p}{\partial \theta}\right) - \frac{\partial}{\partial z}\left(\alpha \frac{\partial p}{\partial z}\right) = f \quad \text{in} \quad \Omega - I, \tag{2.1}$$

$$p > 0 \quad \text{in} \quad \Omega - I, \tag{2.2}$$

$$p = \frac{\partial p}{\partial \nu} = 0 \quad \text{on} \quad \partial I \cap \Omega, \tag{2.3}$$

$$p(\theta, b) = p(\theta, -b) = 0, \qquad 0 \leq \theta \leq 2\pi,$$

and (2.4)

$$p(2\pi, z) = p(0, z), \qquad -b \leq z \leq b,$$

where

$$\alpha = (1 + \varepsilon \cos \theta)^3 \quad \text{and} \quad f = 6\omega\eta\varepsilon(1 - r_b)^{-2} \sin \theta. \tag{2.5}$$

This description, i.e., (2.3), presumes that $\partial I \cap \Omega$ is smooth, an issue to which we shall return momentarily. The first boundary condition of (2.4) means that the lubricant can flow freely at both ends from reservoirs at atmospheric pressure. The second is an obvious periodicity condition.

Suppose that $\partial I \cap \Omega$ is indeed smooth and (2.3) holds. Then setting $p = 0$ in I extends it as a $C^1(\Omega)$ function in $(\Omega - I) \cup I \subset \bar{\Omega}$. It would seem an intelligent move to replace (2.3) by

$$p = 0 \quad \text{in} \quad I.$$

In this way, $\Omega - I$ becomes precisely the set where p is positive. The conditions (2.1), (2.2), (2.3) may be summarized as

$$p \geq 0 \quad \text{and} \quad p(Ap - f) = 0 \quad \text{in} \quad \Omega.$$

A solution to this problem may be found by solving a variational inequality, more specifically, Problem 2.2 below. It will enjoy the additional property that $Ap - f \geq 0$ in Ω.

Denote by $H^1_*(\Omega)$ the subspace of $H^1(\Omega)$ whose elements v satisfy

$$\begin{aligned} v(0, z) &= v(2\pi, z), & |z| &< b, \\ v(b, \theta) &= v(-b, \theta) = 0, & 0 &\leq \theta \leq 2\pi, \end{aligned} \tag{2.6}$$

and set

$$\mathbb{K} = \{v \in H^1_*(\Omega) : v \geq 0 \text{ in } \Omega\}. \tag{2.7}$$

A solution to Problem 2.1 is given by the solution p of

Problem 2.2. *Find* $p \in \mathbb{K}$: $a(p, v - p) \geq \int_\Omega f(v - p)\, d\theta\, dz$ *for all* $v \in \mathbb{K}$, *where*

$$a(u, v) = \int_\Omega \alpha(u_\theta v_\theta + u_z v_z)\, d\theta\, dz \tag{2.8}$$

and α *and* f *are defined by* (2.5).

It is easy to check that $a(u, v)$ is coercive on $H^1_*(\Omega)$. Therefore by Theorem 2.1 of Chapter II we deduce the existence of a unique solution $p(\theta, z)$ to Problem 2.2. Moreover, p is smooth in $\overline{\Omega}$. To verify this in the most elementary manner, transform Ω into an annulus O in the $x = (x_1, x_2)$ plane according to

$$\begin{aligned} x_1 &= (a + b + z)\cos\theta, \\ x_2 &= (a + b + z)\sin\theta \end{aligned} \tag{2.9}$$

for a fixed $a > 0$. Thus

$$O = \{x \in \mathbb{R}^2 : a < |x| < a + 2b\}$$

and the segments $z = -b$ and $z = b$ are mapped onto the circles $|x| = a$ and $|x| = a + 2b$, respectively, which bound O. The periodicity condition (2.6) merely ensures that (2.9) induces a continuous mapping of $H^1_*(\Omega)$ onto $H^1_0(O)$. In this way $\mathbb{K}$ is transformed to the closed convex set $\widetilde{\mathbb{K}} = \{v \in H^1_0(O) : v \geq 0 \text{ in } O\}$.

For a function $g(\theta, z)$ let us set $\tilde{g}(x) = g(\theta, z)$ and define

$$a_{11}(x) = \frac{\tilde{\alpha}(x)}{|x|}\left(\frac{x_1^2}{|x|^2} + x_2^2\right),$$

$$a_{12}(x) = \frac{\tilde{\alpha}(x)}{|x|}\left(\frac{x_1 x_2}{|x|^2} - x_1 x_2\right) = a_{21}(x),$$

$$a_{22}(x) = \frac{\tilde{\alpha}(x)}{|x|}\left(x_2^2 + \frac{x_1^2}{|x|^2}\right).$$

In lieu of Problem 2.2 we consider the variational inequality

$$u \in \tilde{\mathbb{K}}: \quad \int_O a_{ij} u_{x_i} (v - u)_{x_j} \, dx \geq \int_O \tilde{f}(v - u) \, dx \qquad \text{for} \quad v \in \tilde{\mathbb{K}}, \tag{2.10}$$

whose solution is $u(x) = \tilde{p}(x)$. Now $\tilde{p} \in H^{2,s}(O)$, $1 \leq s < \infty$, by Chapter IV, Theorem 2.3; hence $p \in H^{2,s}(\Omega)$, $1 \leq s < \infty$. To summarize:

Theorem 2.3. *There exists a unique solution $p(\theta, z)$ to Problem 2.2. In addition, $p \in H^{2,s}(\Omega) \cap C^{1,\lambda}(\overline{\Omega})$, $1 \leq s < \infty, 0 \leq \lambda < 1$.*

We briefly describe some properties of the solution p. One might note in particular that regardless of how slowly the inner cylinder Σ_b rotates, cavitation is always present, namely, $I \neq \varnothing$.

Theorem 2.4. *Let $p(\theta, z)$ denote the solution to Problem 2.2 and define*

$$I = \{(\theta, z) \in \Omega : p(\theta, z) = 0\},$$

the set of cavitation. Then

(i) $p(\theta, z) = p(\theta, -z)$ *and*
(ii) $I \neq \varnothing$.

Proof. To prove (i) simply observe that $u(\theta, z) = p(\theta, -z)$ is a solution of Problem 2.2. By uniqueness, we conclude that $p(\theta, z) = p(\theta, -z)$.

To prove (ii) assume $I = \varnothing$. Then p is the unique solution of the Dirichlet problem

$$\begin{aligned} Ap &= f && \text{in} \quad \Omega, \\ p &= 0 && \text{for} \quad z = b \quad \text{and} \quad z = -b \\ p(0, z) &= p(2\pi, z) && \text{for} \quad -b \leq z \leq b. \end{aligned} \tag{2.11}$$

An elementary calculation shows that $u(\theta, z) = -p(2\pi - \theta, z)$ is also a solution of (2.11), whence $u = p$ in Ω. In this event, however, $p \notin \mathbb{K}$ since there are points where it is negative. Hence $I \neq \varnothing$. Q.E.D.

Other interesting features of p are given in the exercises. Let us note especially the dependence of the solution on the eccentricity parameter ε. To this end, we rewrite α and f in (2.5) as

$$\alpha = \alpha_\varepsilon = 1 + \varepsilon\beta(\theta, z, \varepsilon)$$

and (2.12)

$$f = f_\varepsilon = \varepsilon f_0$$

and let p_0 denote the solution of

$$p_0 \in \mathbb{K}: \quad \int_\Omega \left[\frac{\partial}{\partial \theta} p_0 \frac{\partial}{\partial \theta} (v - p_0) + \frac{\partial}{\partial z} p_0 \frac{\partial}{\partial z} (v - p_0) \right] d\theta \, dz$$

$$\geq \int_\Omega f_0(v - p_0) \, d\theta \, dz \qquad \text{for} \quad v \in \mathbb{K}. \tag{2.13}$$

Theorem 2.5. *With p_ε the solution of Problem 2.2 for α_ε and f_ε and p_0 defined in* (2.13),

$$\|p_\varepsilon - \varepsilon p_0\|_{H^1(\Omega)} \leq C\varepsilon^2$$

for a constant $C > 0$.

The function p_0 may be regarded as the first term in an asymptotic expansion for p_ε. The proof is left as an exercise.

3. The Filtration of a Liquid through a Porous Medium

In this section we shall describe a problem of the filtration of a liquid through a porous dam assuming that the geometry of the dam is very simple. A more general problem is studied in Section 7. Consider two reservoirs of water separated by an earthen dam. This dam is to be infinite in extent and of constant cross section consisting of permeable vertical walls and a horizontal impermeable base. We are thus led to a two dimensional problem defined in the rectangle R which is the cross section of the physical configuration. The unknowns of the problem are (the cross section of) the wet portion $\Omega \subset R$ of the dam and the pressure distribution of the water. We shall offer a precise description in terms of a free boundary problem and then transform it to a variational inequality by the introduction of a new unknown function (see Fig. 2).

A mathematical model of this situation leads to

Problem 3.1. *Let a, h, $H \in \mathbb{R}$ satisfy $a > 0$ and $0 < h < H$. Find a decreasing function $y = \varphi(x)$, $0 \leq x \leq a$, with*

$$\varphi(0) = H \qquad \text{and} \qquad \varphi(a) > h$$

and a function $u(x, y)$, $(x, y) \in \overline{\Omega}$, where

$$\Omega = \{(x, y) : 0 < y < \varphi(x), 0 < x < a\},$$

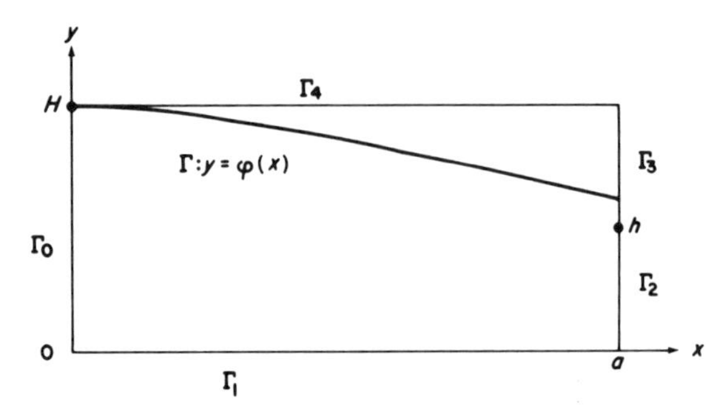

Figure 2. The porous dam in two dimensions.

satisfying

(i) $u(x, y)$ *is harmonic in* Ω *and continuous in* $\overline{\Omega}$,

$$u(0, y) = H \qquad \textit{for} \quad 0 \le y \le H,$$

$$\text{(ii)} \qquad u(a, y) = \begin{cases} h & \textit{for} \quad 0 \le y \le h \\ y & \textit{for} \quad h \le y \le \varphi(a), \end{cases}$$

$$u_y(x, 0) = 0 \qquad \textit{for} \quad 0 < x < a,$$

$$\text{(iii)} \qquad \begin{matrix} u(x, y) = y \\ \dfrac{\partial u}{\partial v}(x, y) = 0 \end{matrix} \qquad \textit{for} \quad y = \varphi(x), \quad 0 < x < a,$$

where v *is the outward normal to the curve* $y = \varphi(x)$.

The relation of u, the piezometric head of the fluid, to the pressure p is given by

$$u(x, y) = y + (1/\gamma)p(x, y), \qquad (x, y) \in \Omega,$$

where the constant of gravitation is suitably normalized and γ is the specific weight of the fluid. It must be remarked that both the domain Ω and the function $u(x, y)$ are unknown in this problem. In the unknown part of $\partial\Omega$, namely,

$$\Gamma: \quad y = \varphi(x), \qquad 0 < x < a, \tag{3.1}$$

the function $u(x, y)$ has to fulfill two requirements, Cauchy conditions, so Γ is a free boundary. Our aim is to reformulate this free boundary problem as a variational inequality. Condition (iii) imposes a smoothness criterion on Γ.

Let us assume first of all that Problem 3.1 admits a solution pair $\{\varphi, u\}$, where φ is smooth and $u \in H^1(\Omega) \cap C^0(\overline{\Omega})$. From (i), (ii), and the second line of (iii) we may prove

Lemma 3.2. *The solution u of Problem 3.1 satisfies*

$$\int_\Omega [u_x\zeta_x + u_y\zeta_y]\,dx\,dy = 0 \tag{3.2}$$

for all $\zeta \in C^1(\overline{\Omega})$ such that $\zeta = 0$ in a neighborhood of the segments $\partial\Omega \cap \{x = 0\}$ and $\partial\Omega \cap \{x = a\}$.

The proof is a consequence of the Green formula

$$-\int_\Omega \Delta u\,\zeta\,dx\,dy = \int_\Omega [u_x\zeta_x + u_y\zeta_y]\,dx\,dy - \int_{\partial\Omega} \frac{\partial u}{\partial \nu}\,\zeta\,ds.$$

We now prove

Lemma 3.3. *Let $u \in H^1(\Omega)$ be a solution of Problem 3.1. Then*

$$u(x, y) \geq y \qquad \text{in} \quad \Omega.$$

Proof. Note that physically this is nothing more than the statement that the pressure of the fluid is nonnegative. Observe first that

$$u(0, y) \geq y, \qquad u(a, y) \geq y,$$

and

$$u(x, y) = y \qquad \text{on} \quad \Gamma.$$

Choose

$$\zeta = \min(u - y, 0) = \begin{cases} 0 & \text{if} \quad u \geq y \\ u - y & \text{if} \quad u \leq y, \end{cases}$$

which is nonpositive and vanishes for $x = 0$ and $x = a$. Moreover,

$$\zeta_x = \begin{cases} 0 & \text{for} \quad u \geq y \\ u_x & \text{for} \quad u \leq y \end{cases}$$

and

$$\zeta_y = \begin{cases} 0 & \text{for} \quad u \geq y \\ u_y - 1 & \text{for} \quad u \leq y. \end{cases}$$

We apply the preceding lemma, observing that $\zeta = 0$ on Γ, so that

$$
\begin{aligned}
0 &= \int_\Omega [u_x\zeta_x + u_y\zeta_y]\,dx\,dy = \int_{\{u\le y\}} (\zeta_x^2 + \zeta_y^2 + \zeta_y)\,dx\,dy \\
&= \int_\Omega (\zeta_x^2 + \zeta_y^2 + \zeta_y)\,dx\,dy \\
&= \int_\Omega (\zeta_x^2 + \zeta_y^2)\,dx\,dy + \int_0^a [\zeta(x, \varphi(x)) - \zeta(x, 0)]\,dx \\
&= \int_\Omega (\zeta_x^2 + \zeta_y^2)\,dx\,dy - \int_0^a \zeta(x, 0)\,dx \\
&\ge \int_\Omega (\zeta_x^2 + \zeta_y^2)\,dx\,dy
\end{aligned}
$$

using here that $\zeta \le 0$. It follows that $\zeta = 0$ in Ω, so that $u \ge y$ in Ω. Q.E.D.

For a fixed $x \in (0, a)$, the flux of the dam, or its discharge, is given by

$$-\int_0^{\varphi(x)} u_x(x, t)\,dt.$$

Lemma 3.4. *Let u be a solution of Problem* 3.1. *Then*

$$-\int_0^{\varphi(x)} u_x(x, t)\,dt = \frac{1}{2a}(H^2 - h^2) \qquad \text{for} \quad 0 < x < a. \tag{3.3}$$

Proof. In fact, choosing $\zeta = \zeta(x)$, a function of x alone which vanishes at $x = 0$ and $x = a$, we have by Lemma 3.2

$$0 = \int_0^a \zeta'(x)\left[\int_0^{\varphi(x)} u_x(x, t)\,dt\right] dx.$$

Thus

$$\int_0^{\varphi(x)} u_x(x, t)\,dt = \text{const} = k,$$

but

$$k = \frac{1}{a}\int_0^a \int_0^{\varphi(x)} u_x(x, t)\,dt\,dx = \frac{1}{a}\int_{\partial\Omega} u(x, t)\,dt = \frac{1}{2a}(h^2 - H^2). \quad \text{Q.E.D.}$$

Unlike the solution $p(\theta, z)$ of the lubrication problem, $u(x, y)$ does not itself resolve a variational inequality. What properties can it enjoy in the rectangle

$$R = \{(x, y) : 0 < x < a, 0 < y < H\}? \tag{3.4}$$

The function $v(x, y) = u(x, y) - y$ is continuous in $\bar{\Omega}$ and vanishes on Γ. Let us define v continuously $R - \Omega$ by setting it equal to 0 there. The extended v is in $H^1(R)$, so for $\zeta \in C_0^\infty(R)$

$$\begin{aligned}\int_R [v_x \zeta_x + v_y \zeta_y]\, dx\, dy &= \int_\Omega [v_x \zeta_x + v_y \zeta_y]\, dx\, dy \\ &= \int_\Omega [u_x \zeta_x + u_y \zeta_y]\, dx\, dy - \int_\Omega \zeta_y\, dx\, dy \\ &= -\int_\Omega \zeta_y\, dx\, dy = -\int_R I_\Omega \zeta_y\, dx\, dy\end{aligned}$$

by Lemma 3.2, where I_Ω denotes the characteristic function of Ω. What we have calculated is that

$$-\Delta(u - y) = \frac{\partial}{\partial y} I_\Omega \quad \text{in} \quad R \tag{3.5}$$

in the sense of distributions. This suggests introducing a new function $w(x, y)$, the solution of the Cauchy problem

$$\begin{aligned} w_y &= y - u \quad &&\text{in} \quad \Omega, \\ w &= 0 &&\text{on} \quad \Gamma. \end{aligned} \tag{3.6}$$

The problem may be integrated explicitly, so

$$w(x, y) = \int_y^{\varphi(x)} [u(x, t) - t]\, dt, \qquad (x, y) \in \Omega. \tag{3.7}$$

From (3.6) and the assumed smoothness of Γ, $w_x = 0$ on Γ, so the function

$$\tilde{w}(x, y) = \begin{cases} w(x, y), & (x, y) \in \Omega \\ 0, & (x, y) \in R \backslash \Omega \end{cases} \tag{3.8}$$

is of class $C^1(R)$ and satisfies

$$\tilde{w} = |\text{grad}\, \tilde{w}| = 0 \qquad \text{on} \quad R - \Omega.$$

This property suggests that $\tilde{w}$ might be the solution of a variational inequality. Continuing this formal line of argument, (3.5) leads us to suspect that

$$-\Delta \tilde{w} = -I_\Omega \qquad \text{in} \quad R,$$

so that

$$-\Delta\tilde{w}(v - \tilde{w}) \geq -(v - \tilde{w}) \qquad \text{a.e. in} \quad R$$

whenever $v \geq 0$ in R. This is a compelling statement, for we have identified the convex set $\mathbb{K}$ of competing functions for our variational inequality, namely, nonnegative v, modulo the appropriate boundary conditions. Let us identify these boundary values and demonstrate precisely that $\tilde{w}$ is the solution of a variational inequality.

From (3.6) or (3.7),

$$\begin{aligned} w(0, y) &= \tfrac{1}{2}(H - y)^2, & 0 \leq y \leq H,& \\ w(a, y) &= \begin{cases} \tfrac{1}{2}(h - y)^2, & 0 \leq y \leq h \\ 0, & h \leq y \leq \varphi(a). \end{cases} \end{aligned} \tag{3.9}$$

By (iii) and Lemma 3.4,

$$\begin{aligned} w_x(x, 0) &= \int_0^{\varphi(x)} u_x(x, t)\, dt + [u(x, \varphi(x)) - \varphi(x)]\varphi'(x) \\ &= \int_0^{\varphi(x)} u_x(x, t)\, dt = \frac{1}{2a}(h^2 - H^2), \end{aligned}$$

so

$$w(x, 0) = \tfrac{1}{2}H^2 + [1/(2a)](h^2 - H^2)x, \qquad 0 \leq x \leq a. \tag{3.10}$$

Let us denote by $g(x, y)$ the function defining the boundary values of $w(x, y)$ in Ω and extend it to an $H^{2,\infty}(R)$ function, still called g, by

$$g(x, y) = \begin{cases} \tfrac{1}{2}(H - y)^2 + [x/(2a)][(h - y)^2 - (H - y)^2], & 0 \leq y \leq h \\ \tfrac{1}{2}(H - y)^2 - [x/(2a)](H - y)^2, & h \leq y \leq H. \end{cases} \tag{3.11}$$

Note that $\tilde{w} \in H^1(\Omega)$ since $u \in H^1(\Omega)$ by assumption. Also $w_y = y - u \leq 0$ in Ω, so

$$\tilde{w} \geq 0 \quad \text{in} \quad R. \tag{3.12}$$

Lemma 3.5. *Let $\tilde{w}$ be defined by (3.8), where w denotes the solution of the Cauchy problem (3.6). Then, with I_Ω the characteristic function of Ω,*

$$\int_R [\tilde{w}_x \zeta_x + \tilde{w}_y \zeta_y]\, dx\, dy + \int_R I_\Omega \zeta\, dx\, dy = 0$$

for all $\zeta \in C^\infty(R)$ which vanish near $x = 0$, $x = a$, and $y = 0$.

Proof. Set $\zeta = \chi_y$, where χ and χ_x vanish on the bottom and on the lateral sides of R; for instance, set

$$\chi(x, y) = \int_0^y \zeta(x, t)\, dt.$$

Then

$$\begin{aligned}
\int_R [\tilde{w}_x \zeta_x + \tilde{w}_y \zeta_y]\, dx\, dy &= \int_\Omega [w_x \zeta_x + w_y \zeta_y]\, dx\, dy \\
&= \int_\Omega [w_x \chi_{xy} + w_y \chi_{yy}]\, dx\, dy \\
&= \int_\Omega [-w_{xy} \chi_x + w_y \chi_{yy}]\, dx\, dy + \int_{\partial\Omega} w_x \chi_x v_y\, ds \\
&= \int_\Omega [w_y(\chi_{xx} + \chi_{yy})]\, dx\, dy - \int_{\partial\Omega} w_y \chi_x v_x\, ds \\
&= \int_\Omega w_y\, \Delta\chi\, dx\, dy
\end{aligned}$$

after two integrations by parts, using that at least one of the two factors involved vanishes at each point of $\partial\Omega$. Continuing, using Lemma 3.2 and (3.6),

$$\begin{aligned}
\int_R [\tilde{w}_x \zeta_x + \tilde{w}_y \zeta_y]\, dx\, dy &= \int_\Omega (y - u)\, \Delta\chi\, dx\, dy \\
&= -\int_\Omega \operatorname{grad} y \cdot \operatorname{grad} \chi\, dx\, dy \\
&\quad + \int_\Omega \operatorname{grad} u \cdot \operatorname{grad} \chi\, dx\, dy \\
&= -\int_\Omega \chi_y\, dx\, dy = -\int_\Omega \zeta\, dx\, dy.
\end{aligned}$$

This proves the lemma. Q.E.D.

Keeping in mind that Ω is an unknown of our formulation, the characterization of w just given cannot be used to solve Problem 3.1. We have shown, however, that

$$-\Delta\tilde{w} = I_\Omega \quad \text{in} \quad R.$$

Now set

$$\mathbb{K} = \{v \in H^1(R) : v \geq 0 \text{ in } R \text{ and } v = g \text{ on } \partial R\}, \tag{3.13}$$

a nonvoid closed subset of $H^1(R)$ whether or not we assume the existence of u because $g \in \mathbb{K}$. By (3.9)–(3.12), $\tilde{w} \in \mathbb{K}$ also. For any $v \in \mathbb{K}$, $v - \tilde{w}$ can be approximated by functions ζ satisfying the hypotheses of Lemma 3.5; hence

$$\begin{aligned}
\int_R [\tilde{w}_x(v - \tilde{w})_x + \tilde{w}_y(v - \tilde{w})_y]\,dx\,dy &= -\int_\Omega (v - \tilde{w})\,dx\,dy \\
&= -\int_R (v - \tilde{w})\,dx\,dy \\
&\quad + \int_{R-\Omega} (v - \tilde{w})\,dx\,dy \\
&= -\int_R (v - \tilde{w})\,dx\,dy + \int_{R-\Omega} v\,dx\,dy \\
&\geq -\int_R (v - \tilde{w})\,dx\,dy
\end{aligned}$$

since $v \geq 0$ in R. We have proved

Theorem 3.6. *Let $\{\varphi, u\}$ be a solution of Problem 3.1 with $u \in H^1(\Omega) \cap C^0(\overline{\Omega})$ and φ smooth. Let w be the solution of the Cauchy problem*

$$\begin{aligned}
w_y &= y - u \quad &&\text{in} \quad \Omega, \\
w &= 0 \quad &&\text{on} \quad \Gamma:\ y = \varphi(x), \quad 0 < x < a,
\end{aligned}$$

and define

$$\tilde{w} = \begin{cases} w & \text{in } \Omega \\ 0 & \text{in } R - \Omega. \end{cases}$$

Then $\tilde{w}$ satisfies the variational inequality

$$\begin{aligned}
\tilde{w} \in \mathbb{K}:\ &\int_R [\tilde{w}_x(v - \tilde{w})_x + \tilde{w}_y(v - \tilde{w}_y)]\,dx\,dy \\
&\geq -\int_R (v - \tilde{w})\,dx\,dy \qquad \text{for all} \quad v \in \mathbb{K},
\end{aligned} \tag{3.14}$$

where $\mathbb{K}$ is defined by (3.13). Moreover,

$$\Omega = \{(x, y) : w(x, y) > 0\}.$$

A consequence is

Corollary 3.7. *If* $\{\varphi, u\}$ *is a solution of Problem* 3.1 *with* $u \in H^1(\Omega) \cap C^0(\overline{\Omega})$ *and* φ *smooth, then the solution of Problem* 3.1 *is unique.*

For otherwise there would be more than one solution of the variational inequality (3.14).

4. The Resolution of the Filtration Problem by Variational Inequalities

From this point, our object will be the discovery of a solution to our original problem by studying the variational inequality (3.14). The bilinear form associated to the Dirichlet integral

$$a(u, v) = \int_R [u_x v_x + u_y v_y]\, dx\, dy, \qquad u, v \in H^1(R),$$

which we term for brevity the Dirichlet form, is coercive on $H_0^1(R)$. So we deduce from Chapter II, Theorem 2.1 the existence of a unique solution of (3.14), which we call $w(x, y)$. In what follows, we shall prove that $u = y - w_y$ and the set $\Gamma = \partial\Omega \cap R$, $\Omega = \{(x, y) : w(x, y) > 0\}$, constitute a solution to Problem 3.1. A study of the free boundary Γ itself will be a part of this project. The proofs we give here are two dimensional in nature. Their higher dimensional generalizations appear in Section 6.

Our initial concern is the smoothness of the solution w. To apply the regularity theory of Chapter IV we must prove that a solution of the Dirichlet problem

$$\begin{aligned} -\Delta u &= f \quad \text{in} \quad R, \\ u &= h \quad \text{on} \quad \partial R \end{aligned} \tag{4.1}$$

satisfies $u \in H^{2,p}(R)$ whenever $f \in L^p(R)$ and $h \in H^{2,p}(R)$ for $2 \leq p < \infty$. This result is not immediate in a neighborhood of a vertex of R since ∂R lacks smoothness there. This problem admits a unique solution $u \in H^1(R)$, for example, by Chapter II, Theorem 2.1. Replacing u by $u - h$ we may take $h = 0$.

Extend u to the region $R^* = \{(x, y) : 0 < x < a, -H < y < 0\}$ by setting

$$\tilde{u}(x, y) = \begin{cases} u(x, y), & (x, y) \in R \\ -u(x, -y), & (x, y) \in R^*. \end{cases}$$

With $\Gamma_1 = \{(x, 0) : 0 < x < a\}$, the base of the dam, we have that $\tilde{u} \in H^1(R \cup \Gamma_1 \cup R^*)$ by the H^1 matching lemma (Chapter II, Lemma A.8). Moreover, for any x_0, $0 < x_0 < a$,

$$\tilde{u} \in H^2(R \cap B_\varepsilon(x_0, 0)) \cap H^2(R^* \cap B_\varepsilon(x_0, 0))$$

for ε small by the elliptic regularity theory. Note that

$$-\Delta\tilde{u} = \tilde{f} \qquad \text{a.e. in} \quad R \cup R^*, \tag{4.2}$$

where

$$\tilde{f}(x, y) = \begin{cases} f(x, y), & (x, y) \in R \\ -f(x, -y), & (x, y) \in R^*. \end{cases}$$

We claim that (4.2) is valid in the sense of distributions. To check this, select $\zeta \in C_0^\infty(B_\varepsilon(x_0, 0))$, $0 < x_0 < a$ and $\varepsilon > 0$ small, write $B_\varepsilon^+ = R \cap B_\varepsilon(x, 0)$, $B_\varepsilon^- = R^* \cap B_\varepsilon(x_0, 0)$, and calculate that

$$\begin{aligned}
\int_{B_\varepsilon} [\tilde{u}_x\zeta_x + \tilde{u}_y\zeta_y]\,dx\,dy &= \int_{B_\varepsilon^+} [\tilde{u}_x\zeta_x + \tilde{u}_y\zeta_y]\,dx\,dy \\
&\quad + \int_{B_\varepsilon^-} [\tilde{u}_x\zeta_x + \tilde{u}_y\zeta_y]\,dx\,dy \\
&= \int_{B_\varepsilon^+} f\zeta\,dx\,dy - \int_{|x - x_0| < \varepsilon} u_y(x, 0)\zeta(x, 0)\,dx \\
&\quad + \int_{B_\varepsilon^-} -f(x, -y)\zeta(x, y)\,dx\,dy \\
&\quad + \int_{|x - x_0| < \varepsilon} u_y(x, 0)\zeta(x, 0)\,dx \\
&= \int_{B_\varepsilon} \tilde{f}\zeta\,dx\,dy.
\end{aligned}$$

Here we have used that the traces of u_y on $\Gamma_1 \cap B_\varepsilon(x_0, 0)$ are well defined since $u \in H^2(B_\varepsilon \cap R) \cap H^2(B_\varepsilon \cap R^*)$. Hence (4.2) is valid in the sense of distributions, so $\tilde{u}$ is a solution of

$$\begin{aligned}
-\Delta\tilde{u} &= \tilde{f} \qquad \text{in} \quad R \cup R^*, \\
\tilde{u} &= 0 \qquad \text{on} \quad x = 0.
\end{aligned}$$

Now $x = 0$ is a smooth surface, so the regularity theory now informs us that $\tilde{u} \in H^{2,p}(B_\varepsilon(0, 0) \cap (R \cup R^*))$. Thus our criterion is established at $(x, y) =$

(0, 0). The treatment at the other vertices of R is identical. This ensures that the solution w to (3.14) satisfies

$$w \in H^{2,p}(R) \cap C^{1,\lambda}(\overline{R}) \qquad \text{for} \quad 1 \le p < \infty, \quad 0 < \lambda < 1.$$

We define

$$\Omega = \{(x, y) \in R : w(x, y) > 0\} \tag{4.3}$$

and observe that

$$\Delta w = 1 \quad \text{in} \quad \Omega \qquad \text{and} \qquad w = w_x = w_y = 0 \quad \text{in} \quad R - \Omega.$$

To facilitate our description of Ω, we set

$$\Gamma_0 = \{(0, y) : 0 < y < H\}, \qquad \Gamma_1 = \{(x, 0) : 0 < x < a\},$$
$$\Gamma_2 = \{(a, y) : 0 < y < h\}, \qquad \Gamma_3 = \{(a, y) : h < y < H\},$$

and

$$\Gamma_4 = \{(x, H) : 0 < x < a\}.$$

The first lemma follows by direct calculation and by noting that w assumes its minimum on $\Gamma_3 \cup \Gamma_4$.

Lemma 4.1. *With the preceding notations*

$$\begin{aligned} w_x &\le 0 \quad \text{on} \quad \Gamma_3, \qquad & w_y &\le 0 \quad \text{on} \quad \Gamma_4,\\ w_x &= 0 \quad \text{on} \quad \Gamma_4, \qquad & w_x &\le 0 \quad \text{on} \quad \Gamma_1,\\ w_y &= 0 \quad \text{on} \quad \Gamma_3, \qquad & w_y &\le 0 \quad \text{on} \quad \Gamma_2, \end{aligned} \tag{4.4}$$

and

$$w_y \le 0 \quad \text{on} \quad \Gamma_0,$$

where w denotes the solution of (3.14).

Next, we prove

Lemma 4.2. *The solution w of* (3.14) *is continuous together with its second derivatives in a neighborhood of $\Gamma_0 \cup \Gamma_1 \cup \Gamma_2$ in R and*

$$w_{xx} = 0 \quad \text{on} \quad \Gamma_0 \cup \Gamma_2 \qquad \text{and} \qquad w_{yy} = 1 \quad \text{on} \quad \Gamma_1. \tag{4.5}$$

Proof. By continuity of w and the positivity of g on $\Gamma_0 \cup \Gamma_1 \cup \Gamma_2$, there is a neighborhood in Ω of $\Gamma_0 \cup \Gamma_1 \cup \Gamma_2$. Recalling that $\Delta w = 1$ in Ω and $g \in C^{2,\lambda}(\Gamma_0 \cup \Gamma_1 \cup \Gamma_2)$, the regularity theory permits us to conclude that there is a ball $B_\varepsilon(x_0, y_0)$, ε sufficiently small, for each $(x_0, y_0) \in \Gamma_0 \cup \Gamma_1 \cup \Gamma_2$, such that $w \in C^{2,\lambda}(B_\varepsilon(x_0, y_0) \cap \overline{\Omega})$, $0 < \lambda < 1$. Hence $w_{xx} = 1 - w_{yy} = 1 - g_{yy} = 0$ on $\Gamma_0 \cup \Gamma_2$ while $w_{yy} = 1 - w_{xx} = 1 - g_{xx} = 1$ on Γ_1. Q.E.D.

The next lemma leads to intriguing conclusions about the free boundary Γ.

Lemma 4.3. *The solution w to (3.14) satisfies*

$$w_x \le 0 \quad \textit{and} \quad w_y \le 0 \quad \textit{in} \quad R. \tag{4.6}$$

Proof. We shall rely on the maximum principle. We know from our regularity theorem that $w_x, w_y \in C^{0,\lambda}(\overline{R})$, $0 < \lambda < 1$, and $w_x = w_y = 0$ in $R - \Omega$. Since $\Delta w_x = \Delta w_y = 0$ in Ω,

$$w_x \le \sup_{\partial\Omega} w_x \qquad \text{and} \qquad w_y \le \sup_{\partial\Omega} w_y.$$

Consider w_x. By the preceding lemma, $w_{xx} = 0$ on $\Gamma_0 \cup \Gamma_2$, so w_x does not achieve its maximum there in view of the Hopf maximum principle. Therefore

$$w_x \le \sup_{\Gamma \cup \overline{\Gamma}_1 \cup \overline{\Gamma}_3} w_x, \qquad \Gamma = \partial\Omega - (\Gamma_0 \cup \Gamma_1 \cup \Gamma_2 \cup \Gamma_3 \cup \Gamma_4).$$

By continuity, $w_x = 0$ on Γ whereas $w_x \le 0$ on $\overline{\Gamma}_1 \cup \overline{\Gamma}_3 \cup \overline{\Gamma}_4$. Hence $w_x \le 0$ in Ω.

Turning to consideration of w_y, its maximum cannot lie on Γ_1 since $(w_y)_y = 1 > 0$ on Γ_1. It is easily seen that $w_y \le 0$ at other points of $\partial\Omega$. Q.E.D.

For any point $P_0 = (x_0, y_0) \in R$, set

$$Q_{P_0}^+ = \{(x, y) \in \overline{R} : x > x_0, y > y_0\}$$

and

$$Q_{P_0}^- = \{(x, y) \in R : x < x_0, y < y_0\}.$$

Lemma 4.4. *If $P \in R - \Omega$, then $Q_P^+ \subset \overline{R} - \overline{\Omega}$, and if $P \in R \cap \partial\Omega$, then $Q_P^- \subset \Omega$.*

Proof. Let $P_0 \in R - \Omega$, so that $w(x_0, y_0) = 0$. Since $w_x \le 0$ and $w_y \le 0$ in R by Lemma 4.3, $w(x, y) = 0$ for $x \ge x_0$ or $y \ge y_0$, namely, $\overline{Q}_P^+ \subset \overline{R} - \Omega$. Hence $Q_P^+ \subset \overline{R} - \overline{\Omega}$.

If $P \in R \cap \partial\Omega$ and Q_P^- were not contained in Ω, there would be a point $P_0 \in Q_P^-$ where $w(x_0, y_0) = 0$. From the above $Q_{P_0}^+ \subset R - \overline{\Omega}$. But $(x, y) \in Q_{P_0}^+$ which is disjoint from $R \cap \partial\Omega$. This is a contradiction. Q.E.D.

Here is our final preparatory lemma.

Lemma 4.5. *The solution* w *of* (3.14) *satisfies*

$$w_y = 0 \qquad on \quad \Gamma_4.$$

Moreover, $\partial\Omega \cap \Gamma_4 = \varnothing$ *and* $\partial\Omega \cap R \neq \varnothing$.

Proof. Let $0 < x_0 < a$ and choose λ small enough that $0 < x_0 < x_0 + \lambda < a$. Then

$$\begin{aligned} w_y(x_0 + \lambda, H) &- w_y(x_0, H) \\ &= -\lim_{\eta \to 0^+} \frac{1}{\eta} \{[w(x_0 + \lambda, H - \eta) - w(x_0 + \lambda, H)] \\ &\quad - [w(x_0, H - \eta) - w(x_0, H)]\} \\ &= -\lim_{\eta \to 0^+} \frac{1}{\eta} \{[w(x_0 + \lambda, H - \eta) - g(x_0 + \lambda, H)] \\ &\quad - [w(x_0, H - \eta) - g(x_0, H)]\} \\ &= -\lim_{\eta \to 0^+} \frac{1}{\eta} [w(x_0 + \lambda, H - \eta) - w(x_0, H - \eta)] \\ &= -\lambda \lim_{\eta \to 0^+} \frac{1}{\eta} w_x(x_0 + \theta\lambda, H - \eta) \geq 0 \end{aligned}$$

since $w_x \leq 0$ in $\bar{R}$. Hence $w_y(x, H)$ is a nondecreasing function of x. Since it vanishes at $x = 0$ and $x = a$, necessarily, $w_y(x, H) = 0$ for $0 < x < a$.

A different proof of this statement may be adapted from Lemma 6.3.

Now if $(x_0, H) \in \partial\Omega$, then necessarily the interval

$$\sigma = \{(x, H) : 0 < x < x_0\} \subset \partial\Omega$$

for otherwise $(x_0, H) \in Q_P^+$ for some $P \in R \cap \partial\Omega$. For $(x_1, H) \in \sigma$, $\Delta w_x = 0$ near (x_1, H) and w_x assumes its minimum value 0 at (x_1, H). Hence $w_{xy}(x_1, H) < 0$. But from the first statement of the lemma, $w_{xy}(x_1, H) = w_{yx}(x_1, H) = 0$. So $\partial\Omega \cap \Gamma_4 = \varnothing$, whence $\partial\Omega \cap R \neq \varnothing$. Q.E.D.

At this stage we are in a position to define

$$\begin{gathered} \varphi(x) = \inf\{y : (x, y) \in R - \Omega\}, \qquad 0 < x < a, \\ \varphi(0) = \lim_{x \to 0^+} \varphi(x), \qquad \text{and} \qquad \varphi(a) = \lim_{x \to a^-} \varphi(x). \end{gathered} \tag{4.7}$$

The point $(x, \varphi(x)) \in \partial\Omega$, $0 < x < a$. From Lemma 4.4 $\partial\Omega \cap Q_P^+ = \varnothing$ for $P = (x, \varphi(x)) \in R - \Omega$, so φ is a nonincreasing function. In particular, the limits above exist. Finally,

$$\Omega = \{(x, y): 0 < y < \varphi(x), 0 < x < a\},$$

where we recall that Ω is *defined* by (4.3), because $w_y \le 0$.

Lemma 4.6. *The set $\partial\Omega \cap R$ does not contain segments parallel to the x or y axes. Hence φ is continuous and strictly decreasing.*

Proof. For a proof by contradiction, assume there exists a segment σ parallel to the y axis in $\partial\Omega \cap R$. Then there is an open neighborhood $U \subset \Omega$ with $\sigma \subset \partial U$ such that $w \in C^\infty(U \cup \sigma)$ and

$$\begin{aligned} \Delta w_y &= 0 && \text{in} \quad U, \\ w_y &\le 0 && \text{in} \quad U, \\ w = w_x = w_y &= 0 && \text{on} \quad \sigma \end{aligned}$$

by Lemma 4.3.

Consequently the nonpositive harmonic w_y attains its maximum on σ so, availing ourselves of the Hopf maximum principle, we deduce

$$(w_y)_x < 0 \qquad \text{on} \quad \sigma.$$

However, $(w_y)_x = w_{yx} = w_{xy} = 0$ on σ by the conditions above and the smoothness of w. Similarly, there are no segments parallel to the x axis in $\partial\Omega \cap R$. Q.E.D.

So the set $\Gamma = \partial\Omega \cap R = \partial\Omega - (\Gamma_0 \cup \Gamma_1 \cup \Gamma_2 \cup \Gamma_3 \cup \Gamma_4)$ may be described as follows:

$$\Gamma: \quad y = \varphi(x), \qquad 0 < x < a,$$

with φ defined by (4.7).

Theorem 4.7. *Let w denote the solution of* (3.14), *and set*

$$\Omega = \{(x, y) \in R : w(x, y) > 0\}.$$

Then $\Gamma = \partial\Omega \cap R$ is an analytic curve.

This theorem is a direct application of Chapter V, Theorem 1.1, with $u = w - \frac{1}{2}|z|^2$ and $\psi = -\frac{1}{2}|z|^2$, $z = x + iy$. We conclude our discussion of Problem 3.1 with

Theorem 4.8. *Problem* 3.1 *admits a unique solution pair* $\{u, \varphi\}$. *In particular, let w denote the solution of the variational inequality*

$$w \in \mathbb{K}: \quad \int_R [w_x(v - w)_x + w_y(v - w)_y]\, dx\, dy$$

$$\geq -\int_R (v - w)\, dx\, dy \qquad \text{for all} \quad v \in \mathbb{K},$$

where $\mathbb{K} = \{v \in H^1(R) : v \geq 0$ *in* R *and* $v = g$ *on* $\partial R\}$, *with* g *defined in* (3.11), *and set*

$$u = y - w_y.$$

Define $\varphi(x)$, $0 < x < a$, *by* (4.7). *Then* $\{u, \varphi\}$ *is the solution of Problem* 3.1. *Moreover*, $u \in C^{0,\lambda}(\overline{\Omega})$ *and the curve* $y = \varphi(x)$, $0 < x < a$, *admits an analytic parametrization.*

Proof. It remains to verify (3.1), from which (i), (ii), and (iii) easily follow. Let ζ be a smooth function in R which vanishes near $x = 0$ and $x = a$. Then

$$\begin{aligned}
\int_\Omega (u_x \zeta_x + u_y \zeta_y)\, dx\, dy &= \int_\Omega [-w_{xy}\zeta_x + (1 - w_{yy})\zeta_y]\, dx\, dy \\
&= \int_\Omega (-w_{xy}\zeta_x + w_{xx}\zeta_y)\, dx\, dy \\
&= \int_R (-w_{xy}\zeta_x + w_{xx}\zeta_y)\, dx\, dy \\
&= \int_R \left[\frac{\partial}{\partial x}(w_x\zeta_y) - \frac{\partial}{\partial y}(w_x\zeta_x)\right] dx\, dy \\
&= \int_{\partial R} w_x\zeta_x\, dx + w_x\zeta_y\, dy \\
&= \int_{\partial R} w_x\zeta_x\, dx = \frac{h^2 - H^2}{2a}\int_0^a \zeta_x\, dx = 0.
\end{aligned}$$

We shall not verify that $\varphi(a) > h$. Q.E.D.

5. The Filtration of a Liquid through a Porous Medium with Variable Cross Section

We renew our attention to the filtration problem by considering a question which requires three independent variables to formulate owing to the complex geometry of the dam. Specifically, we shall assume the dam to be fashioned of vertical walls of variable thickness. The major difficulty we shall encounter in posing the variational inequality is determining the boundary values at the base of the dam of the elements of the convex set of competing functions. In the previous case they were linear, but this will no longer be the case. To resolve the problem requires a deeper investigation of the free boundary.

To begin we describe the classical problem (see Figs. 3 and 4). Consider a dam M given by

$$M = \{x \in \mathbb{R}^3 : g_1(x_1) < x_2 < g_2(x_1), 0 < x_1 < a, 0 < x_3 < H\},$$

where $a > 0$, $H > 0$, and $g_j \in C^{2,\lambda}([0, a])$ for some $\lambda > 0$ satisfy

$$\begin{aligned} g_1'(0) = g_2'(0) = g_1'(a) = g_2'(a) = 0, \\ g_1(x_1) < g_2(x_1), \qquad 0 \leq x_1 \leq a. \end{aligned} \tag{5.1}$$

Let B be the base and T the top of M:

$$B = \{x : 0 < x_1 < a, g_1(x_1) < x_2 < g_2(x_1), x_3 = 0\},$$

$$T = \{x : 0 < x_1 < a, g_1(x_1) < x_2 < g_2(x_1), x_3 = H\}.$$

Assume given h, $0 < h < H$, and let $\varphi: B \to [0, H]$ be continuous and satisfy

$$\varphi(x_1, g_1(x_1)) = H \qquad \text{and} \qquad \varphi(x_1, g_2(x_1)) \geq h, \quad 0 \leq x_1 \leq a. \tag{5.2}$$

For any such φ we set

$$\Omega = \{x \in M : 0 < x_3 < \varphi(x'), x' = (x_1, x_2)\};$$

it will correspond to the wet part of the dam. We adopt these notations for the portions of its boundary:

$$G_1 = \{x \in \overline{M} : x_2 = g_1(x_1), 0 < x_1 < a, 0 < x_3 < H\},$$

$$G_2 = \{x \in \overline{M} : x_2 = g_2(x_1), 0 < x_1 < a, 0 < x_3 < h\},$$

$$G_2^+ = \{x \in \overline{M} : x_2 = g_2(x_1), h < x_3 < \varphi(x'), 0 < x_1 < a\},$$

$$S^- = \{x \in \overline{M} : x_1 = 0, g_1(0) < x_2 < g_2(0), 0 < x_3 < \varphi(0, x_2)\},$$

$$S^+ = \{x \in \overline{M} : x_1 = a, g_1(a) < x_2 < g_2(a), 0 < x_3 < \varphi(a, x_2)\},$$

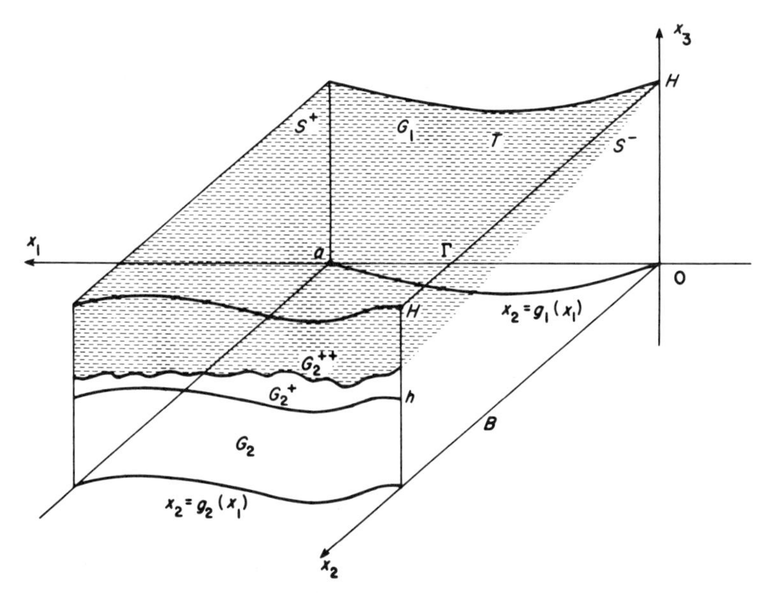

Figure 3. The three dimensional dam M.

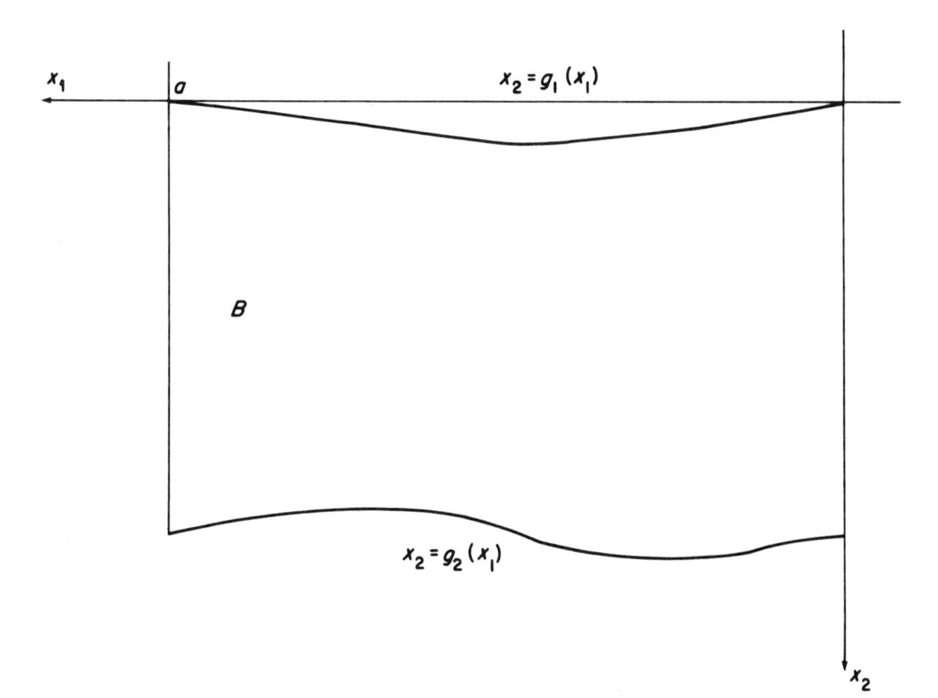

Figure 4. The base of the three dimensional dam, $M = B \times (0, H)$.

and finally,

$$\Gamma = \{x \in \overline{M} : x_3 = \varphi(x'),\ x' \in B\}.$$

Now we can state our problem.

Problem 5.1. *Find a function* $\varphi(x')$, $x' \in B$, *satisfying* (5.2) *and a function* u *defined in* $\overline{\Omega}$ *such that*

(i) $$\Delta u = 0 \quad \text{in} \quad \Omega \quad \text{and} \quad u \in C(\overline{\Omega}),$$

$$u = H \quad \text{on} \quad G_1,$$

(ii) $$u = \begin{cases} h & \text{on} \quad G_2 \\ x_3 & \text{on} \quad G_2^+, \end{cases}$$

$$\frac{\partial u}{\partial \nu} = 0 \quad \text{on} \quad B \cup S^+ \cup S^-,$$

and

(iii) $$u = x_3 \quad \text{and} \quad \frac{\partial u}{\partial \nu} = 0 \quad \text{on} \quad \Gamma.$$

As is customary, in the statement of the problem we suppose that φ is smooth enough that (iii) makes sense. We shall rephrase this problem as a variational inequality in which φ does not appear explicitly, analogous to our treatment of Problem 3.1. Indeed, just as in (3.6), we assume the existence of a solution $u \in H^1(\Omega)$ of Problem 5.1 and define a new function $w(x)$, the solution of the Cauchy problem

$$w_{x_3}(x) = x_3 - u(x), \quad x \in \Omega; \qquad w(x) = 0, \quad x \in \Gamma. \tag{5.3}$$

Although $w(x)$ may be expressed as an integral, its explicit form is not required. We extend w to all of M by setting it equal to zero in $M - \Omega$. Our assumption that Γ is smooth implies that the resulting $w \in C^1(\overline{M})$ and

$$w = |\text{grad}\ w| = 0 \quad \text{in} \quad M - \Omega.$$

Before proceeding it will be helpful to review a maximum principle for a mixed boundary value problem. Let $\Omega \subset \mathbb{R}^N$ be a bounded domain with a Lipschitz boundary $\partial\Omega$ and suppose that Γ_1 and Γ_2 are open subsets of $\partial\Omega$ satisfying

$$\Gamma_1 \cap \Gamma_2 = \varnothing, \qquad \overline{\Gamma}_1 \cup \Gamma_2 = \partial\Omega. \tag{5.4,i}$$

Finally, suppose that Γ_1 is "sufficiently large" that the Dirichlet form is coercive on the functions which vanish on Γ_1, namely,

$$\|\zeta\|_{H^1(\Omega)} \le C\|\zeta_x\|_{L^2(\Omega)} \qquad \text{for} \quad \zeta = 0 \quad \text{on} \quad \Gamma_1, \tag{5.4,ii}$$

$\zeta \in H^1(\Omega)$, for a constant $C > 0$ (independent of ζ).

It is easy to verify the existence of a unique solution $u \in H^1(\Omega)$ of the problem

$$\begin{aligned} -\Delta u &= f && \text{in} \quad \Omega, \\ u &= g && \text{on} \quad \Gamma_1, \\ \frac{\partial u}{\partial \nu} &= h && \text{on} \quad \Gamma_2 \end{aligned} \tag{5.5}$$

given $f \in L^2(\Omega)$, $g \in H^1(\Omega)$, and $h \in L^2(\Gamma_2)$, where ν denotes the outward normal of Γ_2. Indeed, with V_g and V_0 the subspaces of $H^1(\Omega)$ functions v such that $v = g$ on Γ_1 and $v = 0$ on Γ_1, respectively, u is the solution of the weak problem

$$u \in V_g\colon \quad \int_\Omega u_{x_i}\zeta_{x_i}\, dx = \int_\Omega f\zeta\, dx + \int_{\Gamma_2} h\zeta\, d\sigma \qquad \text{for} \quad \zeta \in V_0. \tag{5.6}$$

Existence is a consequence of Chapter II, Theorem 2.1 in view of (5.4,ii).

Proposition 5.2. *Assume that Ω satisfies* (5.4) *and let u be the solution of* (5.5), *where $f \in L^2(\Omega)$, $g \in H^1(\Omega)$, and $h \in L^2(\Gamma_2)$. If $f \le 0$ in Ω and $h \le 0$ on Γ_2, then*

$$\max_\Omega u \le \max_{\Gamma_1} g.$$

The proof of the proposition follows upon taking $\zeta = \max(u - k, 0) \in V_0$ in (5.6), where $k = \max_{\Gamma_1} g$.

Lemma 5.3. *Let w be defined by* (5.3). *Then $w \in H^1(M)$ and $w \ge 0$ and $w_{x_3} \le 0$ in M.*

Proof. It is obvious that $w \in H^1(M)$. Now $w_{x_3} = x_3 - u$ is in $H^1(\Omega)$ and satisfies

$$\begin{aligned} \Delta w_{x_3} &= 0 && \text{in} \quad \Omega, \\ w_{x_3} &= x_3 - H \le 0 && \text{on} \quad \bar{G}_1 \\ w_{x_3} &= x_3 - h \le 0 && \text{on} \quad \bar{G}_2 \\ w_{x_3} &= 0 && \text{on} \quad \bar{G}_2^+ \cup \Gamma \end{aligned}$$

whereas

$$\frac{\partial}{\partial \nu} w_{x_3} = 0 \quad \text{on} \quad S^+ \cup S^-$$

and

$$\frac{\partial}{\partial \nu} w_{x_3} = -1 \qquad \text{on} \quad B.$$

So $w_{x_3} \leq 0$ in Ω by the proposition. The lemma follows. Q.E.D.

To delineate the convex set of admissible functions which will appear in the statement of the variational inequality, we investigate the boundary conditions which pertain to w. It is no difficulty to verify that

$$\begin{aligned} w(x) &= \tfrac{1}{2}(H - x_3)^2, \qquad && x \in G_1, \\ w(x) &= \tfrac{1}{2}(h - x_3)^2, && x \in G_2, \\ w(x) &= 0, && x \in G_2^+, \\ w_{x_1}(x) &= 0, && x \in \partial M, \quad x_1 = 0 \quad \text{or} \quad x_1 = a. \end{aligned} \tag{5.7}$$

In addition we prove

Lemma 5.4. *Let $\alpha_0(x')$, $x' \in \overline{B}$, denote the solution of the mixed problem*

$$\begin{aligned} -\Delta\alpha_0 &= 0 && \textit{in} \quad B, \\ \alpha_0 &= \tfrac{1}{2}H^2 && \textit{for} \quad x_2 = g_1(x_1), \quad 0 \leq x_1 \leq a, \\ \alpha_0 &= \tfrac{1}{2}h^2 && \textit{for} \quad x_2 = g_2(x_1), \quad 0 \leq x_1 \leq a, \\ \frac{\partial \alpha_0}{\partial x_1} &= 0 && \textit{for} \quad x_1 = 0, \quad g_1(0) < x_2 < g_2(0), \\ & && \qquad\;\; x_1 = a, \quad g_1(a) < x_2 < g_2(a). \end{aligned} \tag{5.8}$$

Then

$$\tfrac{1}{2}h^2 \leq \alpha_0 \leq \tfrac{1}{2}H^2 \qquad \textit{in} \quad B \tag{5.9}$$

and the solution w of (5.3) satisfies $w = \alpha_0$ in B.

Proof. First note that there is a Lipschitz function in B which attains the boundary values asked of α_0 on the curves $x_2 = g_1(x_1)$ and $x_2 = g_2(x_1)$, $0 \leq x_1 \leq a$. Hence (5.8) admits a unique solution $\alpha_0 \in H^1(B)$. The inequality (5.9) follows by applying Proposition 5.2 to α_0 and to $\alpha_0 - \frac{1}{2}h^2$.

Observe that

$$w_{x_1 x_3} = -u_{x_1} \qquad \text{and} \qquad w_{x_2 x_3} = -u_{x_2},$$

whence

$$w_{x_j}(x', 0) = -\int_0^{\varphi(x')} u_{x_j}(x', x_3)\, dx_3, \qquad j = 1, 2.$$

Let $\psi(x') \in C^\infty(B)$ vanish near the curves $x_2 = g_1(x_1)$ and $x_2 = g_2(x_1)$, $0 \le x_1 \le a$. Then with $\zeta(x) = \psi(x')$,

$$\begin{aligned}
&\int_B [w_{x_1}(x', 0)\psi_{x_1}(x') + w_{x_2}(x', 0)\psi_{x_2}(x')]\, dx' \\
&\quad = -\int_B \int_0^{\varphi(x')} [u_{x_1}(x', x_3)\psi_{x_1}(x') + u_{x_2}(x', x_3)\psi_{x_2}(x')]\, dx_3\, dx' \\
&\quad = -\int_B \int_0^{\varphi(x')} u_{x_j}(x)\zeta_{x_j}(x)\, dx \\
&\quad = -\int_\Omega u_{x_j}(x)\zeta_{x_j}(x)\, dx = 0
\end{aligned}$$

by Fubini's theorem and because u is a solution of Problem 5.1. So $w(x', 0)$ is a solution of (5.8). The conclusion now follows since (5.8) admits a unique solution. Q.E.D.

By an abuse of notation, let us henceforth adopt the symbols S^+ and S^- to mean

$$S^- = \{x \in \mathbb{R}^3 : x_1 = 0, g_1(0) < x_2 < g_2(0), 0 < x_3 < H\} \subset \partial M$$

and

$$S^+ = \{x \in \mathbb{R}^3 : x_1 = a, g_1(a) < x_2 < g_2(a), 0 < x_3 < H\} \subset \partial M.$$

The proof of the next statement is similar to that of Lemma 3.6 and is left to the reader.

Lemma 5.5. *Let w be defined by* (5.3). *Then*

$$\int_M w_{x_j}\zeta_{x_j}\, dx + \int_M I_\Omega \zeta\, dx = 0,$$

where I_Ω is the characteristic function of Ω, whenever $\zeta \in C^\infty(\overline{M})$ and vanishes near $\partial M - (S^+ \cup S^-)$.

Set

$$G_2^{++} = \{x : x_2 = g_2(x_1), 0 < x_1 < a, h < x_3 < H\}$$

and define

$$\alpha(x) = \begin{cases} \alpha_0(x') & \text{on} \quad \overline{B} \\ \frac{1}{2}(H - x_3)^2 & \text{on} \quad \overline{G}_1 \\ \frac{1}{2}(h - x_3)^2 & \text{on} \quad \overline{G}_2 \\ 0 & \text{on} \quad \overline{G_2^{++}} \\ 0 & \text{on} \quad \overline{T}. \end{cases} \tag{5.10}$$

The set

$$\mathbb{K} = \{v \in H^1(M) : v \geq 0 \text{ in } M \text{ and } v = \alpha \text{ in } \partial M - (S^+ \cup S^-)\} \tag{5.11}$$

is a closed convex nonempty subset of $H^1(M)$. To see that it is not void, merely extend α to a nonnegative $H^1(M)$ function. For example, let $\theta(x') \in C^\infty(\mathbb{R}^2)$ satisfy $0 \leq \theta \leq 1$, $\theta = 1$ near the curve $x_2 = g_1(x_1)$, $0 \leq x_1 \leq a$, and $\theta = 0$ near the curve $x_2 = g_2(x_1)$, $0 \leq x_1 \leq a$. Let $\psi(x_3) \in C^\infty(\mathbb{R})$ satisfy $0 \leq \psi \leq 1$, $\psi(0) = 1$, and $\psi = 0$ for $x_3 \geq h$. Now extend α to M by

$$\begin{aligned} \alpha(x) = {} & \tfrac{1}{2}[\alpha_0(x') - 2Hx_3 + x_3^2]\theta(x') \\ & + [1 - \theta(x')]\{\tfrac{1}{2}[\alpha_0(x') - 2hx_3 + x_3^2]\psi(x_3) \\ & + [\max(h - x_3, 0)]^2[1 - \psi(x_3)]\}. \end{aligned} \tag{5.11$'$}$$

Such an extension has the advantage that $\alpha \in H^{2,s}(M)$ if $\alpha_0 \in H^{2,s}(B)$, a property which will assist us in our regularity considerations. Also, $\mathbb{K} \neq \varnothing$ whether or not u exists.

Theorem 5.6. *Let $\{u(x), \varphi(x')\}$ be a solution to Problem 5.1 with $u \in H^1(\Omega) \cap C(\overline{\Omega})$ and $\varphi(x')$ smooth. Define $w(x)$ by*

$$\begin{aligned} w_{x_3} &= x_3 - u && \text{in} \quad \Omega, \\ w &= 0 && \text{on} \quad \Gamma : x_3 = \varphi(x'), \quad x' \in B, \end{aligned} \tag{5.3$'$}$$

and extend w by zero in $M - \Omega$. Then

$$w \in \mathbb{K}: \quad \int_M w_{x_i}(v - w)_{x_i}\, dx \geq -\int_M (v - w)\, dx \qquad \text{for all} \quad v \in \mathbb{K}, \tag{5.12}$$

where $\mathbb{K}$ is defined by (5.11). Moreover,

$$\Omega = \{x \in M : w(x) > 0\}.$$

Proof. By Lemmas 5.3 and 5.4, $w \in \mathbb{K}$. Now let $v \in \mathbb{K}$. From Lemma 5.5,

$$\begin{aligned}
&\int_M w_{x_i}(v - w)_{x_i}\, dx + \int_M (v - w)\, dx \\
&\quad = \int_M w_{x_i}(v - w)_{x_i}\, dx + \int_M I_\Omega(v - w)\, dx + \int_{M-\Omega} (v - w)\, dx \\
&\quad = \int_{M-\Omega} v\, dx \geq 0
\end{aligned}$$

since $v - w \in H^1(M)$ and vanishes on $\partial M - (S^+ \cup S^-)$. This proves the theorem. Q.E.D.

Corollary 5.7. *Problem* 5.1 *admits at most one solution.*

This is because, of course, the variational inequality (5.12) has a unique solution.

6. The Resolution of the Filtration Problem in Three Dimensions

As in Section 4, we reverse our strategy to seek the solution of Problem 5.1 by examining the variational inequality (5.12). Properties of the free boundary will also be discussed. Once again, our first step is to affirm the smoothness of $w(x)$.

Theorem 6.1. *There exists a unique solution $w(x)$ to the variational inequality* (5.12). *Moreover, $w \in H^{2,s}(M) \cap C^{1,\lambda}(\overline{M})$ for $1 \leq s < \infty$ and $0 < \lambda < 1$.*

The existence and uniqueness of w is a consequence of Chapter II, Theorem 2.1. To assure the smoothness of w we must verify that the solution of the problem

$$\begin{aligned}
-\Delta u &= f && \text{in} \quad M, \\
u &= h && \text{in} \quad \partial M - (S^+ \cup S^-), \\
\frac{\partial u}{\partial \nu} &= 0 && \text{on} \quad S^+ \cup S^-
\end{aligned} \tag{6.1}$$

satisfies $u \in H^{2,s}(M)$ whenever $f \in L^s(M)$ and $h \in H^{2,s}(M)$ in addition to verifying that out choice of boundary values $\alpha(x)$ defined by (5.10) is the restriction of an $H^{2,s}(M)$ function. The proofs of both of these statements are analogous to our discussion of (4.1): the difficulties occur at the edges of M and the vertices of B, but in both cases the functions may be extended to be solutions of equations in $H^{2,\infty}$ domains. Here (5.1) plays an influential role. This is sufficient to apply the elliptic regularity theory. We state these facts as lemmas, the proofs of which are left as exercises.

Lemma 6.2. *Let $\alpha_0(x')$, $x' \in B$, be the solution of (5.8). Then $\alpha_0 \in H^{2,s}(B)$ and hence $\alpha \in H^{2,s}(M)$ for α defined by (5.11').*

Lemma 6.3. *Let u be the solution of (6.1) for $f \in L^s(M)$ and $h \in H^{2,s}(M)$, $1 < s < \infty$. Then $u \in H^{2,s}(M)$.*

We define, for w the solution of (5.12),

$$\begin{aligned} u(x) &= x_3 - w_{x_3}(x), \qquad x \in M, \\ \Omega &= \{x \in M : w(x) > 0\}; \end{aligned} \tag{6.2}$$

thus $u \in C^{0,\lambda}(\overline{\Omega})$ and $\Delta u = 0$ in Ω.

Lemma 6.4. *Let w be the solution of (5.12). Then*

$$w_{x_3}(x) \le 0 \qquad \textit{for} \quad x \in M,$$

and consequently,

$$w_{x_3}(x) < 0 \qquad \textit{and} \qquad u(x) > x_3 \qquad \textit{for} \quad x \in \Omega.$$

Proof. It is convenient to observe that, since $\alpha_0 > 0$ in $\overline{B}$ and w is continuous, there is a "slab" $B \times (0, \delta) \subset \Omega$ for some $\delta > 0$. Thus $B \subset \partial\Omega$.

In Ω, w_{x_3} is harmonic. For $x \in \partial\Omega - M$, $w(x) = |\text{grad } w(x)| = 0$, so $w_{x_3}(x) = 0$. Now suppose $x \in \partial\Omega \cap M$. If $x \in \overline{G}_1 \cup \overline{G}_2 \cup \overline{G}_2^{++} \cup T$, then $w_{x_3}(x) \le 0$ either by direct evaluation or, in the case that $x \in T$, because $w \ge 0$ in M and $w(x) = 0$. If $x \in B \subset \partial\Omega$, then

$$w_{x_3x_3}(x) = 1 - [w_{x_1x_1}(x) + w_{x_2x_2}(x)] = 1 - \Delta\alpha_0(x) = 1 > 0.$$

Thus x cannot be a point where w_{x_3} achieves its maximum.

Finally, suppose that $x \in S^+ \cup S^-$, say $x \in S^-$, and $w_{x_3}(x) \ne 0$. Then, by continuity of w_{x_3}, there is a neighborhood of x in M where $w_{x_3} \ne 0$ and *per forza* a neighborhood of x in M where $w > 0$. Say that $\varepsilon > 0$ is small enough that $B_\varepsilon(x) \cap M \subset \Omega$, so $B_\varepsilon(x) \cap S^- \subset \partial\Omega$. Now $\Delta w = 1$ in $B_\varepsilon(x) \cap M$ and $w_{x_1} = 0$ on $B_\varepsilon(x) \cap S^-$, so w is smooth in $B_\varepsilon(x) \cap \overline{M}$. We may calculate now that $w_{x_1x_3} = 0$ in $B_\varepsilon(x) \cap S^-$, so, in particular, x cannot be a point where w_{x_3} attains an extremum. Therefore

$$\max_{\partial\Omega} w_{x_3} = \max_{G_1 \cup G_2 \cup G_2{}^{++} \cup T} w_{x_3} \leq 0.$$

By the maximum principle, $w_{x_3}(x) < 0$ in Ω. The lemma follows. Q.E.D.

From the lemma, w is decreasing along each line parallel to the x_3 axis. Introduce

$$\begin{aligned} \varphi(x') &= \inf\{x_3 : x = (x', x_3) \in \overline{M} - \Omega\}, \qquad x' \in B, \\ &= \inf\{x_3 : w(x', x_3) = 0\}, \qquad x' \in B. \end{aligned} \tag{6.3}$$

Continuity of w implies that

$$\varphi(x_0') \leq \liminf_{x' \to x_0'} \varphi(x'), \qquad x_0' \in B.$$

Define the point set Γ by

$$\Gamma: \quad x_3 = \varphi(x'), \qquad x' \in B. \tag{6.4}$$

Note that Ω admits the description

$$\Omega = \{x \in M : x_3 < \varphi(x'), x' \in B\}. \tag{6.5}$$

Lemma 6.5. *Let w denote the solution of* (5.12). *Then*

$$w_{x_3}(x) = 0 \qquad \textit{for} \quad x \in T.$$

Proof. Our proof must be based on an idea different from that used in Lemma 4.5. Let $v(x) = \frac{1}{2}(H - x_3)^2$. We claim that $v - w \geq 0$ in Ω. Since this difference is harmonic in Ω, we need only check its values on $\partial\Omega$. Thus

$$\begin{aligned} v(x) &= w(x) = 0 && \text{if} \quad x \in T, \\ v(x) &\geq 0 = w(x) && \text{if} \quad x \in M \cap \partial\Omega, \\ v(x) &\geq w(x) && \text{if} \quad x \in \overline{G}_1 \cup \overline{G}_2 \cup \overline{G}_2^{++}, \end{aligned} \tag{6.6}$$

and

$$v(x) = \tfrac{1}{2}H^2 \geq \alpha_0(x) = w(x) \qquad \text{if} \quad x \in B \subset \partial\Omega$$

recalling (5.9).

Finally consider $(S^+ \cup S^-) \cap \partial\Omega$. We argue similarly to the last lemma. In view of the description of Ω given in (6.5), for each $x \in (S^+ \cup S^-) \cap \partial\Omega$, we may find a ball B_δ in $\overline{\Omega}$ with $x \in \partial B_\delta$, unless $x \in \overline{(\partial\Omega \cap M)}$, where (6.6) holds by continuity. From the Hopf maximum principle, at any such x where $v - w$ attains an extremum, its normal derivative does not vanish. But since w is a solution of (5.12),

$$\frac{\partial}{\partial \nu}(v - w) = \pm \frac{\partial}{\partial x_1}(v - w) = 0 \qquad \text{on} \quad S^+ \cup S^-.$$

Hence $v - w \geq 0$ in Ω.

For $x \in M - \Omega$, $v(x) - w(x) = v(x) \geq 0$. So

$$w(x) \leq v(x) \qquad \text{in} \quad M$$

and

$$w(x) = v(x) = 0 \qquad \text{on} \quad T,$$

whence

$$0 = \frac{\partial v}{\partial x_3}(x) \leq \frac{\partial w}{\partial x_3}(x) \leq 0, \qquad x \in T. \quad \text{Q.E.D.}$$

We cannot yet show that $\{u, \varphi\}$ is a solution to our problem, but we are able to achieve this intermediate step.

Theorem 6.6. *The pair $\{u, \varphi\}$ defined in (6.2), (6.3) is a weak solution of Problem 5.1. Namely,*

$$\int_\Omega u_{x_i} \zeta_{x_i}\, dx = 0$$

for all $\zeta \in H^1(M)$ which vanish near $G_1 \cup G_2 \cup G_2^{++}$ and

$$u = H \qquad \text{on} \quad G_1,$$

$$u = \begin{cases} h & \text{on} \quad G_2 \\ x_3 & \text{on} \quad G_2^{++} \cap \bar{\Omega}, \end{cases}$$

and

$$u = x_3 \qquad \text{on} \quad \Gamma.$$

Proof. In fact, for any ζ which vanishes near $G_1 \cup G_2 \cup G_2^{++}$ we calculate that

$$\begin{aligned}
\int_\Omega u_{x_i}\zeta_{x_i}\,dx &= \int_\Omega [-w_{x_1x_3}\zeta_{x_1} - w_{x_2x_3}\zeta_{x_2} + (1 - w_{x_3x_3})\zeta_{x_3}]\,dx \\
&= \int_\Omega (-w_{x_1x_3}\zeta_{x_1} - w_{x_2x_3}\zeta_{x_2} + w_{x_1x_1}\zeta_{x_3} + w_{x_2x_2}\zeta_{x_3})\,dx \\
&= \int_M (-w_{x_1x_3}\zeta_{x_3} - w_{x_2x_3}\zeta_{x_2} + w_{x_1x_1}\zeta_{x_3} + w_{x_2x_2}\zeta_{x_3})\,dx \\
&= \int_M (w_{x_1}\zeta_{x_1x_3} + w_{x_2}\zeta_{x_2x_3} - w_{x_1}\zeta_{x_1x_3} - w_{x_2}\zeta_{x_2x_3})\,dx \\
&\quad - \int_{\partial M} (w_{x_1}\zeta_{x_1} + w_{x_2}\zeta_{x_2})\,dx_1\,dx_2 \\
&\quad + \int_{\partial M} w_{x_1}\zeta_{x_3}\,dx_2\,dx_3 + \int_{\partial M} w_{x_2}\zeta_{x_3}\,dx_1\,dx_3 \\
&= -\int_{B\cup T} (w_{x_1}\zeta_{x_1} + w_{x_2}\zeta_{x_2})\,dx_1\,dx_2 \\
&= 0
\end{aligned}$$

since $w|_B = \alpha_0$ is harmonic and $w = 0$ on T. Recall that $w_{x_1} = 0$ on $S^+ \cup S^-$ and ζ vanishes near $G_1 \cup G_2 \cup G_2^{++}$.

The other properties of u are immediate. Note especially that

$$\begin{aligned}
u(x) &> x_3, \qquad x \in \Omega, \\
u(x) &= x_3, \qquad x \in \Gamma. \quad \text{Q.E.D.}
\end{aligned}$$

To extend the last theorem, we must show that $\{\varphi, u\}$ is a classical solution to Problem 5.1, at least insofar as the smoothness of Γ is concerned. We begin our analysis with

Theorem 6.7. *Let w be the solution of* (5.12) *and define the free boundary Γ by*

$$\Gamma: \quad x_3 = \varphi(x'), \qquad x \in B.$$

Then φ is a Lipschitz function in B.

Knowing only the definition of Γ, it is not even certain that $\Gamma \cap M$ is closed in M. The theorem tells us that $\partial\Omega \cap M \subset \Gamma$. That $\Gamma = \partial\Omega \cap$ M, or alternatively that $\varphi(x') < H$ for $x \in B'$, will be a corollary of the smoothness of φ. To prove the theorem we first discuss several lemmas.

Lemma 6.8. *Extend w to $B \times (0, 2H)$ by setting*

$$\tilde{w}(x) = \begin{cases} w(x), & x \in M \\ 0, & x \notin M. \end{cases}$$

Then $\tilde{w} \in H^{2,\infty}_{\text{loc}}(B \times (0, 2H))$.

Proof. We know that $w \in H^{2,\infty}_{\text{loc}}(M)$. Our object is to show that $w_{x_i x_j} \in L^\infty$ near T, the top of M. The proof of this fact is quite easy because of Lemma 6.5. Indeed, we assert that $\tilde{w}$ is the solution of the variational inequality

$$\tilde{w} \in \tilde{\mathbb{K}}: \quad \int_D \tilde{w}_{x_i}(v - \tilde{w})_{x_i}\, dx \geq -\int_D (v - \tilde{w})\, dx \qquad \text{for} \quad v \in \tilde{\mathbb{K}},$$

where $D = B \times (0, 2H)$ and

$$\tilde{\mathbb{K}} = \{v \in H^1(D) : v \geq 0 \text{ in } D,\, v = \alpha \text{ on } \partial M - (S^+ \cup S^- \cup T),\\ v = 0 \text{ on } \partial D - \partial M\},$$

where α is defined by (5.10) as usual.

For $v \in \tilde{\mathbb{K}}$, the product

$$\frac{\partial \tilde{w}}{\partial \nu}(v - \tilde{w}) = \frac{\partial w}{\partial \nu}(v - w) = 0 \qquad \text{on} \quad \partial M.$$

This follows from the boundary conditions on $\partial M - T$, but $w_\nu(v - w) = w_{x_3}(v - w) = 0$ on T by Lemma 6.5. Therefore

$$\begin{aligned}
\int_D \tilde{w}_{x_i}(v - \tilde{w})_{x_i}\, dx &= \int_M w_{x_i}(v - w)_{x_i}\, dx \\
&= -\int_M \Delta w(v - w)\, dx + \int_{\partial M} \frac{\partial w}{\partial \nu}(v - w)\, dx \\
&= -\int_\Omega (v - w)\, dx \\
&\geq -\int_\Omega (v - w)\, dx - \int_{D-\Omega} v\, dx = -\int_D (v - w)\, dx.
\end{aligned}$$

As we suggested at the beginning of the proof, $\tilde{w} \in H^{2,\infty}_{\text{loc}}(D)$ by Chapter IV, Theorem 6.3. Q.E.D.

Lemma 6.9. *Let $x_0 \in \Gamma$ and $B_r(x_0) \subset B \times (0, 2H)$. Then there is a cone $\Lambda_r \subset \mathbb{R}^3_+ = \{x \in \mathbb{R}^3 : x_3 > 0\}$ such that*

$$\frac{\partial w}{\partial \xi}(x) = w_x(x) \cdot \xi \leq 0 \qquad \text{for} \quad x \in B_{r/2}(x_0)$$

and

$$\frac{\partial w}{\partial \xi}(x) < 0 \qquad \text{for} \quad x \in \Omega \cap B_{r/2}(x_0)$$

whenever $\xi \in \Lambda_r \cap S^2$.

Proof. Define $\zeta(x) \in C^{1,1}(\overline{B_r(x_0)}) = H^{2,\infty}(B_r(x_0))$ by

$$\zeta(x) = \begin{cases} 0, & \rho \leq r/2, \\ (4/r^2)(\rho - \frac{1}{2}r)^2, & r/2 \leq \rho \leq r, \end{cases} \qquad \rho = |x - x_0|,$$

and for $0 \neq \xi \in \mathbb{R}^3$ and $\varepsilon > 0$ set

$$v(x) = \xi \cdot w_x(x) + w(x) - \varepsilon\zeta(x), \qquad x \in B_r(x_0).$$

Since $\Delta v = \Delta w - \varepsilon\, \Delta\zeta = 1 - \varepsilon\, \Delta\zeta \geq 0$ in $\Omega \cap B_r(x_0)$ for $\varepsilon = \varepsilon(r)$ sufficiently small,

$$v(x) \leq \max_{\partial(\Omega \cap B_r(x_0))} v, \qquad x \in B_r(x_0) \cap \Omega.$$

Suppose $x \in \Gamma$. Then if $x \in M$, $w(x) = |\text{grad } w(x)| = 0$ whereas if $x \in T$, then $w(x) = |\text{grad } w(x)| = 0$ by Lemma 6.5. Thus $v \leq 0$, $x \in \Gamma$. For any other $x \in \partial(\Omega \cap B_r(0))$ we must have $x \in \partial B_r(0) \cap \Omega$. Consider two cases. First suppose that $w(x) \leq \varepsilon/2$. Since $w_{x_j}(x)$ is Lipschitz in $B_r(x_0)$ and vanishes at $x = x_0$, applying here the preceding lemma,

$$v(x) = \xi_1 w_{x_1}(x) + \xi_2 w_{x_2}(x) + \xi_3 w_{x_3}(x) + w(x) - \varepsilon\zeta(x)$$

$$\leq C_1\sqrt{\xi_1^2 + \xi_2^2}\, r + \xi_3 w_{x_3}(x) + \varepsilon/2 - \varepsilon.$$

When $\xi_3 \geq 0$, $\xi_3 w_{x_3}(x) \leq 0$ by Lemma 6.4. Choosing $\xi_1^2 + \xi_2^2$ sufficiently small, namely,

$$C_1\sqrt{\xi_1^2 + \xi_2^2}\, r \leq \varepsilon/2,$$

we obtain

$$v(x) \leq 0 \qquad \text{for} \quad \rho = r \qquad \text{and} \qquad w(x) \leq \varepsilon/2.$$

On the other hand, if $w(x) \geq \varepsilon/2 > 0$, then $w_{x_3}(x) < 0$ so that

$$\begin{aligned} v(x) &\leq \varepsilon/2 + \xi_3 w_{x_3}(x) + w(x) - \varepsilon \\ &\leq -\varepsilon/2 + \xi_3 \sup_{\{w \geq \varepsilon/2\} \cap B_r(x_0)} w_{x_3} + \sup_{B_r(x_0)} w \\ &\leq 0 \qquad \text{for} \quad \rho = r, \quad w(x) \geq \varepsilon/2 \end{aligned}$$

when ξ_3 is sufficiently large. Therefore

$$v(x) \leq 0 \qquad \text{in} \quad \Omega \cap B_r(x_0).$$

In particular, $\zeta(x) = 0$ for $\rho \leq r/2$, so

$$\xi \cdot w_x \leq \xi \cdot w_x + w \leq 0 \qquad \text{in} \quad B_{r/2}(x_0). \quad \text{Q.E.D.}$$

Proof of Theorem 6.7. Given $x_0 \in \Gamma$, choose $r > 0$ with $B_r(x_0) \subset B \times (0, 2H)$ as in the previous lemma and let $x \in \Gamma \cap B_{r/4}(x_0)$. Then w is decreasing on any ray $x + t\xi$, $0 \leq t \leq r/4$, $\xi \in \Lambda_r \cap S^2$. We may take Λ_r to be open in $\mathbb{R}^3$, so, since $w(x) = 0$, the set

$$x + \{t\xi : 0 < t < r/4, \xi \in \Lambda_r \cap S^2\} \subset M - \overline{\Omega}.$$

Similarly, suppose that $w(y) = 0$ and

$$y \in x + \{t\xi : -r/4 < t < 0, \xi \in \Lambda_r \cap S^2\}.$$

Then

$$x \in y + \{t\xi : 0 < t < r/4, \xi \in \Lambda_r \cap S^2\} \subset M - \overline{\Omega},$$

which contradicts $x \in \Gamma$. Hence

$$x + \{t\xi : -r/4 < t < 0, \xi \in \Lambda_r \cap S^2\} \subset \Omega.$$

This shows that φ is Lipschitz. Q.E.D.

Corollary 6.10. *Let w denote the solution of* (5.12) *and define u, φ by* (6.2), (6.3) *and Γ by* (6.4). *Then Γ is an analytic surface and $\{u, \varphi\}$ is a classical solution of Problem* 5.1.

Proof. The criterion Chapter VI, 2.20 applies since Γ is Lipschitz. Hence Γ is of class C^1 and $w_{x_i x_j} \in C(\Omega \cup \Gamma)$. The conclusion follows from Chapter VI, Theorem 2.1. Q.E.D.

Corollary 6.11. *With the previous notations,*

$$\varphi(x') < H \qquad \text{for} \quad x' \in B.$$

Proof. Suppose $\varphi(x_0') = H$ so that $(x_0', H) \in \Gamma \cap T$. Once again we compare $\frac{1}{2}(H - x_3)^2$ with w in Ω but now, since φ is smooth, the Hopf boundary point lemma may be applied at (x_0', H). Hence

$$\frac{\partial}{\partial x_3}\left(\frac{1}{2}(H - x_3)^2\right) - \frac{\partial}{\partial x_3} w \neq 0 \qquad \text{at} \quad (x_0', H).$$

But each term vanishes. This contradicts that $\varphi(x_0') = H$. Q.E.D.

7. Flow past a Given Profile: The Problem in the Physical Plane

The determination of the flow of a fluid past a given profile consists in ascertaining the fluid velocity at each point of the region subject to various conditions at the profile and at infinity. We illustrate here how an especially simple case may be regarded as a variational inequality. This case is the motion of a steady, irrotational, and incompressible fluid about a symmetric convex profile in $\mathbb{R}^2$. The unknown function in the variational inequality will be a transform of the original stream function, similar to the situation encountered in our study of the filtration problem, but its domain of definition will be the hodograph plane of the flow. The parameters which occur in the variational inequality will depend only on the given profile, the given fluid speed at infinity, and any estimate for the maximum of the fluid speed in the entire region. This last, in turn, may be estimated a priori (cf. Theorem 8.7).

To give a mathematical formulation of the physical problem, let there be given a closed convex profile $\mathscr{P}$ in $\mathbb{R}^2$ which is symmetric with respect to the x axis and contains the origin in its interior. Let $G = \mathbb{R}^2 - \mathscr{P}$, an open set, and let $\mathbf{v}$ be the outward normal to $\mathscr{P}$, which we assume defined everywhere on $\partial\mathscr{P}$ except perhaps at the two points A and B where $y = 0$.

Problem 7.1. *Find the velocity* $\mathbf{q} = (q_1, q_2)$ *defined in* $\bar{G}$ *and satisfying*

(i) $\mathbf{q}$ *is of class* C^1 *in* G *and continuous in* $\bar{G}$,

(ii) $\operatorname{div} \mathbf{q} = \dfrac{\partial q_1}{\partial x} + \dfrac{\partial q_2}{\partial y} = 0$ *in* G,

(iii) $\operatorname{curl} \mathbf{q} = \dfrac{\partial q_2}{\partial x} - \dfrac{\partial q_1}{\partial y} = 0$ *in* G,

(iv) $\mathbf{q} \cdot \mathbf{v} = 0$ *on* ∂G,

(v) $q_1(x, y) \to q_\infty$ *and* $q_2(x, y) \to 0$ *as* $(x, y) \to \infty$,

(vi) $q_1(x, y) = q_1(x, -y)$ *and* $q_2(x, y) = -q_2(x, -y)$ *(symmetry)*,

where $q_\infty > 0$ *is given.*

Assume for the present that a solution $\mathbf{q}$ of Problem 7.1 exists. Then there exist locally in G two functions φ and ψ such that

$$\varphi_x = \psi_y = q_1 \qquad \text{and} \qquad \varphi_y = -\psi_x = q_2 \tag{7.1}$$

from (ii) and (iii). These are called the velocity potential and the stream function, respectively, and the relations above are the statement that they satisfy the Cauchy–Riemann equations. In particular, the function of a complex variable

$$V(z) = q_1(x, y) - iq_2(x, y), \qquad z = x + iy \in G, \tag{7.2}$$

is defined and holomorphic in all of G.

Let us analyze some of the properties of V, for they lead to the existence of a solution to Problem 7.1. From (v) we see that $V(z) \to q_\infty$ as $|z| \to \infty$, hence V admits an expansion

$$V(z) = q_\infty + \sum_1^\infty \frac{a_n}{z^n} \qquad \text{near} \quad z = \infty.$$

Because of the symmetry condition (vi),

$$V(\bar{z}) = \overline{V(z)}, \tag{7.3}$$

so that $a_n \in \mathbb{R}$ for $n \geq 1$. Choosing a counterclockwise orientation for $\partial\mathscr{P} = \partial G$, we have by Cauchy's theorem

$$\int_{\partial G} V(z)\, dz = 2\pi i a_1.$$

Also,

$$\int_{\partial G} V(z)\, dz = \int_{\partial G} q_1\, dx + q_2\, dy + i \int_{\partial G} q_1\, dy - q_2\, dx.$$

The boundary condition (iv) tells us that $q_1\, dy - q_2\, dx = 0$ on ∂G, so the integral is real. This reveals that

$$a_1 = 0,$$

which is the statement that symmetric flows have no circulation about $\mathscr{P}$. Consequently, there is a single valued holomorphic function $\Phi(z)$ in G such that

$$\Phi'(z) = V(z)$$

and with the expansion

$$\Phi(z) = q_\infty z - \sum_{n=1}^\infty \frac{a_{n+1}}{nz^n} \qquad \text{near} \quad z = \infty. \tag{7.4}$$

The functions $\varphi = \operatorname{Re}\Phi$ and $\psi = \operatorname{Im}\Phi$ are the velocity potential and the stream function, respectively. They are harmonic in G and in $C^1(\bar{G})$ by virtue of the continuity of $\mathbf{q}$. In addition, $\Phi(\bar{z}) = \overline{\Phi(z)}$ in view of (7.3); thus

$$\psi(x, -y) = -\psi(x, y) \qquad \text{for} \quad z \in G$$

and, in particular,

$$\psi(x, 0) = 0, \qquad z = x \in G.$$

Also, (7.1) holds in G. Moreover, (iv) asserts that the tangential derivative of ψ vanishes on $\partial\mathscr{P}$, so that

$$\psi = 0 \qquad \text{on} \quad \partial\mathscr{P}.$$

Theorem 7.2. *Let $\partial\mathscr{P}_+ = \partial\mathscr{P} \cap \{z : \operatorname{Im} z \geq 0\}$ and assume that $\partial\mathscr{P}_+$ is of class $C^{2,\lambda}$ up to and including its end points. Then there exists a unique solution to Problem* 7.1.

We are assuming, of course, that $\mathscr{P}$ is a convex profile.

Proof. To check the uniqueness it suffices to check that ψ is unique. Note that for any two stream functions ψ and $\tilde{\psi}$, the difference $\psi - \tilde{\psi}$ is harmonic in G, vanishes on $\partial\mathscr{P}$, and tends to 0 at ∞. By the maximum principle $\psi - \tilde{\psi} = 0$.

Existence is a consequence of the Riemann mapping theorem. Let f be a conformal mapping from G onto $D = \{\zeta : |\zeta| > 1\}$ such that $f(\infty) = \infty$ and $f'(\infty) > 0$, so f is a 1 : 1 holomorphic mapping of G, $f'(z) \neq 0$, $|f(z)| \to \infty$ as $|z| \to \infty$, and $f'(z)$ converges to a positive limit as $|z| \to \infty$. Also f maps $\bar{G}$ continuously onto $\bar{D}$.

Since G is symmetric and f thus specified is unique, $f(\bar{z}) = \overline{f(z)}$. Now define

$$\Phi(z) = \frac{q_\infty}{f'(\infty)}\left[1 - \frac{1}{f(z)^2}\right], \qquad z \in G,$$

and set $V(z) = \Phi'(z)$. Then V is holomorphic in G and $V(\bar{z}) = \overline{V(z)}$ so (ii), (iii), and (vi) hold. Since

$$V(z) = \frac{q_\infty}{f'(\infty)} f'(z)\left[1 - \frac{1}{f(z)^2}\right], \qquad z \in G,$$

$V(z) \to q_\infty$ as $|z| \to \infty$. Moreover

$$\psi(z) = \operatorname{Im}\Phi(z) = [q_\infty/f'(\infty)](1 - |f(z)|^2)\operatorname{Im} f(z); \tag{7.5}$$

hence $\psi = 0$ on $\partial\mathscr{P}$.

It remains to show that $\mathbf{q} = (\psi_y, -\psi_x)$ is continuous in $\bar{G}$. Once this is known, (iv) follows from the vanishing of ψ on ∂G. We leave this detail of the behavior of conformal mappings to the reader. Q.E.D.

Let A and B denote the end points of $\partial\mathscr{P}_+$, defined in the statement of the theorem. Note that

$$\mathbf{q}(A) = \mathbf{q}(B) = 0$$

because $q_2(x, 0) = 0$ and

$$\lim_{\substack{z \to A \\ z \in \partial\mathscr{P}_+}} \mathbf{q} \cdot \mathbf{v} = 0$$

but the one-sided limits of $\mathbf{v}$ as $z \to A$ are different from $(0, \pm 1)$ by hypothesis.

We establish some properties of ψ which will be useful later.

Proposition 7.3. *Let ψ denote the stream function of Problem* 7.1. *Then $\psi(z) > 0$ for $z \in G_+ = \{z \in G : \operatorname{Im} z > 0\}$.*

Proof. Inspecting (7.5), we see that $\psi(z)$ and $\operatorname{Im} f(z)$ have the same sign. Now f maps $G \cap \{z : \operatorname{Im} z = 0\}$ onto $D \cap \{\zeta : \operatorname{Im} \zeta = 0\}$ with $f'(\infty) > 0$. It follows that $f(G_+) = \{\zeta \in D : \operatorname{Im} \zeta > 0\}$, namely, $\operatorname{Im} f(z) > 0$ for $z \in G_+$. Q.E.D.

Proposition 7.4. *Let ψ denote the stream function of Problem* 7.1. *Then $\psi_y = \operatorname{Re} V > 0$ in $G \cup \partial G$ except at $z = A, B$ (where $V = 0$).*

Proof. The harmonic ψ achieves its minimum on ∂G_+, where $\psi = 0$. Hence by the maximum principle, $\partial\psi/\partial v > 0$ on ∂G_+ except at $z = A, B$ where it is not defined. Hence $\psi_y(x, 0) > 0$ for $(x, 0) \in \partial G_+$, $(x, 0) \neq A, B$, and since $\psi = 0$ on $\partial\mathscr{P}_+$, it follows that $\psi_y > 0$ on $\partial\mathscr{P}_+$, $(x, 0) \neq A, B$. Now $\psi_y \geq 0$ on ∂G_+ and harmonic in G_+, so $\psi_y > 0$ in G_+. Q.E.D.

8. Flow past a Given Profile: Resolution by Variational Inequalities

We wish to reinterpret Problem 7.1 in terms of new variables and new unknown functions. The new independent variables will be obtained via a hodograph transform and, as one would suspect, the new unknown will be

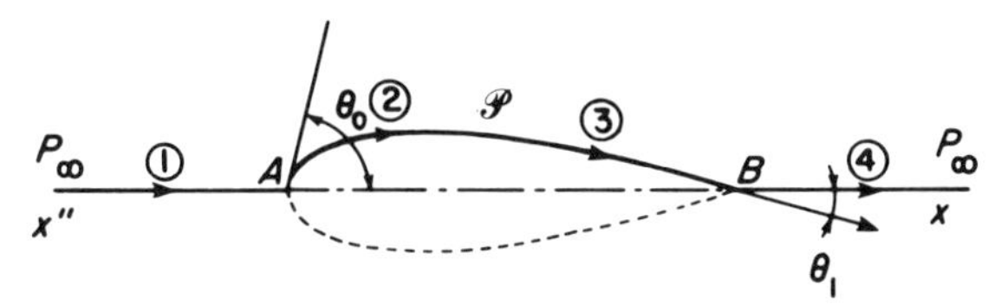

Figure 5. Physical plane.

related to its Legendre transform. Consider the hodograph mapping $z \to V(z)$. We state the theorem below, whose proof is reserved for the exercises.

Theorem 8.1. (a) *The function $q_1(x, 0)$ decreases from q_∞ to 0 as x increases from $-\infty$ to A and increases from 0 to q_∞ as x increases from B to ∞.*

(b) *The set $V(\partial\mathscr{P}_+)$ is a simple closed curve. If Σ_+ denotes the bounded open component of $\mathbb{R}^2 - V(\partial\mathscr{P}_+)$, then $(0, q_\infty] \subset \Sigma_+$ and the hodograph mapping $z \to V(z)$ is 1 : 1 from G_+ onto $\Sigma_+ - (0, q_\infty]$.*

Refer to Figs. 5 and 6. Slightly different hodograph variables will be more convenient. Let $\theta_A \in (0, \frac{1}{2}\pi]$ ($\theta_B \in [-\frac{1}{2}\pi, 0)$) be the angle determined by the x axis and the tangent to $\partial\mathscr{P}_+$ at A (B). Assuming $\partial\mathscr{P}_+$ to be strictly convex and smooth, for each $\theta \in (\theta_B, \theta_A)$, there is a unique point $P \in \partial\mathscr{P}_+$ where the tangent to $\partial\mathscr{P}_+$ at P makes an angle θ with the positive x axis. We denote the coordinates of P by $(X(\theta), Y(\theta))$, $C^{1,\lambda}$ functions of $\theta \in [\theta_B, \theta_A]$ under our hypothesis that $\partial\mathscr{P}_+$ is of class $C^{2,\lambda}$. In terms of θ, the radius of curvature $R(\theta)$ of $\partial\mathscr{P}_+$ at $(X(\theta), Y(\theta))$ is given by

$$R(\theta) = -\sqrt{X'(\theta)^2 + Y'(\theta)^2} < \infty. \tag{8.1}$$

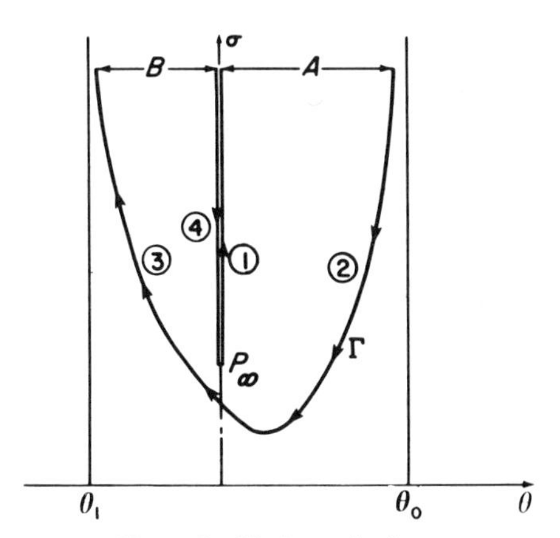

Figure 6. Hodograph plane.

For $z \in \bar{G}_+$, $z \neq A, B$, we define

$$\theta + i\sigma = W(z) = -i \log \overline{V(z)} = -\arg V(z) - i \log |V(z)|, \tag{8.2}$$

where the range of the argument is taken so that $-\theta_A < \arg V(z) < -\theta_B$. For a point $z = X(\theta) + iY(\theta) \in \partial\mathscr{P}_+$,

$$V(z) = |V(z)| e^{-i\theta} \tag{8.3}$$

in view of (iv); thus

$$W(z) = \theta - i \log |V(z)|, \qquad z \in \partial\mathscr{P}_+.$$

The curve Γ defined by $\Gamma = W(\partial\mathscr{P}_+)$ thereby admits the representation

$$\Gamma : \sigma = l(\theta), \qquad \theta_B < \theta < \theta_A, \tag{8.4}$$

where $l(\theta) = -\log |V(X(\theta) + iY(\theta))|$. Observe that $l(\theta) \in C^{1,\lambda}(\theta_B, \theta_A)$ and

$$\lim_{\theta \to \theta_B} l(\theta) = \lim_{\theta \to \theta_A} l(\theta) = +\infty.$$

It follows from Theorem 8.1 that the mapping $z \to W(z)$ is 1 : 1 from G_+ onto an open domain $\mathscr{D}$ defined by

$$\mathscr{D} = \{\theta + i\sigma : \sigma > l(\theta), \theta_B < \theta < \theta_A\} - \{i\sigma : \sigma \geq \sigma_\infty\},$$

with $\sigma_\infty = -\log q_\infty$.

We define the stream function in $\mathscr{D}$ in terms of our new variables:

$$\tilde{\psi}(\theta, \sigma) = \psi(x, y), \qquad \theta + i\sigma \in \mathscr{D}.$$

Since ψ is harmonic and W is antiholomorphic, $\tilde{\psi}$ is harmonic in $\mathscr{D}$. In addition, $\tilde{\psi}(\theta, \sigma) = 0$ on Γ, $\tilde{\psi}(0, \sigma) = 0$ for $\sigma > \sigma_\infty$, and $\tilde{\psi} \to 0$ as $\theta + i\sigma \to \infty$. At $i\sigma_\infty$, however $\tilde{\psi}$ exhibits a singularity, so it does not vanish identically. To understand this more clearly, we refer to Exercise 9, but this viewpoint is not convenient for us. Indeed, it is possible to calculate $\tilde{\psi}_\sigma$ and $\tilde{\psi}_\theta$ on Γ, which is evidence that the pair $\tilde{\psi}$, Γ is the solution of a free boundary problem, specifically,

$$\begin{aligned}
\Delta\tilde{\psi} &= 0 && \text{in } \mathscr{D}, \\
\tilde{\psi} &= 0 && \text{on } \partial\mathscr{D} - \{i\sigma_\infty\}, \\
\tilde{\psi} &= 0 && \\
\frac{\partial\tilde{\psi}}{\partial\sigma} &= \frac{R(\theta)e^{-\sigma}}{1 + (dl/d\theta)^2} && \text{on } \Gamma.
\end{aligned}$$

Although we also eschew this approach, we do wish to emphasize that the solution of the free boundary problem provides the velocity distribution of the fluid along the profile $\partial\mathscr{P}_+$ by (8.3) and (8.4). Namely,

$$l(\theta) = -\log|V(z)| = -\log|\mathbf{q}|, \quad \theta_B < \theta < \theta_A,$$

so $q_1 = e^{-l(\theta)} \cos\theta$ and $q_2 = e^{-l(\theta)} \sin\theta$. This is in contradistinction to our previous obstacle problems and physical applications, where the free boundary always represented a transition in the behavior of the solution. Here instead it gives the velocity distribution along the profile.

Analogous to our treatment of the filtration problem, we would prefer a new dependent variable which arose as the solution of a Cauchy problem. This is a schizophrenic situation given that the use of hodograph variables tempts us to introduce a Legendre transform. More precisely, let us discuss

Method I. *Introduce $u(\theta, \sigma)$, the solution of*

$$\begin{aligned} u_\sigma + u &= \tilde{\psi} && \text{in} \quad \mathscr{D}, \\ u = u_\theta &= u_\sigma = 0 && \text{on} \quad \Gamma \end{aligned} \tag{8.5}$$

and

Method II. *Introduce $u(\theta, \sigma)$, a modified Legendre transform of ψ by*

$$u(\theta, \sigma) = \psi(x, y) - x\psi_x(x, y) - y\psi_y(x, y) - e^{-\sigma}[X(\theta)\sin\theta - Y(\theta)\cos\theta]. \tag{8.6}$$

Theorem 8.2. *There is a unique solution $u(\theta, \sigma)$ to the Cauchy problem* (8.5) *and it is the Legendre transform* (8.6). *In addition,*

$$\begin{cases} -\Delta u = e^{-\sigma}R(\theta) & \text{(8.7)} \\ u > 0 & \text{(8.8)} \end{cases} \quad \text{in} \quad \mathscr{D},$$

where $R(\theta)$ is the radius of curvature (8.1).

Consequently, the two methods are equivalent. Assume the theorem for the moment and choose any $m < \min_{\theta_B < \theta < \theta_A} l(\theta)$ and let $\Omega = (\theta_B, \theta_A) \times (m, \infty)$. By (8.5), extending u as 0 for $\theta + i\sigma \in \Omega - \overline{\mathscr{D}}$ gives rise to a function in $C^1(\Omega - \{i\sigma : \sigma \geq \sigma_\infty\})$ which fulfills the relation

$$\int_\Omega [u_\theta(v - u)_\theta + u_\sigma(v - u)_\sigma]\, d\theta\, d\sigma \geq \int_\Omega e^{-\sigma}R(\theta)(v - u)\, d\theta\, d\sigma$$

for $v \geq 0$ in Ω, vanishing near $\{i\sigma : \sigma \geq \sigma_\infty\}$, and having compact support.

(8.9)

We conclude that our new u is a likely candidate to be the solution of a variational inequality. We now turn to the Proof of Theorem 8.2.

Proof. First observe that if $v(\theta, \sigma)$ is any solution of the equation

$$v_\sigma + v = f \qquad \text{in} \quad \mathscr{D},$$

then $(\partial/\partial\sigma)(e^\sigma v) = e^\sigma f$. So if $v = u(\theta, \sigma)$ and $f = \tilde{\psi} > 0$, it follows that $e^\sigma u(\theta, \sigma)$ is a strictly increasing function of σ for each θ. Since $e^\sigma u(\theta, \sigma) = 0$ on Γ, $u(\theta, \sigma) > 0$ in $\mathscr{D}$. On the other hand, if v is the difference of two solutions of (8.5) so $f = 0$, then $e^\sigma v(\theta, \sigma) = \zeta(\theta)$, a function of θ alone, and may be evaluated on Γ, where $\zeta(\theta) = 0$. Hence (8.5) admits a unique solution.

We now show that this solution is obtained from (8.6). From the definition of θ, σ given by (8.2),

$$\psi_y + i\psi_x = e^{-\sigma}(\cos\theta - i\sin\theta)$$

or

$$\psi_x = -e^{-\sigma}\sin\theta \qquad \text{and} \qquad \psi_y = e^{-\sigma}\cos\theta, \qquad \theta + i\sigma \in \mathscr{D}.$$

For the moment, set $u_0 = \psi - x\psi_x - y\psi_y$. Then

$$du_0 = -x\,d\psi_x - y\,d\psi_y$$

or

$$du_0 = e^{-\sigma}(x\cos\theta + y\sin\theta)\,d\theta + e^{-\sigma}(y\cos\theta - x\sin\theta)\,d\sigma. \tag{8.10}$$

Now $u = u_0 - e^{-\sigma}(X(\theta)\sin\theta - Y(\theta)\cos\theta)$, so by (8.10), which displays $\partial u_0/\partial\sigma$,

$$u_\sigma + u = \tilde{\psi} \qquad \text{in} \quad \mathscr{D}. \tag{8.5$'$}$$

Recalling that $x = X(\theta)$ and $y = Y(\theta)$ on Γ, it is evident that $u = 0$ on Γ. Hence $u_\sigma = 0$ on Γ since $\tilde{\psi}$ vanishes there and (8.5′) holds. Finally

$$\frac{\partial u}{\partial\theta} = e^{-\sigma}[X(\theta)\cos\theta + Y(\theta)\sin\theta] - e^{-\sigma}[X(\theta)\cos\theta + Y(\theta)\sin\theta]$$

$$- e^{-\sigma}[X'(\theta)\sin\theta - Y'(\theta)\cos\theta] = 0 \qquad \text{for} \quad \theta + i\sigma \in \Gamma$$

since $\tan\theta = Y'(\theta)/X'(\theta)$ was the defining property of θ.

The function $\psi - x\psi_x - y\psi_y$ is harmonic in G_+ and $z \to \theta + i\sigma$ is anticonformal, so $\psi - x\psi_x - y\psi_y = u_0$ is a harmonic function of θ, σ in $\mathscr{D}$. Consequently,

$$\begin{aligned} \Delta u &= -\Delta(e^{-\sigma}[X(\theta)\sin\theta - Y(\theta)\cos\theta]) \qquad \text{in} \quad \mathscr{D} \\ &= -e^{-\sigma}R(\theta) \qquad \text{in} \quad \mathscr{D}, \end{aligned}$$

keeping in mind that $X'(\theta) = R(\theta)\cos\theta$ and $Y'(\theta) = R(\theta)\sin\theta$. Q.E.D.

To complete our project we shall provide a legitimate meaning for (8.9) by determining the conditions satisfied by u "at infinity" and on the ray $\{i\sigma : \sigma \geq \sigma_\infty\}$ and by showing that u has the requisite integrability properties.

Theorem 8.3. *The function $u(\theta, \sigma)$ defined by (8.5) or (8.6) is continuous in $\overline{\mathscr{D}}$ provided we define*

$$u(0, \sigma) = He^{-\sigma}, \qquad \sigma \geq \sigma_\infty,$$

where $H = Y(0)$ is the height of the profile $\partial\mathscr{P}_+$.

Proof. Recalling the function $\Phi(z)$ from (7.4),

$$\psi - x\psi_x - y\psi_y = \mathrm{Im}(\Phi(z) - z\Phi'(z)),$$

$$\left.\begin{aligned} \Phi(z) &= q_\infty z - \sum_1^\infty \frac{a_{n+1}}{nz^n}, \\ z\Phi'(z) &= q_\infty z + \sum_1^\infty \frac{a_{n+1}}{z^n}, \end{aligned}\right\} \qquad |z| \text{ large.}$$

As $\theta + i\sigma \to i\sigma_\infty$, the corresponding $z \in G_+$ tends to ∞; thus

$$\begin{aligned} u(\theta, \sigma) &= \mathrm{Im}(\Phi(z) - z\Phi'(z)) - e^{-\sigma}[X(\theta)\sin\theta - Y(\theta)\cos\theta] \\ &\to Y(0)e^{-\sigma_\infty} = He^{-\sigma_\infty} \qquad \text{as} \quad \theta + i\sigma \to i\sigma_\infty. \end{aligned}$$

Now consider a point $i\sigma_1$ with $\sigma_1 > \sigma_\infty$. It has two preimages in ∂G_+, $(x_1, 0)$ with $x_1 > 0$ and $(x_2, 0)$ with $x_2 < 0$. In either case, when $\theta + i\sigma$ is in a neighborhood of $i\sigma_1$, the corresponding (x, y) is in a neighborhood of $(x_1, 0)$ or $(x_2, 0)$ so from (8.6),

$$u(\theta, \sigma) \to He^{-\sigma_1},$$

recalling here that $\psi_x = -e^{-\sigma}\sin\theta$. Q.E.D.

Corollary 8.4. *Let u be defined by (8.5) or (8.6) and set $H = Y(0)$, the height of the profile $\mathscr{P}$. Then*

$$0 \leq u(\theta, \sigma) \leq He^{-\sigma} \qquad \textit{for} \quad \theta + i\sigma \in \Omega, \tag{8.11}$$

and in particular, $u \in L^2(\Omega)$.

Proof. We have to show that (8.11) holds in $\mathscr{D}$. Set $v(\theta, \sigma) = He^{-\sigma}\cdot\cos\theta$, which is harmonic and positive in Ω since $-\pi/2 \leq \theta_B < \theta_A \leq \pi/2$. Thus

$$-\Delta(u - v) = -\Delta u = e^{-\sigma}R(\theta) \leq 0 \qquad \text{in} \quad \mathscr{D};$$

hence

$$\sup_{\mathscr{D}}(u - v) \le \sup_{\partial\mathscr{D}}(u - v)$$

according to the maximum principle, provided u is bounded. Reserving this detail for the end of the proof, note that

$$u - v = 0 \qquad \text{on} \quad i\sigma, \quad \sigma \ge \sigma_\infty,$$

whereas

$$u - v = -v \le 0 \qquad \text{on} \quad \Gamma.$$

Consequently (8.11) holds once we verify that u is bounded.

By continuity of u, that is, Theorem 8.3,

$$0 \le u(\theta, \sigma) \le K \qquad \text{for} \quad \theta + i\sigma \in \mathscr{D}, \quad \sigma \le 2\sigma_\infty.$$

Now for some $M > 0$,

$$\tilde{\psi}(\theta, \sigma) \le M \qquad \text{for} \quad \theta + i\sigma \in \mathscr{D}, \quad \sigma \ge 2\sigma_\infty,$$

since the points $\theta + i\sigma$ with $\sigma \ge 2\sigma_\infty$ refer to points z with $|z|$ bounded. From (8.5),

$$\frac{\partial}{\partial\sigma}(e^\sigma u) = e^\sigma \tilde{\psi} \qquad \text{in} \quad \mathscr{D},$$

so

$$0 \le u(\theta, \sigma) \le M(1 - e^{2\sigma_\infty - \sigma}) \qquad \text{in} \quad \mathscr{D}, \quad \sigma \ge 2\sigma_\infty. \quad \text{Q.E.D.}$$

Lemma 8.5. *Let u be defined by* (8.5) *or* (8.6). *Then $u \in H_0^1(\Omega)$.*

Proof. We only need to verify that $\operatorname{grad} u \in L^2(\Omega)$. To this end, let $v \in C^\infty(\Omega) \cap H_0^1(\Omega)$ be any function satisfying

$$v(0, \sigma) = He^{-\sigma} \qquad \text{for} \quad \sigma \ge \sigma_\infty,$$

$$|v(\theta, \sigma)| \le 2He^{-\sigma} \qquad \text{in} \quad \mathscr{D},$$

$$\Delta v \in L^\infty(\Omega),$$

and

$$\operatorname{supp} v \subset \mathscr{D} \cup \{i\sigma : \sigma \ge \sigma_\infty\};$$

this last condition has been imposed for convenience.

Let $\xi_0(\sigma) \in C^\infty(\mathbb{R})$ satisfy $0 \le \xi_0 \le 1$ and

$$\xi_0(\sigma) = \begin{cases} 1, & \sigma \le 0 \\ 0, & \sigma \ge 1 \end{cases}$$

and set $\xi_n(\sigma) = \xi_0(\sigma - n)$. Also let $\eta_0 \in C^\infty(\mathbb{R}^2)$ satisfy $0 \le \eta_0 \le 1$ and

$$\eta_0(\theta, \sigma) = \begin{cases} 1 & \text{for} \quad |\theta| + |\sigma| > 1 \\ 0 & \text{for} \quad |\theta| + |\sigma| < \frac{1}{2}, \end{cases}$$

and set $\eta_n(\theta, \sigma) = \eta_0(n\theta, n(\sigma - \sigma_\infty))$. Finally, set $\zeta_n(\theta, \sigma) = \xi_n(\theta, \sigma) \cdot \eta_n(\theta, \sigma)$.

After an integration by parts, we obtain

$$\int_\Omega |\operatorname{grad}(u - v)|^2 \zeta_n \, d\theta \, d\sigma = \int_{\mathscr{D}} |\operatorname{grad}(u - v)|^2 \zeta_n \, d\theta \, d\sigma$$

$$= -\int_{\mathscr{D}} \Delta(u - v)(u - v)\zeta_n \, d\theta \, d\sigma - \int_{\mathscr{D}} \operatorname{grad}(u - v) \operatorname{grad} \zeta_n (u - v) \, d\theta \, d\sigma.$$

Also we calculate that

$$\int_{\mathscr{D}} \operatorname{grad}(u - v) \operatorname{grad} \zeta_n (u - v) \, d\theta \, d\sigma = \frac{1}{2} \int_{\mathscr{D}} \operatorname{grad}(u - v)^2 \operatorname{grad} \zeta_n \, d\theta \, d\sigma$$

$$= -\frac{1}{2} \int_{\mathscr{D}} (u - v)^2 \, \Delta\zeta_n \, d\theta \, d\sigma.$$

Combining these two we have that

$$\int_\Omega |\operatorname{grad}(u - v)|^2 \zeta_n \, d\theta \, d\sigma$$

$$= -\int_{\mathscr{D}} (u - v) \, \Delta(u - v)\zeta_n \, d\theta \, d\sigma + \frac{1}{2} \int_{\mathscr{D}} (u - v)^2 \, \Delta\zeta_n \, d\theta \, d\sigma$$

$$\le \|u - v\|_{L^\infty(\Omega)} \|\Delta(u - v)\|_{L^\infty(\Omega)} + \frac{1}{2} \|u - v\|^2_{L^\infty(\Omega)} \int_{\mathscr{D}} |\Delta\zeta_n| \, d\theta \, d\sigma.$$

The right-hand side is finite by (8.11).

But $\int_{\mathscr{D}} \Delta\zeta_n \, d\theta \, d\sigma$ remains bounded as $n \to \infty$. Indeed, for n sufficiently large, the supports of grad ξ_n and grad η_n are disjoint, so

$$\Delta\zeta_n = n^2 \, \Delta\eta_0(n\theta, n(\sigma - \sigma_\infty)) + \xi_0''(\sigma - n),$$

which leads us to the estimate

$$\int_{\mathscr{D}} |\Delta\zeta_n|\, d\theta\, d\sigma \le n^2 \int_{1/(2n)<|\theta|+|\sigma-\sigma_\infty|<1/n} |\Delta\eta_0(n\theta, n(\sigma-\sigma_\infty))|\, d\theta\, d\sigma$$

$$+ \int_{n<\sigma<n+1} |\xi_0''(\sigma-n)|\, d\theta\, d\sigma$$

$$\le \pi\|\Delta\eta_0\|_{L^\infty(\mathbb{R}^2)} + \pi\|\xi_0''\|_{L^\infty(\mathbb{R})}. \quad \text{Q.E.D.}$$

We are now at the final stage of our investigation. Set

$$\mathbb{K} = \{v \in H_0^1(\Omega) : v \ge 0 \text{ in } \Omega \text{ and } v(0,\sigma) = He^{-\sigma} \text{ for } \sigma \ge \sigma_\infty\}, \quad (8.12)$$

which is a closed convex subset of $H_0^1(\Omega)$ by virtue of the continuity of the trace (cf. Chapter II, Appendix A).

Theorem 8.6. *The function u defined by* (8.5) *or* (8.6) *satisfies*

$$u \in \mathbb{K}: \quad \int_\Omega [u_\theta(v-u)_\theta + u_\sigma(v-u)_\sigma]\, d\theta\, d\sigma \ge \int_\Omega R(\theta)e^{-\sigma}(v-u)\, d\theta\, d\sigma \tag{8.13}$$

for all $v \in \mathbb{K}$.

Proof. By Theorem 8.2 and Lemma 8.5 we know that $u \in \mathbb{K}$. It is sufficient to show that (8.13) holds for $v \in \mathbb{K} \cap L^\infty(\Omega)$; the general case follows by a truncation argument. For any $v \in \mathbb{K} \cap L^\infty(\Omega)$,

$$\int_\Omega \operatorname{grad} u \operatorname{grad}(v-u)\zeta_n\, d\theta\, d\sigma$$

$$= \int_{\mathscr{D}} \operatorname{grad} u \operatorname{grad}(v-u)\zeta_n\, d\theta\, d\sigma$$

$$= -\int_{\mathscr{D}} \Delta u(v-u)\zeta_n\, d\theta\, d\sigma - \int_{\mathscr{D}} \operatorname{grad} u \operatorname{grad} \zeta_n(v-u)\, d\theta\, d\sigma$$

$$= \int_{\mathscr{D}} R(\theta)e^{-\sigma}(v-u)\zeta_n\, d\theta\, d\sigma - \int_{\mathscr{D}} \operatorname{grad} u \operatorname{grad} \zeta_n(v-u)\, d\theta\, d\sigma.$$

Now

$$\int_{\mathscr{D}} R(\theta)e^{-\sigma}(v-u)\zeta_n\, d\theta\, d\sigma \ge \int_\Omega R(\theta)e^{-\sigma}(v-u)\zeta_n\, d\theta\, d\sigma$$

since $R(\theta) < 0$, $u = 0$ on $\Omega - \mathscr{D}$, and $v \ge 0$ in Ω.

Thus to prove (8.13), it suffices to demonstrate that

$$\int_{\mathscr{D}} \operatorname{grad} u \operatorname{grad} \zeta_n(v - u)\, d\theta\, d\sigma \to 0 \qquad \text{as} \quad n \to \infty. \tag{8.14}$$

For n sufficiently large, grad ζ_n = grad ξ_n + grad η_n. Considering the contribution from ξ_n first,

$$\left| \int_{\mathscr{D}} \operatorname{grad} u \operatorname{grad} \xi_n(v - u)\, d\theta\, d\sigma \right|$$

$$= \left| \int_{\mathscr{D} \cap \{\theta + i\sigma : n \le \sigma \le n+1\}} \operatorname{grad} u \operatorname{grad} \xi_n(v - u)\, d\theta\, d\sigma \right|$$

$$\le \sqrt{\pi} \|v - u\|_{L^\infty(\Omega)} \|\xi_0'\|_{L^\infty(\Omega)} \left(\int_{n \le \sigma \le n+1} |\operatorname{grad} u|^2\, d\theta\, d\sigma \right)^{1/2} \to 0$$

as $n \to \infty$ since $|\operatorname{grad} u| \in L^2(\Omega)$.

Now for η_n we have

$$\left| \int_{\mathscr{D}} \operatorname{grad} u \operatorname{grad} \eta_n(v - u)\, d\theta\, d\sigma \right|$$

$$= \left| \int_{\mathscr{D} \cap \{\theta + i\sigma : |\theta| + |\sigma - \sigma_\infty| < 1/n\}} \operatorname{grad} u \operatorname{grad} \eta_n(v - u)\, d\theta\, d\sigma \right|$$

$$\le \sqrt{\pi}\, \|v - u\|_{L^\infty(\Omega)} n \|\operatorname{grad} \eta_0\|_{L^\infty(\Omega)} \frac{1}{n} \left(\int_{|\theta| + |\sigma - \sigma_\infty| < 1/n} |\operatorname{grad} u|^2\, d\theta\, d\sigma \right)^{1/2}$$

$$\to 0$$

as $n \to \infty$, again because $|\operatorname{grad} u| \in L^2(\Omega)$. This proves (8.14) and concludes the demonstration of the theorem. Q.E.D.

We may now consider the variational inequality (8.14), in agreement with our usual policy, which may be formulated once the parameters Ω, σ_∞, H, and $R(\theta)$ are known. Of these, H and $R(\theta)$ are functions of the profile $\mathscr{P}$ and $\sigma_\infty = -\log q_\infty$ is given. The set $\Omega = (\theta_B, \theta_A) \times (m, \infty)$ depends on $\mathscr{P}$ through θ_A and θ_B and on $m \le \min l(\theta)$. Recall that

$$l(\theta) = -\log|\mathbf{q}| \ge -\log\left(\max_{\partial\mathscr{P}_+} |\mathbf{q}| \right) > -\infty \tag{8.15}$$

because the speed of the fluid is bounded in G, and in particular, along the profile $\partial\mathscr{P}$. A careful estimate for m may be useful in computing the solution.

The variational inequality (8.13) may be solved to yield the desired solution of Problem 7.1. To this end observe that the Dirichlet form

$$a(u, v) = \int_{\Omega} [u_\theta v_\theta + u_\sigma v_\sigma]\, d\theta\, d\sigma$$

is coercive on $H_0^1(\Omega)$.

In conclusion, we observe that an estimate for $\max|\mathbf{q}|$ occurring in (8.15) may be found a priori.

Theorem 8.7. *Let* $r = \min_{\theta_B \le \theta \le \theta_A} |R(\theta)| > 0$. *Then*

$$\max_{\partial \mathscr{P}_+} |\mathbf{q}| \le q_\infty e^{-\sigma_0},$$

where $\sigma_0 > 0$ *is the unique solution of*

$$(1 + \sigma)e^{-\sigma} = 1 - (H/r).$$

The proof of this is left to the reader.

9. The Deflection of a Simply Supported Beam

A variational inequality of higher order concerns the small deflections of a stiff beam constrained to lie above a given obstacle. We shall investigate this problem restricting ourselves to the case of a one dimensional beam. Given α, $0 < \alpha < \infty$, set $\Omega = (-\alpha, \alpha) \subset \mathbb{R}^1$ and let ψ be a smooth function in $\overline{\Omega}$ satisfying

$$\max_{\Omega} \psi > 0 \qquad \text{and} \qquad \psi < 0 \qquad \text{on} \quad \partial\Omega. \tag{9.1}$$

Let $V = H^2(\Omega) \cap H_0^1(\Omega)$ and set

$$\mathbb{K} = \{v \in V : v \ge \psi \text{ in } \Omega\}, \tag{9.2}$$

a closed convex subset of V.

Define the bilinear form

$$a(v, w) = \int_{\Omega} v''(t) w''(t)\, dt, \qquad v, w \in V. \tag{9.3}$$

It is easy to check that $a(v, w)$ is coercive on V. The norm on V is given by $\|v'\|_{L^2(\Omega)} + \|v''\|_{L^2(\Omega)}$, so checking the coerciveness means showing that

$$\|v'\|_{L^2(\Omega)} \le C \|v''\|_{L^2(\Omega)}, \qquad v \in V,$$

for some $C > 0$. Indeed, for $v \in V$, $v \in C^1(\overline{\Omega})$ and $v(\alpha) = v(-\alpha) = 0$. Suppose $y \in \Omega$ is a point where $v'(y) = 0$. Then

$$v'(x)^2 = \int_y^x \frac{d}{dt} v'(t)^2 \, dt = 2 \int_y^x v'(t) v''(t) \, dt$$

$$\leq 2 \int_{-\alpha}^{\alpha} |v'(t)| \, |v''(t)| \, dt.$$

Integrating over Ω and applying Schwarz' Inequality yields

$$\|v'\|_{L^2(\Omega)} \leq 4\alpha \|v''\|_{L^2(\Omega)}. \tag{9.4}$$

Our project is to study

Problem 9.1. *Given $f \in V'$, find*

$$u \in \mathbb{K}: \quad a(u, v - u) \geq \langle f, v - u \rangle \qquad \text{for all} \quad v \in \mathbb{K}.$$

Theorem 9.2. *There exists a unique solution to Problem 9.1.*

This is a direct application of Chapter II, Theorem 2.1. Note that $u \in C^{1,1/2}(\overline{\Omega})$ since $\Omega \subset \mathbb{R}^1$. We examine some properties of the solution. For $\zeta \in C_0^\infty(\Omega)$, $\zeta \geq 0$, the function $u + \zeta \in \mathbb{K}$, so the mapping

$$\zeta \to a(u, \zeta) - \langle f, \zeta \rangle$$

is a nonnegative distribution. By the Schwarz theorem, there is a measure $\mu \geq 0$ such that

$$a(u, \zeta) = \langle f, \zeta \rangle + \int_\Omega \zeta(t) \, d\mu(t), \qquad \zeta \in C_0^\infty(\Omega), \tag{9.5}$$

and moreover,

$$\operatorname{supp} \mu \subset \{t : u(t) = \psi(t)\}.$$

From this characterization we infer some regularity properties of u. For example:

Theorem 9.3. *Let u denote the solution of Problem 9.1 with $f = 0$. Then $(d^3/dt^3)u$ is a function of bounded variation and $(d^2/dt^2)u$ is continuous in Ω.*

To prove this observe that the distribution $(d^4/dt^4)u = d\mu$ by (9.5), where $\mu \geq 0$. Thus we may find an increasing function $m(t)$ such that

$$\frac{d^3}{dt^3} u = m(t) = \mu((0, t)) + c.$$

The result follows. The astute reader will observe that u'' is actually Lipschitz in Ω.

We now direct our attention to a means of deriving smoothness properties of u by penalization. The particular penalization will involve u itself. Define $u_\varepsilon(t)$ to be the solution of the second order problem

$$\begin{aligned} u_\varepsilon - \varepsilon u_\varepsilon'' &= u \qquad \text{in} \quad \Omega \\ u_\varepsilon(\alpha) &= u_\varepsilon(-\alpha) = 0 \end{aligned} \tag{9.6}$$

where u is the solution of Problem 9.1.

Lemma 9.4. *Let u_ε be the solution of* (9.6). *Then*

$$u_\varepsilon \in \mathbb{K} \cap H^4(\Omega) \qquad \text{and} \quad u_\varepsilon''(\alpha) = u_\varepsilon''(-\alpha) = 0.$$

Proof. Since $u \in H^2(\Omega)$, $u_\varepsilon \in H^4(\Omega)$. In particular, $u_\varepsilon \in C^3(\Omega)$. At a point $x \in \Omega$ where $u_\varepsilon(x) = \min u_\varepsilon(t)$, $u_\varepsilon''(x) \geq 0$, so

$$u_\varepsilon(x) - \psi(x) = u(x) - \psi(x) + \varepsilon u_\varepsilon''(x) \geq u(x) - \psi(x) \geq 0.$$

It follows that $u_\varepsilon \in \mathbb{K}$. Q.E.D.

Lemma 9.5. *Suppose that $f \in H^{-1}(\Omega)$ and u_ε is defined by* (9.6). *Then for a constant $C > 0$*

$$\|u_\varepsilon\|_{H^3(\Omega)} \leq C\|f\|_{H^{-1}(\Omega)} \qquad \text{for} \quad 0 < \varepsilon \leq 1.$$

Proof. Write $f = f_0 - (d/dt)f_1, f_0, f_1 \in L^2(\Omega)$. Then

$$a(u, v - u) \geq \langle f, v - u\rangle = \int_\Omega [f_0(v - u) + f_1(v - u)']\, dt$$

or, invoking Minty's lemma (Chapter III, Lemma 1.5),

$$a(v, v - u) \geq \int_\Omega [f_0(v - u) + f_1(v - u)']\, dt \qquad \text{for all} \quad v \in \mathbb{K}.$$

Choosing $v = u_\varepsilon$, we have $v - u = \varepsilon u_\varepsilon''$, so

$$a(u_\varepsilon, u_\varepsilon'') \geq \int_\Omega [f_0 u_\varepsilon'' + f_1 u_\varepsilon''']\, dt.$$

Integrating by parts in the first term gives

$$a(u_\varepsilon, u_\varepsilon'') = -\int_\Omega (u_\varepsilon''')^2\, dt$$

since $u_\varepsilon''(\alpha) = u_\varepsilon''(-\alpha) = 0$. *Therefore*

$$\|u_\varepsilon'''\|_{L^2(\Omega)}^2 \leq \|f_0\|_{L^2(\Omega)}\|u_\varepsilon''\|_{L^2(\Omega)} + \|f_1\|_{L^2(\Omega)}\|u_\varepsilon'''\|_{L^2(\Omega)}.$$

Inasmuch as $u_\varepsilon'' \in H_0^1(\Omega)$, we may appeal to Poincare's lemma, whence

$$\|u_\varepsilon''\|_{L^2(\Omega)} \leq 4\alpha\|u_\varepsilon'''\|_{L^2(\Omega)}.$$

Therefore

$$\|u_\varepsilon'''\|_{L^2(\Omega)} \leq 4\alpha\|f_0\|_{L^2(\Omega)} + \|f_1\|_{L^2(\Omega)}.$$

The lemma follows observing that $u_\varepsilon \in \mathrm{V}$. Q.E.D.

The theorem below has essentially the same content as Theorem 9.1 in this one dimensional case. However, the method of proof (Lemmas 9.4 and 9.5) are available for use in higher dimensions too.

Theorem 9.6. *Let u be the solution of Problem* 9.1 *with* $f \in H^{-1}(\Omega)$. *Then* $u \in H^3(\Omega) \cap C^{2,1/2}(\overline{\Omega})$.

Proof. We need only show that $u_\varepsilon \to u$ in some sense. Indeed, by the previous lemma, $\|u_\varepsilon''\|_{L^2(\Omega)} \leq C < \infty$ for $0 < \varepsilon \leq 1$, so by (9.6)

$$\|u_\varepsilon - u\|_{L^2(\Omega)} \leq \varepsilon\|u_\varepsilon''\|_{L^2(\Omega)} \leq \varepsilon C \to 0 \quad \text{as} \quad \varepsilon \to 0. \tag{9.7}$$

There exists a subsequence u_{ε_k} which converges weakly in $H^3(\Omega)$ to some function which by (9.7) must be u. Hence $u \in H^3(\Omega)$. Q.E.D.

Comments and Bibliographical Notes

The problem of hydrodynamic lubrication was discussed by O. Reynolds in 1886 and by A. Sommerfeld in 1904. Its formulation as a variational inequality is due to Cimatti [1].

The asymptotic expansion of Theorem 2.5 also appears there and its extension to an asymptotic series is derived in Capriz and G. Cimatti [1].

The filtration problem has an extensive literature both before and after the advent of variational inequalities. An interesting idea about its solution was proposed by Renato Caccioppoli (see the article by C. Miranda [1]). The introduction of the transformation (3.7) is due to C. Baiocchi [1]. This has stimulated much research in the subject. We refer to the papers of Baiocchi [2, 4]. For extensions and additional literature, there is the recent book of

Baiocchi and A. Capelo [1]. The existence of the seepage interval $(h, \varphi(a))$ with $\varphi(a) > h$ was shown by A. Friedman and R. Jensen [1].

The higher dimensional problem of Sections 5 and 6 was formulated in Stampacchia [6]. The analysis of the free boundary in this case is due to Caffarelli [1]. The proof of Theorem 6.7 is adapted from Alt [2].

There are other methods to approach the filtration problem. We refer to Alt [1] and Brezis, Kinderlehrer, and Stampacchia [1].

A general bibliography about fluid flow may be found in the book of Bers [1]. Here we have followed Brezis and Stampacchia [2]. See also Stampacchia [8]. Although the present method is limited to symmetric flow, it applies to compressible fluids (Brezis and Stampacchia [3]) cavitation (Brezis and Duvaut [1]), and flow in a channel (Tomarelli [1]) (cf. the exercises.)

We have just barely touched on the theory of higher order variational inequalities. For examples of work in this general field we cite P. Villaggio [1, 2], Frehse [2] Caffarelli and Friedman [1], and Brezis and Stampacchia [4] and Stampacchia [7].

Among the applications we have not had space to consider we note the problem of elastoplastic torsion. Here the convex set of competing functions has the form $\mathbb{K} = \{v \in H_0^1(\Omega): |v_x| \le 1 \text{ a.e. in } \Omega\}$. The reader may wish to consult Brezis [2], Brezis and Stampacchia [1], Caffarelli and Riviere [4], and Ting [1]. We have briefly touched on the problem of the confined plasma and its formulation as a variational problem (see Chapter VI, Section 4 and Exercise 15). In addition to the references mentioned there we cite Berestycki and Brezis [1]. There are questions in steady vortex flow which lead to similar considerations; see Fraenkel and Berger [1].

Exercises

1. Let $\Omega \subset \mathbb{R}^N$ be a domain and $\psi \in C^1(\bar{\Omega})$ an obstacle, namely, $\max_\Omega \psi > 0$ and $\psi < 0$ on $\partial\Omega$. With $\mathbb{K} = \{v \in H_0^1(\Omega): v \ge \psi \text{ in } \Omega\}$ consider the variational inequality

$$u \in \mathbb{K}: \quad \int_\Omega u_{x_i}(v - u)_{x_i}\, dx \ge \int_\Omega f(v - u)\, dx \qquad \text{for} \quad v \in \mathbb{K},$$

where $f \in L^\infty(\Omega)$ is given. Illustrate that this problem is not necessarily equivalent to the free boundary problem:

Find a closed subset $I \subset \Omega$ such that

$$-\Delta u = f \qquad \text{in} \quad \Omega - I,$$
$$u = \psi \qquad \text{in} \quad I,$$

and

$$u \in C^1(\overline{\Omega}).$$

Consider a one dimensional case. When are the two problems the same? What implications does this have for the lubrication of a journal bearing treated in Section 2?

2. Prove Theorem 2.4. [*Hint*: First take $v = 0$ in (2.8) to establish that $\|\text{grad } p_\varepsilon\|_{L^2(\Omega)} \leq C\varepsilon$. Then multiply (2.13) by ε^2 and set $v = p_\varepsilon \in \mathbb{K}$ and add the result to (2.8) with $v = \varepsilon p_0$.]

3. For the solution of Problem 2.1 show that

$$\frac{\partial}{\partial z} p(\theta, z) \leq 0 \qquad \text{for} \quad 0 \leq z < b,$$

$$\frac{\partial}{\partial z} p(\theta, z) \geq 0 \qquad \text{for} \quad -b < z \leq 0.$$

(*Hint*: $(\partial/\partial z)p$ obeys a maximum principle in $\Omega - I$.)

4. Let $M \subset \mathbb{R}^N$ be a domain and let $\xi \in S^{N-1} \subset \mathbb{R}^N$, and finally; let f be smooth in $\overline{M}$. Consider the problem: Find a pair $\{\Omega, u\}$, $\Omega \subset M$ and $u \in C(\overline{\Omega})$ such that

$$-\Delta u = f \qquad \text{in} \quad \Omega,$$
$$u = 0$$
$$\frac{\partial u}{\partial v} = \xi \cdot v \qquad \text{on} \quad \partial\Omega \cap M.$$

Introduce a new unknown and write this problem as "nearly a variational inequality," by which we mean that boundary conditions of elements of $\mathbb{K}$ are left unspecified. Under what additional hypotheses about u on $\partial M \cap \partial\Omega$ can the problem be rewritten as a variational inequality? To begin an investigation, write the problem in weak form.

5. Let w be the solution of the variational inequality (3.14) and assume that $\varphi(a) > h$. Show that $w \notin H^{2,\infty}(R)$.

6. Consider Problem 3.1 for a dam of variable porosity. More specifically, suppose that the piezometric head $u(x, y) \in C(\overline{\Omega})$ satisfies the equation

$$\frac{\partial}{\partial x}(k(x, y)u_x(x, y)) + \frac{\partial}{\partial y}(k(x, y)u_y(x, y)) = 0 \qquad \text{in} \quad \Omega$$

and the boundary conditions of Problem 3.1, where

$$k(x, y) = e^{f(x)+g(y)} > 0 \quad \text{in} \quad R,$$

and

$$g'(y) \geq 0 \qquad \text{for} \quad h \leq y \leq H.$$

Introduce the new unknown $w(x, y)$ as the solution of the Cauchy problem

$$w_y = -e^{g(y)}(u - y) \quad \text{in} \quad \Omega,$$
$$w = 0 \quad \text{on} \quad \Gamma.$$

(Benci [1].)

7. Prove Lemma 6.2.

8. Prove Lemma 6.3.

9. (Example of Fluid Flow) Suppose the profile in Problem 7.1 is $\mathscr{P} = \{z : |z| \leq 1\}$ and $q_\infty = 1$. Then the conformal mapping f of Theorem 7.2 reduces to $f(z) = z$. Calculate $\Phi(z)$, $\psi(z)$, $V(z)$, and the solution $u(\theta, \sigma)$ of the variational inequality. In particular, study the singularity of $\tilde{\psi}(\theta, \sigma)$ at $\theta + i\sigma = i\sigma_\infty = 0$.

10. Suppose the profile in Problem 7.1 is the ellipse

$$\mathscr{P} = \{z : (x^2/a^2) + (y^2/b^2) \leq 1\}$$

and $q_\infty = 1$. Find the solution to Problem 7.1 explicitly.

11. Study the steady motion of a compressible flow about a symmetric convex profile. In this case there is a given density function $\rho(x, y)$ such that

(ii′) $\operatorname{div}(\rho \mathbf{q}) = 0$, and
(iii′) $\operatorname{curl} \mathbf{q} = 0$

in G together with (i), (iv), (v) and (vi) of Problem 7.1. One assumes that $\rho = h(|\mathbf{q}|)$. There are a velocity potential φ and a stream function ψ which satisfy

$$\mathbf{q} = \operatorname{grad} \varphi, \qquad \rho q_1 = \psi_y, \qquad \text{and} \qquad \rho q_2 = -\psi_x$$

(Brezis and Stampacchia [3] and Stampacchia [8]).

12. Suppose in Problem 7.1 that the fluid is in motion about the same convex profile situated in the middle of a channel. Find a variational inequality for this problem (Tomarelli [1]).

13. Study the possibility of cavitation in Problem 7.1. (Brezis and Duvaut [1]).

14. (Detail of the Proof of Theorem 7.2) Let L: $z = x(s) + iy(s)$,

$0 \le s \le 1$, be a closed arc of class $C^{1,\lambda}$, some $\lambda > 0$, in the second quadrant of the $z = x + iy$ plane with $z(0) = 0$, and for $\varepsilon > 0$ small set

$$U = \{z : y > y(s) \text{ if } x \le 0 \text{ and } y > 0 \text{ if } x \ge 0\} \cap \{z : |z| < \varepsilon\}.$$

Suppose that $\psi \in C(\overline{U})$ satisfies

$$\Delta\psi = 0 \quad \text{in} \quad U$$

$$\psi = 0 \quad \text{for} \quad z \in L \quad \text{and for} \quad 0 \le z \le \varepsilon.$$

Then for some $\mu > 0$, $\psi \in C^{1,\mu}(U \cap \{z : |z| < \delta\})$ for every $\delta < \varepsilon$. [*Hint*: Suppose the tangent to L at $z = 0$ meets the positive x axis at an angle α, $\pi/2 \le \alpha \le \pi$, and consider the mapping $z \to z^{\pi/\alpha}$; (Brezis and Stampacchia [2]).

15. Prove Theorem 8.1. (*Hint*: Use the principle of the argument, i.e., the index of the mapping $z \to V(z)$; (Brezis and Stampacchia [2]).

16. Prove Theorem 8.7. [*Hint*: Let

$$h(\sigma) = \begin{cases} re^{-\sigma} + re^{-\sigma_\infty - \sigma_0}(\sigma - \sigma \quad) + He^{-\sigma_\infty} - re^{-\sigma_\infty}, & \sigma \ge \sigma_\infty + \sigma_0 \\ 0, & \sigma < \sigma_\infty + \sigma_0 \end{cases}$$

and prove that the solution $u(\theta, \sigma)$ of the variational inequality satisfies $u \le h$. Hence $u = 0$ for $\sigma < \sigma_\infty + \sigma_0$, so $l(\theta) \ge \sigma_\infty + \sigma_0$, $\theta_B < \theta < \theta_A$]

17. Solve Problem 9.1 for $V = \{v \in H^2(\Omega) : v = v' = 0 \text{ on } \partial\Omega\}$.

18. Consider Problem 9.1 for dimension $N > 1$. Show a solution u exists and that $u \in H^3(\Omega)$.

19. Suppose in Problem 9.1, $\mathbb{K} = \{v \in H^2(\Omega) \cap H_0'(\Omega) : |v''(t)| \le \beta\}$. Show that a solution exists and, moreover, it is in $H^{4,s}(\Omega)$.

20. Describe the dual space V' for Problem 9.1 and its variants above.

21. Suppose $N > 1$ and analyze formally the free boundary of Problem 9.1 in the spirit of Chapter VI. Although the equations are of higher order, they may be written as second order systems when $a(u, v) = \int_\Omega \Delta u \, \Delta v \, dx$.

22. Let $\Gamma : y = \varphi(x)$, $0 < x < a$, be the free boundary of Problem 3.1. Check that $\varphi(x)$ is an analytic function of x. (Refer to Theorem 6.7).

CHAPTER **VIII**

A One Phase Stefan Problem

1. Introduction

A one phase Stefan problem is the description, typically, of the melting of a body of ice maintained at 0°C in contact with a region of water. In general, the Stefan problem refers to the change of phase in a thermoconductive medium when energy, that is, heat, is contributed to the system. Unknown are (i) the temperature distribution of the water as a function of space and time and (ii) the free boundary consisting of the ice–water interface. The temperature distribution is required to solve the heat equation in the aqueous region and energy is conserved across the interface. By limiting ourselves to the case of one space dimension we shall be able to explore many of the features of this problem while avoiding much of the technical complication. This will be especially true in our investigation of the free boundary by means of the Legendre transform. Let $T > 0$ and $s_0 > 0$ be given. This is the classical Stefan problem.

Problem 1.1. *To find a function* $\Theta(x, t)$ *and a curve* $\Gamma : t = s(x)$, $x > s_0$, *such that*

$$-\Theta_{xx} + \Theta_t = 0 \qquad \textit{in} \quad \{(x, t) : s(t) < t < T\},$$

$$\begin{cases} \Theta = 0 \\ \Theta_x s'(x) = -k \end{cases} \qquad \textit{for} \quad (x, t) \in \Gamma,$$

$$\Theta(x, 0) = h(x) \qquad \textit{for} \quad 0 < x < s_0,$$

$$\Theta(0, t) = g(t) \qquad \textit{for} \quad 0 < t < T,$$

where $h(x) > 0$ *is the initial temperature,* $g(t) \geq 0$ *is the heat contributed at time* t*, and* $k > 0$ *is the heat of fusion.*

The function $\Theta(x, t)$ is interpreted as the temperature of the water at the point x at time t and the curve Γ represents the interface between the ice and the water. At time t the water occupies the subset defined by $\{x : s(x) < t\}$, where we understand that $s(x) = 0$ when $x \leq s_0$. Note that $\Theta(x, t) = \min \Theta = 0$ for $(x, t) \in \Gamma_1$ so $\Theta_x \leq 0$ on Γ. Hence

$$s'(x) \geq 0, \tag{1.1}$$

which means that *the curve* Γ *is monotone.* This is the property of the one phase problem which permits its transformation without difficulty to a variational inequality.

At this point we transform Problem 1.1 to a variational inequality. Let $R > s_0 > 0$. Later we shall exhibit this R as a function of the initial data k, and so on. Set $D = (0, R) \times (0, T)$. Our variational inequality will be for $u(x, t)$ defined by

$$\begin{aligned}
u(x, t) &= \int_{s(x)}^{t} \Theta(x, \tau)\, d\tau, && s(x) \leq t \leq T, \quad s_0 \leq x \leq R, \\
u(x, t) &= 0, && 0 \leq t \leq s(x), \quad s_0 \leq x \leq R \\
u(x, t) &= \int_0^t \Theta(x, \tau)\, d\tau, && 0 \leq t \leq T, \quad 0 \leq x \leq s_0.
\end{aligned} \tag{1.2}$$

We compute the differential equation satisfied by u assuming that Θ and s are smooth:

$$\begin{aligned}
u_x(x, t) &= \int_{s(x)}^{t} \Theta_x(x, \tau)\, d\tau - s'(x)\Theta(x, s(x)) \\
&= \int_{s(x)}^{t} \Theta_x(x, \tau)\, d\tau, \qquad s(x) < t < T, \quad s_0 < x < R.
\end{aligned}$$

Also

$$\begin{aligned}
u_{xx}(x, t) &= \int_{s(x)}^{t} \Theta_{xx}(x, \tau)\, d\tau - s'(x)\Theta_x(x, s(x)) = \int_{s(x)}^{t} \Theta_t(x, \tau)\, d\tau + k \\
&= \Theta(x, t) + k = u_t(x, t) + k, \qquad s(x) < t < T, \quad s_0 < x < R.
\end{aligned}$$

In the same manner,

$$u_{xx}(x, t) = u_t(x, t) - h(x), \qquad 0 < t < T, \quad 0 < x < s_0.$$

So let us define

$$f(x) = \begin{cases} h(x), & 0 \le x < s_0 \\ -k, & s_0 \le x \le R \end{cases} \tag{1.3}$$

and

$$\psi(t) = \int_0^t g(\tau)\, d\tau, \qquad 0 \le t \le T, \tag{1.4}$$

where we recall that $k > 0$ and g and h are smooth functions positive in $[0, T]$ and $[0, s_0)$, respectively. With $\mathbb{K}$ the set of nonnegative functions in $L^2(D)$, the conditions on u may be expressed as the alternatives

$$\left.\begin{aligned} (-u_{xx} + u_t)(v - u) - f(v - u) &= 0 \\ u &> 0 \end{aligned}\right\} \quad \text{for} \quad v \in \mathbb{K}$$

or

$$\left.\begin{aligned} (-u_{xx} + u_t)(v - u) - f(v - u) &= kv \ge 0 \\ u &= 0 \end{aligned}\right\} \quad \text{for} \quad v \in \mathbb{K}.$$

This leads to the variational inequality, for $T > 0$ given:

Problem 1.2. *To find $u \in L^2(0, T; H^2(0, R)) \cap \mathbb{K}$ such that*

$$\begin{aligned} u_t &\in \mathbb{K}, && \\ (-u_{xx} + u_t)(v - u) &\ge f(v - u) && \textit{a.e. for} \quad v \in \mathbb{K}, \\ u &= \psi && \textit{for} \quad 0 \le t \le T, \quad x = 0, \\ u &= 0 && \textit{for} \quad 0 \le t \le T, \quad x = R, \\ u &= 0 && \textit{for} \quad t = 0, \quad 0 \le x \le R. \end{aligned}$$

Here we have used the notation $L^2(0, T; H^2(0, R))$ to denote the functions $u(x, t)$ satisfying

$$\int_0^T \int_0^R (u^2 + u_x^2 + u_{xx}^2)\, dx\, dt < \infty.$$

2. Existence and Uniqueness of the Solution

A solution to Problem 1.2 will be found by approximation with a suitable penalty function. The particular choice of penalty is intended to simplify the existence and regularity proofs. For $\varepsilon > 0$ we choose $\beta_\varepsilon(t) \in C^\infty(\mathbb{R})$ with the properties

$$\begin{aligned} &\beta_\varepsilon(t) = 0 \qquad \text{if} \quad t \geq \varepsilon \\ &\beta_\varepsilon(0) = -1 \\ &\beta_\varepsilon'(t) > 0 \qquad \text{and} \quad \beta_\varepsilon''(t) \leq 0, \quad -\infty < t < \varepsilon \end{aligned} \tag{2.1}$$

Next choose a sequence $f_\varepsilon(x)$ of smooth functions in $[0, R]$ which are uniformly bounded and decrease to $f(x)$ defined by (1.3) as $\varepsilon \to 0$. Finally choose $\eta(x) \in C_0^\infty(\mathbb{R})$ to satisfy

$$\eta(x) = \begin{cases} 1 & \text{for} \quad 0 \leq x \leq \frac{1}{3}s_0 \\ 0 & \text{for} \quad \frac{2}{3}s_0 \leq x \end{cases}$$

and $0 \leq \eta(x) \leq 1$, $x \in \mathbb{R}$. For $T > 0$ and $\varepsilon > 0$ given, consider the initial boundary value problem

Problem 2.1. *To find $u(x, t)$, $(x, t) \in \bar{D}$, such that*

$$\begin{aligned} -u_{xx} + u_t + k\beta_\varepsilon(u) &= f_\varepsilon && \text{in} \quad D, \\ u &= \varepsilon\eta, && t = 0, \quad 0 \leq x \leq R, \end{aligned} \tag{2.2}$$

$$\begin{aligned} u &= \psi + \varepsilon, && 0 \leq t \leq T, \quad x = 0, \\ u &= 0, && 0 \leq t \leq T, \quad x = R, \end{aligned} \tag{2.3}$$

where $k > 0$ is the constant of Problem 1.1 *and ψ is defined by* (1.4).

It is well known that Problem 2.1 admits a classical solution $u = u_\varepsilon$; cf. Friedman [2].

Lemma 2.2. *Let u_ε, $\varepsilon > 0$, denote the solution to Problem* 2.1 *in D. Then there is an $\varepsilon_0 > 0$ such that*

$$0 \leq \frac{\partial u_\varepsilon}{\partial t}(x, t) \leq K \qquad \text{in} \quad D,$$

where $K > 0$ is independent of ε, $0 < \varepsilon \leq \varepsilon_0$.

Proof. We differentiate (2.2) with respect to t and apply the maximum principle. For $w = \partial u_\varepsilon/\partial t$, we obtain

$$\begin{aligned} -w_{xx} + w_t + k\beta_\varepsilon'(u_\varepsilon)w &= 0 && \text{in} \quad D, \\ w &= f_\varepsilon + \varepsilon\eta_{xx} - k\beta(u_\varepsilon), && t = 0, \\ w &= \psi', && x = 0, \\ w &= 0, && x = R. \end{aligned} \tag{2.4}$$

Since $\beta_\varepsilon'(u_\varepsilon) \geq 0$ in D we infer from the maximum principle that

$$\min\left(\min_{\partial_p D} w, 0\right) \leq w(x, t) \leq \max\left(\max_{\partial_p D} w, 0\right) \quad \text{in} \quad D, \tag{2.5}$$

where $\partial_p D$ denotes the parabolic boundary of D, i.e.,

$$\partial D - \{(x, T): 0 < x < R\}.$$

First we evaluate the left-hand member of this expression. Recalling that $f_\varepsilon \geq f$ and

$$0 \leq u_\varepsilon(x, 0) = \varepsilon\eta(x) \leq \varepsilon \qquad \text{for} \quad 0 \leq x \leq \tfrac{2}{3}s_0,$$

we deduce that

$$\begin{aligned} w(x, 0) &\geq f(x) + \varepsilon\eta_{xx}(x), && 0 < x \leq \tfrac{2}{3}s_0 \\ &= h(x) + \varepsilon\eta_{xx}(x) \geq 0, && 0 < x \leq \tfrac{2}{3}s_0 \end{aligned}$$

for $\varepsilon \leq \varepsilon_0$, ε_0 sufficiently small, because $h(x)$ is positive in $[0, s_0)$. On the other hand

$$w(x, 0) = f_\varepsilon(x) - k\beta_\varepsilon(0) \geq h(x) + k \geq 0 \qquad \text{for} \quad \tfrac{2}{3}s_0 \leq x \leq s_0$$

and

$$w(x, 0) = f_\varepsilon(x) - k\beta_\varepsilon(0) \geq k - k = 0 \qquad \text{for} \quad s_0 < x < R.$$

Hence

$$w(x, 0) \geq 0 \qquad \text{for} \quad 0 < x < R.$$

On the vertical sides of the boundary,

$$w(0, t) = \psi'(t) = g(t) > 0 \qquad \text{for} \quad 0 < t < T$$

and

$$w(R, t) = 0 \qquad \text{for} \quad 0 < t < T.$$

Consequently, $\min(\min_{\partial_p D} w, 0) = 0$.

We now inspect the right-hand inequality of (2.5). We need only note that

$$|f_\varepsilon + \varepsilon\eta_{xx} - k\beta_\varepsilon(u_\varepsilon)| \le \sup|f_\varepsilon| + \varepsilon\sup|\eta_{xx}| + k \le K_1,$$

when $t = 0$, for a K_1 independent of $\varepsilon \le \varepsilon_0$ to show that

$$\max\left(\max_{\partial_p D} w, 0\right) \le \max(\max g, K_1) = K.$$

The assertion of the lemma now follows from (2.5). Q.E.D.

Lemma 2.3. *Let u_ε, $0 < \varepsilon \le \varepsilon_0$, denote the solution to Problem* 2.1. *Then*

$$\|\beta_\varepsilon(u_\varepsilon)\|_{L^\infty(D)} \le 1 \qquad \textit{for} \quad 0 < \varepsilon < \varepsilon_0.$$

Proof. From the previous lemma, $\partial u_\varepsilon/\partial t \ge 0$ in D so $u_\varepsilon(x, t) \ge 0$ in D by (2.3). Recalling our choice of β_ε in (2.1),

$$-1 \le \beta_\varepsilon(u_\varepsilon(x, t)) \le 0 \qquad \text{if} \quad (x, t) \in D. \quad \text{Q.E.D.}$$

At this point we address the existence and uniqueness of a solution to the variational inequality.

Theorem 2.4. *There exists a unique solution u to Problem* 1.2. *It enjoys the properties that*

$$u, u_x, u_t, u_{xx} \in L^\infty(D)$$

and

$$u \ge 0, \qquad u_t \ge 0 \qquad \textit{in} \quad D.$$

Let u_ε denote the solution to Problem 2.1. *Then as $\varepsilon \to 0$*

$$u_\varepsilon \to u \qquad \textit{weakly in} \quad H^{1,p}(D), \qquad 1 < p < \infty,$$

and

$$u_\varepsilon \to u \qquad \textit{weakly in} \quad H^{2,p}(0, R), \quad 1 < p < \infty,$$

for each t, $0 < t < T_0$, and hence $u_\varepsilon \to u$ uniformly in D and $u_{\varepsilon x} \to u_x$ uniformly in $(0, R)$ for each $t \in (0, T)$.

Proof. In view of the preceding lemmas, the solution u_ε, $0 < \varepsilon \le \varepsilon_0$, of Problem 2.1 satisfies

$$\begin{aligned}\|u_{\varepsilon xx}\|_{L^\infty(D)} &= \|f_\varepsilon - u_{\varepsilon t} - k\beta(u_\varepsilon)\|_{L^\infty(D)} \\ &\le \|f_\varepsilon\| + K + k \le C_1, \qquad 0 < \varepsilon < \varepsilon_0.\end{aligned}$$

Hence $u_{\varepsilon x}(x, t)$ is a family of bounded functions, because the u_ε are uniformly bounded by Lemma 2.2. Therefore

$$\|u_\varepsilon\|^p_{H^{1,p}(D)} = \iint_D |u_\varepsilon|^p \, dx\, dt + \iint_D |u_{\varepsilon x}|^p \, dx\, dt + \iint_D |u_{\varepsilon t}|^p \, dx\, dt \le C_2$$

independent of ε, $0 < \varepsilon < \varepsilon_0$. This implies the existence of a sequence $\varepsilon' \to 0$ and a function $u \in H^{1,p}(D)$, $1 \le p < \infty$, such that

$$u_{\varepsilon'} \to u \qquad \text{in} \quad H^{1,p}(D) \tag{2.6}$$

and uniformly in $\bar{D}$. In particular, $u \ge 0$ and satisfies the boundary conditions of Problem 1.2.

In addition, a subsequence of the $u_{\varepsilon'}$ converges weakly to u in $H^{2,p}(0, R)$, $1 \le p < \infty$, for each fixed t, $0 < t < T$. Since, however, $u_{\varepsilon'} \to u$ uniformly in D, it follows that the entire sequence $u_{\varepsilon'} \to u$ weakly in $H^{2,p}(0, R)$, $0 < t < T$. In particular, $u_{xx} \in L^\infty(D)$.

To complete the proof that u is a solution to Problem 1.2, let $v \in L^\infty(D)$ satisfy $v \ge \delta > 0$. Multiplying (2.2) by $v - u_\varepsilon$ and noting that $\beta_\varepsilon(v) = 0$ if $\varepsilon < \delta$, we see that

$$(-u_{\varepsilon xx} + u_{\varepsilon t})(v - u_\varepsilon) - k[\beta_\varepsilon(v) - \beta_\varepsilon(u_\varepsilon)](v - u_\varepsilon) = f_\varepsilon(v - u_\varepsilon), \qquad \varepsilon < \delta.$$

Integrating with respect to $x \in [0, R]$ and observing that

$$[\beta_\varepsilon(v) - \beta_\varepsilon(u_\varepsilon)](v - u_\varepsilon) \ge 0,$$

we obtain

$$\int_0^R (-u_{\varepsilon xx} + u_{\varepsilon t})(v - u_\varepsilon)\, dx \ge \int_0^R f_\varepsilon(v - u_\varepsilon)\, dx.$$

Taking $\varepsilon \to 0$, owing to the weak convergence we get

$$\int_0^R (-u_{xx} + u_t)(v - u)\, dx \ge \int_0^R f(v - u)\, dx. \tag{2.7}$$

By approximation, (2.7) holds for all $v \ge 0$, $v \in L^2(0, R)$. It follows that u is a solution to Problem 1.2.

We can easily check that the solution is unique in $L^2(0, T; H^2(0, R))$. This implies, in particular, that the entire sequence u_ε converges weakly to u in $H^{1,p}(D)$ and in $H^{2,p}(0, R)$ for each t, $0 < t < T$. That $u_t \ge 0$ follows from Lemma 2.2. Q.E.D.

Corollary 2.5. *Let u denote the solution to Problem* (1.2) *and define*

$$\Omega(t) = \{x \in (0, R) : u(x, t) > 0\}, \qquad 0 \le t \le T.$$

Then $\Omega(t) \subset \Omega(t')$ for $t < t'$, $t, t' \in [0, T]$.

Proof. This is obvious because u is Lipschitz with $u_t \geq 0$. Q.E.D.

We shall now show that given $T > 0$, $R > 0$ may be chosen so that $u = 0$ in a neighborhood of $x = R$. This implies the existence of a free boundary. To begin we prove a comparison theorem analogous to our weak maximum principle of Chapter II.

Lemma 2.6. *Suppose that $f \leq \hat{f}$ and $\psi \leq \hat{\psi}$ and that u, $\hat{u}$ are solutions of Problem* 1.2 *for f, ψ and $\hat{f}$, $\hat{\psi}$, respectively. Then*

$$u \leq \hat{u} \quad \text{in} \quad D.$$

Proof. For $v \in \mathbb{K}^*$, the set of nonnegative $H^1(D)$ functions obeying the boundary conditions of Problem 1.2, the integrated form of the problem is valid. Namely,

$$\iint_D [\hat{u}_x(v - \hat{u})_x + \hat{u}_t(v - \hat{u})]\,dx\,dt \geq \iint_D \hat{f}(v - \hat{u})\,dx\,dt, \qquad v \in \mathbb{K}^*.$$

In this expression we may take $v = \max(u, \hat{u}) \in \mathbb{K}^*$ since $\psi \leq \hat{\psi}$. Hence

$$\iint_A [\hat{u}_x(u - \hat{u})_x + \hat{u}_t(u - \hat{u})]\,dx\,dt \geq \iint_A \hat{f}(u - \hat{u})\,dx\,dt, \tag{2.8}$$

$$A = \{(x, t) : u(x, t) > \hat{u}(x, t)\}.$$

Further observe that $-u_{xx} + u_t = f$ a.e. in A since $u > \hat{u} \geq 0$ there. Consequently, for $\zeta = \max(u - \hat{u}, 0)$, which is zero on $\partial_p D$,

$$-\iint_D [u_x\zeta_x + u_t\zeta]\,dx\,dt = -\iint_D f\zeta\,dx\,dt. \tag{2.9}$$

Writing (2.8) in terms of ζ, we see that

$$\iint_D [\hat{u}_x\zeta_x + \hat{u}_t\zeta]\,dx\,dt \geq \iint_D \hat{f}\zeta\,dx\,dt. \tag{2.10}$$

Adding (2.9) and (2.10) and again using the definition of ζ, we obtain

$$-\iint_D \zeta_x^2\,dx\,dt - \int_0^R \zeta(x, T)^2\,dx \geq \iint_D (\hat{f} - f)\zeta\,dx\,dt \geq 0.$$

Hence $\zeta = 0$ a.e. in D, i.e., meas $A = 0$. Q.E.D.

We may now construct an explicit solution of the Stefan problem which we employ to dominate our solution u. Denote by N_M the set

$$N_M = \{x : |x| < M(t+1)^{1/2}\},$$

where M is a positive constant to be determined. Set

$$\hat{\Theta}(x, t) = F\left(\frac{x}{(t+1)^{1/2}}\right)$$

and

$$\Phi(x, t) = x - M(t+1)^{1/2}, \qquad x \geq 0, \quad t \geq 0,$$

where

$$F(z) = C \int_z^\infty e^{-\zeta^2/4}\, d\zeta - C'$$

with C, C' satisfying

$$2Ce^{-M^2/4} = k, \tag{2.11}$$

$$F(M) = C \int_M^\infty e^{-\zeta^2/4}\, d\zeta - C' = 0. \tag{2.12}$$

It is easy to check that $\hat{\Theta}(x, t)$ and the curve $\hat{\Gamma}$ defined by $\Phi(x, t) = 0$ is a solution to the classical Stefan Problem 1.1 with

$$\hat{h}(x) = F(x), \qquad 0 \leq x \leq M,$$

and

$$\hat{g}(t) = F(0), \qquad 0 \leq t \leq T.$$

Hence $\hat{\Theta}$ determines a solution $\hat{u}$ to Problem 1.2 via (1.2). Note that the heat of fusion corresponding to $\hat{\Theta}$ and $\hat{\Gamma}$ is the same k as that pertaining to u.

Now

$$C = \frac{k}{2} e^{M^2/4} \to \infty \qquad \text{as} \quad M \to \infty$$

whereas

$$C' = \frac{k}{2} \int_M^\infty e^{(M^2-\zeta^2)/4}\, d\zeta \leq \frac{k}{2} \sum_M^\infty (n+1) e^{(M^2-n^2)/4} < \infty$$

is bounded as $M \to \infty$. Hence we may find M so large that

$$\hat{h}(x) \geq h(x) \qquad \text{and} \qquad \hat{g}(t) \geq g(t).$$

Now we apply Lemma 2.6 to conclude

$$\hat{u} \geq u \quad \text{in} \quad D.$$

Since $\hat{u}(x, t) = 0$ for $x \geq M(t + 1)^{1/2}$,

$$\Omega(t) \subset \{x : 0 < x < M(t + 1)^{1/2}\}, \qquad 0 \leq t \leq T.$$

In sum:

Proposition 2.7. *Given $T > 0$ we may choose $R > 0$ so that the solution u to Problem* 1.2 *vanishes in a neighborhood of $x = R$.*

Proof. We choose M as above and $R > M(T + 1)^{1/2}$. Q.E.D.

It will always be assumed that R has been chosen so large that the conclusion of Proposition 2.7 is satisfied. Set

$$\begin{aligned} \Omega &= \{(x, t) : u(x, t) > 0\}, \\ \Omega(t) &= \{x : (x, t) \in \Omega\}, \qquad 0 < t \leq T, \\ \Gamma &= \partial\Omega \cap D. \end{aligned} \tag{2.13}$$

Theorem 2.8. *Let u be a solution of Problem* 1.2. *Then Γ defined by* (2.13) *admits the representation*

$$\Gamma: \quad x = \sigma(t), \qquad 0 \leq t \leq T,$$

where σ is a continuous increasing function of t with $s_0 = \sigma(0) < \sigma(t)$ for $t > 0$.

Proof. We first show that $\Omega(t)$ is connected, which implies that $\sigma(t)$ is well defined. The interval $(0, s_0) \subset \Omega(t)$ for all t because $u_t \geq 0$. Suppose that (x_1, x_2) is a component of the open set $\Omega(t)$ which does not contain $(0, s_0)$. Then

$$-u_{xx}(x, t) = -k - u_t(x, t) < 0 \qquad \text{in} \quad (x_1, x_2),$$

so, by the maximum principle,

$$u(x, t) \leq \max(u(x_1, t), u(x_2, t)) = 0, \qquad x_1 < x < x_2.$$

This is a contradiction; hence, $\Omega(t)$ is connected.

We define

$$\sigma(t) = \sup\{x : x \in \Omega(t)\}, \qquad 0 < t \leq T,$$

$$\sigma(0) = s_0.$$

Since $\Omega(t) \subset \Omega(t')$ for $t < t'$, it is obvious that σ is monotone. Suppose now that

$$x_1 = \sigma(t) < \lim_{\substack{t' \to t \\ t' > t}} \sigma(t') = x_2$$

for some t, $0 \le t < T$. In this event, the rectangle

$$Q = \{(x, \tau) : x_1 + \varepsilon < x < x_2 - \varepsilon, t - \tau < T\} \subset \Omega,$$

for $\varepsilon > 0$ small, and

$$\begin{aligned} -u_{xx} + u_\tau &= -k \quad &&\text{in} \quad Q, \\ u(x, t) &= 0 &&\text{for} \quad x_1 + \varepsilon < x < x_2 + \varepsilon. \end{aligned}$$

By the regularity theory for parabolic equations, u is smooth in $\bar{Q}$ so

$$u(x, t) = u_x(x, t) = u_{xx}(x, t) = 0 \qquad \text{for} \quad x_1 + \varepsilon < x < x_2 - \varepsilon.$$

From the equation, $u_t(x, t) = -k$ for $x_1 + \varepsilon < x < x_2 - \varepsilon$, whence $u_t(x, \tau) < 0$ in a neighborhood in Q of $(\frac{1}{2}(x_1 + x_2), t)$. This contradicts the terms of our existence theorem, Theorem 2.4. In the same way we may show that u is continuous from below.

Finally we point out that $\sigma(t) > s_0$ for $t > 0$. Suppose not and that $\sigma(t_0) = s_0$ for some $t_0 > 0$. By monotonicity of σ, $\sigma(t) = s_0$ for $0 \le t \le t_0$. Consider the rectangle

$$Q = \{(x, t) : s_0 - \varepsilon < x < s_0, 0 < t < t_0\}$$

for $\varepsilon > 0$, small, and observe that

$$\begin{aligned} -u_{xx} + u_t = f = h &\ge 0 \quad &&\text{in} \quad Q, \\ u &= 0 &&\text{for} \quad x = s_0, \quad 0 < t < t_0, \\ u &> 0 &&\text{for} \quad x = s_0 - \varepsilon, \quad 0 < t < t_0. \end{aligned}$$

Since u_x is continuous in D and u attains its minimum at $x = s_0$,

$$u_x(s_0, t) = 0 \qquad \text{for} \quad 0 < t < t_0. \tag{2.14}$$

Introduce the solution $v(x, t)$ to the Dirichlet problem

$$\begin{aligned} -v_{xx} + v_t &= h \quad &&\text{in} \quad Q, \\ v &= 0 &&\text{on} \quad \partial_p Q, \end{aligned}$$

where $\partial_p Q$ denotes the parabolic boundary of Q. Since $h \ge 0$,

$$v \ge \min_{\partial_p Q} v = 0 \qquad \text{in} \quad Q.$$

Hence by the Hopf–Friedman maximum principle,

$$v_x(s_0, t) < 0, \qquad 0 < t < t_0. \tag{2.15}$$

On the other hand, $u - v$ is a solution to the heat equation in Q with $u - v \geq 0$ on $\partial_p Q$. Hence, since u and v are not identical,

$$(u - v)_x(s_0, t) < 0$$

because $u - v$ attains its minimum there. Combining this with (2.14) and (2.15) we obtain

$$0 = u_x(s_0, t) < v_x(s_0, t) < 0, \qquad 0 < t < t_0,$$

the desired contradiction. Q.E.D.

It is useful to point out that Γ is a Lipschitz curve in the coordinates obtained from (x, t) by rotation through $\pi/4$. Hence a function $\zeta \in H^1(\Omega)$ admits a trace in $L^2(\Gamma)$.

3. Smoothness Properties of the Solution

To discuss the free boundary Γ of our Stefan problem it is useful to know that $u_{xx}(x, t)$ and $u_{xt}(x, t)$ are continuous in Ω near Γ. One feature of the development is the proof that $u_t \in H^1(D \cap \{t \geq t_0 > 0\})$ which proceeds by use of the penalizations. Another is the application of a weak maximum principle and Bernstein's method to show that u_t is Lipschitz in Ω near Γ.

Lemma 3.1. *Let u_ε, $0 < \varepsilon \leq \varepsilon_0$, denote the solution to Problem 2.1. Then*

$$\left| \frac{\partial^2 u_\varepsilon}{\partial x \, \partial t}(x, t) \right| \leq C \qquad \text{for} \qquad 0 \leq x \leq s_0/6, \quad 0 < t < T$$

and

$$\left| \frac{\partial^2 u_\varepsilon}{\partial x^2}(x, t) \right| \leq C \qquad \text{for} \qquad 0 \leq x \leq s_0/6, \quad 0 < t < T,$$

where $C > 0$ is independent of ε, $0 < \varepsilon \leq \varepsilon_0$.

Proof. From the initial conditions (2.4),

$$u_\varepsilon(x, 0) = \varepsilon \qquad \text{for} \quad 0 \leq x \leq s_0/3,$$

so, since $u_{\varepsilon t}(x, t) \geq 0$, in D,

$$u_\varepsilon(x, t) \geq \varepsilon \quad \text{in} \quad 0 \leq x \leq s_0/3, \quad 0 \leq t \leq T.$$

Hence $\beta_\varepsilon(u_\varepsilon(x, t)) = 0$ in $0 \leq x \leq s_0/3$, $0 \leq t \leq T$, and u_t is a solution to the equation

$$-u_{\varepsilon xx} + u_{\varepsilon t} = f_\varepsilon, \qquad 0 < x < s_0/3, \quad 0 < t < T.$$

The conclusion now follows from the standard Schauder theory of the heat equation (Friedman [2]). Q.E.D.

Lemma 3.2. *Let u_ε, $0 < \varepsilon < \varepsilon_0$, denote the solution to Problem 2.1. Then*

$$\iint_D u_{\varepsilon xt}(x, t)^2 \, dx \, dt \leq C$$

where $C > 0$ is independent of ε, $0 < \varepsilon \leq \varepsilon_0$.

Proof. For the duration of the proof, we set $u = u_\varepsilon$, $\beta = \beta_\varepsilon$, and $w = u_t$. Differentiate Eq. (2.3) with respect to t and multiply by w to obtain

$$-ww_{xx} + ww_t + k\beta'(u)w^2 = 0 \qquad \text{in} \quad D,$$

which, integrated over $(0, R)$ for fixed t, $0 < t < T$, yields

$$-\int_0^R ww_{xx} \, dx + \frac{1}{2}\frac{d}{dt}\int_0^R w^2 \, dx + k\int_0^R \beta'(u)w^2 \, dx = 0. \tag{3.1}$$

The first term may be integrated by parts. We observe that

$$\begin{aligned} -\int_0^R ww_{xx} \, dx &= \int_0^R w_x^2 \, dx + w(0, t)w_x(0, t) - w(R, t)w_x(R, t) \\ &= \int_0^R w_x^2 \, dx + w(0, t)w_x(0, t) \end{aligned}$$

because $w(R, t) = u_t(R, t) = 0$. By the previous lemma

$$|w(0, t)w_x(0, t)| \leq |\psi_t(t)u_{\varepsilon xt}(0, t)| \leq \text{const} = C_1.$$

Therefore, from (3.1) and recalling that $\beta'(u) \geq 0$,

$$\int_0^R w_x^2 \, dx + \frac{1}{2}\frac{d}{dt}\int_0^R w^2 \, dx \leq C_1.$$

This, integrated over $(0, T)$, gives

$$\int_0^T \int_0^R w_x^2 \, dx \, dt + \frac{1}{2}\int_0^R [w(x, R)^2 - w(x, 0)^2] \, dx \leq C_1 T.$$

According to Lemma 2.2, $|w(x, t)| \le K$ independent of ε, whence

$$\int_0^T \int_0^R w_x^2 \, dx \, dt \le C_1 T + RK^2. \quad \text{Q.E.D.}$$

Lemma 3.3. *Let u_ε, $0 < \varepsilon \le \varepsilon_0$, denote the solution to Problem 2.1. Then for $0 < \sigma \le t < T$,*

$$\int_0^R u_{\varepsilon xt}(x, t)^2 \, dx + \int_\sigma^t \int_0^R u_{\varepsilon tt}(x, \tau)^2 \, dx \, d\tau \le \frac{C}{\sigma},$$

where $C > 0$ is a constant independent of ε, $0 < \varepsilon \le \varepsilon_0$.

This lemma is somewhat more difficult. Clearly it leads to the integrability criterion for the solution u we mentioned at the beginning of the paragraph. Incidentally, we use here the assumption that $\beta_\varepsilon''(t) \le 0$.

Proof. As before, we set $u = u_\varepsilon$, $\beta = \beta_\varepsilon$, and $w = u_{\varepsilon t}$. Differentiating Eq. (2.2) with respect to t and multiplying by w_t we obtain

$$-w_{xx} w_t + w_t^2 + k\beta'(u) w_t w = 0 \qquad \text{in} \quad D,$$

which, integrated over $(0, R)$, yields

$$-\int_0^R w_{xx} w_t \, dx + \int_0^R w_t^2 \, dx + k \int_0^R \beta'(u) w w_t \, dx = 0. \tag{3.2}$$

We begin as usual with the first term, so,

$$-\int_0^R w_{xx} w_t \, dx = \int_0^R w_x w_{xt} \, dx + w_x(0, t) w_t(0, t), \qquad 0 < t < T,$$

$$= \frac{1}{2} \frac{d}{dt} \int_0^R w_x^2 \, dx + w_x(0, t) w_t(0, t), \qquad 0 < t < T.$$

The boundary term at $x = R$ does not appear because $w = w_t = 0$ there. When $x = 0$, we may use Lemma 3.1 and our assumptions about ψ to conclude

$$|w_x(0, t) w_t(0, t)| \le |w_x(0, t)| \, |\psi_{tt}(t)| \le C_2$$

independent of ε, $0 < \varepsilon \le \varepsilon_0$. Consequently, by (3.2),

$$\frac{1}{2} \frac{d}{dt} \int_0^R w_x^2 \, dx + \int_0^R w_t^2 \, dx + \frac{1}{2} k \int_0^R \beta'(u) \frac{\partial}{\partial t} w^2 \, dx \le C_2. \tag{3.3}$$

To estimate the last term, we write

$$\int_0^R \beta'(u) \frac{\partial}{\partial t} w^2 \, dx = \int_0^R \frac{\partial}{\partial t} [\beta'(u)w^2] \, dx - \int_0^R \beta''(u)w^3 \, dx.$$

At this point we employ Lemma 2.2 in an essential way to show that $w \geq 0$ and $\beta''(u) \leq 0$ imply

$$-\int_0^R \beta''(u)w^3 \, dx \geq 0.$$

Therefore

$$\int_0^R \beta''(u) \frac{\partial}{\partial t} w^2 \, dx \geq \frac{d}{dt} \int_0^R \beta'(u)w^2 \, dx.$$

Using this estimate in (3.3) we obtain

$$\frac{1}{2} \frac{d}{dt} \int_0^R [w_x^2 + k\beta'(u)w^2] \, dx + \int_0^R w_t^2 \, dx \leq C_2.$$

For $\tau < t$ we integrate over (τ, t), which provides us with the estimate

$$\int_0^R [w_x(x, t)^2 + k\beta'(u)w(x, t)^2] \, dx + 2 \int_\tau^t \int_0^R w_t^2 \, dx \, d\tau'$$

$$\leq C_2(t - \tau) + \int_0^R [w_x(x, \tau)^2 + \beta'(u)w(x, \tau)^2] \, dx$$

Now this estimate holds for each $\tau < t$, so we may integrate it with respect to τ over the interval $(0, \sigma)$, $\sigma < t$, say, to obtain

$$\sigma \int_0^R w_x(x, t)^2 \, dx + 2\sigma \int_\sigma^t \int_0^R w_\tau^2 \, dx \, d\tau$$

$$\leq \sigma \int_0^R [w_x(x, t)^2 + k\beta'(u)w(x, t)^2] \, dx + 2\sigma \int_\sigma^t \int_0^R w_\tau^2 \, dx \, d\tau$$

$$\leq C_3 + \int_0^\sigma \int_0^R \left[w_x(x, \tau)^2 + kw(x, \tau) \frac{\partial}{\partial \tau} \beta(u(x, \tau)) \right] dx \, d\tau$$

$$\leq C_3 + \int_0^T \int_0^R w_x(x, \tau)^2 \, dx \, d\tau + Kk \int_0^T \int_0^R \frac{\partial}{\partial \tau} \beta(u(x, \tau)) \, dx \, d\tau$$

$$\leq C_3 + \int_0^T \int_0^R w_x(x, \tau)^2 \, dx \, d\tau + Kk \int_0^R [\beta(u(x, T)) - \beta(u(x, 0))] \, dx$$

$$\leq C_3 + \int_0^T \int_0^R w_x(x, \tau)^2 \, dx \, d\tau + 2RKk.$$

Here we have used the fact that β' and w are nonnegative and Lemma 2.3. In view of Lemma 3.2, our lemma is proved. Q.E.D.

Theorem 3.4. *Let u be the solution to Problem 1.2. Then there is a constant $C > 0$ such that for each $\sigma > 0$*

$$\int_0^R u_{xt}(x, t)^2 \, dx + \int_\sigma^t \int_0^R u_{\tau\tau}(x, \tau)^2 \, dx \, d\tau \leq \frac{C}{\sigma} \qquad \textit{for} \quad \sigma < t < T.$$

Proof. This follows immediately from Lemma 3.3 by the weak convergence of $u_{\varepsilon t}$ to u_t in L^p. In particular, note that $u_t \in H^1 (D \cap \{(x, t) : t \geq \sigma\})$ for each $\sigma > 0$. Q.E.D.

We now embark on the next phase of our program, the Lipschitz continuity of u_t near Γ.

Lemma 3.5. *Let u denote the solution of Problem 1.2 and let Γ denote the free boundary associated to it. Then for $(x_0, t_0) \in \Gamma$, there is a neighborhood U of (x_0, t_0) such that*

$$u_x(x, t) < 0 \qquad \textit{in} \quad U \cap \Omega.$$

Proof. This is elementary. In view of Theorem 2.8, there is a neighborhood U of $(x_0, t_0) \in \Gamma$ such that $U \subset \{(x,t) : x > s_0, t > 0\}$. Hence $f(x) = -k$ in U and

$$u_{xx}(x, t) = u_t(x, t) + k \geq k > 0 \qquad \text{in} \quad U \cap \Omega.$$

Therefore

$$u_x(x, t) = -\int_x^\sigma u_{xx}(y, t) \, dy < 0 \qquad \text{in} \quad U \cap \Omega.$$

Lemma 3.6. *Let Γ denote the free boundary associated to the solution u of Problem 1.2 and let $Q = \{(x, t) : |x - x_0| < \varepsilon, 0 < t_0 - t < \delta\}$ for a given $(x_0, t_0) \in \Gamma$ and $\varepsilon > 0$, $\delta > 0$. Suppose that $w, \Theta \in H^1 (Q \cap \Omega)$ satisfy*

$$-w_{xx} + w_t \geq 0 \qquad \textit{in} \quad Q \cap Q,$$

$$-\Theta_{xx} + \Theta_t = 0 \qquad \textit{in} \quad Q \cap \Omega,$$

$$w \geq \Theta \qquad \textit{on} \quad \partial_p(Q \cap \Omega),$$

where $\partial_p(Q \cap \Omega) = \partial(Q \cap \Omega) - \{(x, t_0) : |x - x_0| < \varepsilon\}$ denotes the parabolic boundary of $Q \cap \Omega$. Then

$$w \geq \Theta \qquad \textit{in} \quad Q \cap \Omega.$$

The proof of this maximum principle is analogous to our weak maximum principle of Chapter II. We relegate it to an exercise.

Lemma 3.7. *Let u denote the solution to Problem 1.2 and Γ its free boundary. Then for each $(x_0, t_0) \in \Gamma$ with $t_0 \geq \delta > 0$,*

$$0 \leq u_t(x, t_0) \leq c_1(x - x_0)^2 - c_2 x u_x(x, t) \qquad \text{for} \quad (x, t) \in Q \cap \Omega,$$

where c_1, c_2, and ε are positive constants independent of $(x_0, t_0) \in \Gamma$, $t_0 \geq \delta > 0$, and

$$Q = \{(x, t) : |x - x_0| < \varepsilon \text{ and } |t - t_0| < \varepsilon\}.$$

Proof. Let us first summarize our knowledge about u_t. We know there is a neighborhood $Q = \{(x, t) : |x - x_0| < \varepsilon, |t - t_0| < \varepsilon\}$ such that $u_t \in H^1(Q)$ by Theorem 3.4. Since Γ is a Lipschitz curve, u_t admits a trace $u_t(\sigma(t), t)$ on $\Gamma \cap Q$. To see that trace is zero simply note that

$$u_t(x, t) = \lim_{\mu \to 0} u_t(x + \mu, t) \qquad \text{in} \quad H^1(Q)$$

and that $u_t = 0$ in $Q \backslash \Omega$. Since the trace is a continuous mapping from $H^1(Q)$ to $L^2(\Gamma \cap Q)$, $\text{trace}_\Gamma u_t = 0$. Also note that

$$-u_{txx} + u_{tt} = 0 \qquad \text{in} \quad Q \cap \Omega. \tag{3.4}$$

Let us set

$$w(x, t) = c_1(x - x_0)^2 - c_2 x u_x(x, t), \qquad (x, t) \in Q \cap \Omega,$$

and choose $c_1 > 0$ so large that

$$c_1(x - x_0)^2 \geq u_t(x, t) \qquad \text{for} \quad (x, t) \in \partial(Q \cap \Omega).$$

This is possible, clearly, because $u_t = 0$ on Γ. Now choose $c_2 > 0$ so large that

$$\begin{aligned} -w_{xx} + w_t &= -2c_1 + c_2 u_{xx} && \text{in} \quad Q \cap \Omega \\ &\geq -2c_1 + 2kc_2 > 0 && \text{in} \quad Q \cap \Omega. \end{aligned} \tag{3.5}$$

According to Lemma 3.5, $w(x, t) \geq 0$ in $Q \cap \Omega$, for suitable Q, so

$$w \geq u_t \qquad \text{on} \quad \partial_p(Q \cap \Omega)$$

By (3.4), (3.5) we may apply our maximum principle Lemma 3.6, whence

$$0 \leq u_t(x, t) \leq w(x, t), \qquad (x, t) \in Q \cap \Omega.$$

The conclusion follows. Q.E.D.

A particular consequence of the lemma is that u_t is continuous.

Theorem 3.8. *Let u denote the solution to Problem* 1.2 *and Γ its free boundary. Then for each $(x_0, t_0) \in \Gamma$, there is a neighborhood U of (x_0, t_0) such that*

$$u_t \in H^{1,\infty}(U).$$

Proof. It suffices to prove that $u_t \in H^{1,\infty}(U \cap \Omega)$. Choose Q as in the previous lemmas so the conclusion of Lemma 3.7 holds, and set $\Theta = u_t$. We shall apply the method of S. Bernstein, but to an approximation of Θ in place of Θ because its smoothness at Γ is in question (Bernstein [1]).

Let $\alpha(\zeta) \in C_0^\infty(\mathbb{R})$ be a mollifier and set

$$\alpha_h(\xi) = (1/h)\alpha(\xi/h) \qquad \text{for} \quad 0 < h < 1,$$

$$\Theta_h(x, t) = \int_0^R \alpha_h(x - \xi)\Theta(\xi, t)\, d\xi, \qquad (x, t) \in Q,$$

and

$$\Omega_h = \{(x, t) : 0 < x < \sigma(t) - h, 0 < t < T\}.$$

Observe that

$$-\Theta_{hxx} + \Theta_{ht} = 0 \qquad \text{in} \quad \Omega_h \cap Q$$

and

$$0 \le \Theta_h(x, t) \le \sup \Theta \qquad \text{in} \quad \Omega_h \cap Q. \tag{3.6}$$

We now transform the estimate of Lemma 3.7 to an estimate for Θ_{hx} on $\Gamma \cap Q$. First note that when $(x, t) \in \partial\Omega_h \cap Q$, there is a y with $|x - y| = h$ and $(y, t) \in \Gamma$. Now by Lemma 3.7

$$\begin{aligned} 0 \le \sup_{|x-\xi|<h} \Theta(\xi, t) &\le \sup_{|y-\xi|<2h} \Theta(\xi, t) \\ &\le \sup_{|y-\xi|<2h} \{c_1(y - \xi)^2 - c_2 \xi u_\xi(\xi, t)\} \le c_1 h, \end{aligned}$$

recalling here that $u_x(x, t)$ is a Lipschitz function of x. So we have established that

$$0 \le \sup_{|x-\xi|<h} \Theta(\xi, t) \le c_1 h \qquad \text{for} \quad (x, t) \in \partial\Omega_h \cap Q. \tag{3.7}$$

We compute that

$$
\begin{aligned}
|\Theta_{hx}(x, t)| &\leq \int_{|x-\xi|<h} |\alpha_{hx}(x-\xi)|\,\Theta(\xi, t)\,d\xi, \qquad (x, t) \in \partial\Omega_h \cap Q,\\
&\leq \sup_{|x-\xi|<h} \Theta(\xi, t) \int_{|x-\xi|<h} |\alpha_{hx}(x-\xi)|\,d\xi,\\
&\leq \frac{1}{h} \sup_{|x-\xi|<h} \Theta(\xi, t) \int_{|\eta|\leq 1} |\alpha_\eta(\eta)|\,d\eta \leq c_2
\end{aligned} \tag{3.8}
$$

in view of (3.7).

Now choose a function $\zeta \in C_0^\infty(Q)$, $0 \leq \zeta \leq 1$, with $\zeta = 1$ near (x_0, t_0) and set

$$
w = w_h = \zeta^2 \Theta_{hx}^2 + \mu \Theta_h^2, \qquad (x, t) \in Q \cap \Omega_h.
$$

Applying the heat operator to w we compute

$$
\begin{aligned}
-\Delta w + w_t &= -2\{\zeta^2 \Theta_{hxx}^2 + 4\zeta\zeta_x \Theta_{hx}\Theta_{hxx} + \Theta_{hx}^2[(\zeta\zeta_x)_x - \zeta_t + \mu]\}\\
&\leq -2\{(1-\varepsilon)\zeta^2\Theta_{hxx}^2 + \Theta_{hx}^2[\mu + (\zeta\zeta_x)_x - \zeta_t - (4/\varepsilon)\zeta_x^2]\}\\
&\leq 0
\end{aligned} \tag{3.9}
$$

for a $\mu \geq 0$ which depends only on ζ. Hence by the maximum principle,

$$
w(x, t) \leq \max_{\partial_p(\Omega_h \cap Q)} w \leq c_2^2 + \mu(\sup \Theta)^2,
$$

keeping in mind (3.8) and (3.6). Restricting ourselves to a neighborhood U of (x_0, t_0) where $\zeta = 1$ and passing to the limit as $h \to 0$ we conclude that

$$
|u_{tx}(x, t)| \leq \text{const} \qquad \text{in} \quad \Omega \cap U.
$$

As a particular consequence of this fact, u_x is a Lipschitz function of t, so, employing the estimate of Lemma 3.7, we deduce that

$$
0 \leq u_t(x, t) \leq \text{const}\,\sqrt{(x-x_0)^2 + (t-t_0)^2}
$$

when $(x_0, t_0) \in \Gamma$ and $(x, t) \in U \cap \Omega$. To complete the proof of the theorem, introduce the mollification Ψ_h of u_t in the time coordinate only and derive a differential inequality analogous to (3.9) for the function

$$
z = z_h = \zeta^2 \Psi_{ht}^2 + \mu \Psi_{hx}^2.
$$

The details are left to the reader. Q.E.D.

We conclude this section with

Theorem 3.9. *Let u denote the solution to Problem* 1.2 *and* Γ *its free* *boundary. Then for each* $(x_0, t_0) \in \Gamma$, *there is a neighborhood* U *of* (x_0, t_0) *such that*

$$u_{xx}, u_{xt} \in C(\overline{\Omega} \cap U).$$

Proof. Choose U as in the previous theorem. Since

$$u_{xx} = u_t + k \qquad \text{for} \quad (x, t) \in U \cap \Omega,$$

u_{xx} is not only continuous but Lipschitz in $U \cap \overline{\Omega}$. Differentiating with respect to t,

$$u_{txx} = u_{tt} \in L^\infty(U \cap \Omega),$$

so

$$|u_{xt}(x, t) - u_{xt}(x', t)| \leq C|x - x'|, \qquad (x, t), (x', t) \in Q \cap \Omega,$$

with C independent of t. Moreover, u_{xt}, a solution of the heat equation in $U \cap \Omega$, is continuous there.

Finally, let $(x, t) \in \Gamma \cap U$ and $(x, t') \in U \cap \Omega$. Given $\varepsilon > 0$ we estimate

$$\begin{aligned} |u_{xt}(x, t) - u_{xt}(x, t')| &\leq |u_{xt}(x, t) - u_{xt}(x - \varepsilon, t)| \\ &\quad + |u_{xt}(x - \varepsilon, t) - u_{xt}(x - \varepsilon, t')| \\ &\quad + |u_{xt}(x - \varepsilon, t') - u_{xt}(x, t')| \\ &\leq 2C\varepsilon + |u_{xt}(x - \varepsilon, t) + u_{xt}(x - \varepsilon, t')|. \end{aligned}$$

Hence

$$\lim_{t' \to t} |u_{xt}(x, t) - u_{xt}(x, t')| \leq 2C\varepsilon,$$

which proves the theorem. Q.E.D.

4. The Legendre Transform

Applying the Legendre transform to the free boundary value problem for u gives rise to a new problem with a smooth "free" boundary but with a highly nonlinear equation, even in the simplest case. Throughout this section, let u denote the solution to Problem 1.2 and Γ its free boundary. Given $(x_0, t_0) \in \Gamma$, let U be a neighborhood of (x_0, t_0) such that $u_x \in C^1(\overline{\Omega} \cap U)$. Since Γ is a Lipschitz curve it is easy to check that $u_{xx}, u_{xt} \in C(\overline{\Omega} \cap U)$ imply $u_x \in C^1(\overline{\Omega} \cap U)$ and indeed that u_x admits a C^1 extension into the neighborhood U.

Now we introduce the transformation

$$\begin{aligned} \xi &= -u_x(x, t), \\ \tau &= t, \end{aligned} \qquad (x, t) \in U. \tag{4.1}$$

This mapping is C^1 with

$$\frac{\partial(\xi, \tau)}{\partial(x, t)} = \begin{pmatrix} -u_{xx}(x, t) & -u_{xt}(x, t) \\ 0 & 1 \end{pmatrix}. \tag{4.2}$$

Since $u_{xx}(x_0, t_0) = k > 0$, $\partial(\xi, \tau)/\partial(x, t)$ is nonsingular at (x_0, t_0) and maps a neighborhood, say $U \cap \Omega$, 1 : 1 onto a region

$$G \subset \{(\xi, \tau) : \xi > 0\}.$$

Recall here that $u_x < 0$ in Ω near Γ. Also, $U \cap \Gamma$ is mapped onto a subset

$$\Sigma \subset \{(\xi, \tau) : \xi = 0\},$$

inasmuch as $u_x = 0$ on Γ.

The Legendre transform of u is

$$v(\xi, \tau) = x\xi + u(x, t), \qquad (\xi, \tau) \in G \cup \Sigma, \quad (x, t) \in \overline{\Omega} \cap U.$$

It enjoys the property that

$$\begin{aligned} dv &= x\, d\xi + \xi\, dx + u_x\, dx + u_t\, dt \\ &= x\, d\xi + u_t\, d\tau, \end{aligned}$$

or $v_\xi = x$ and $v_\tau = u_t$. In particular,

$$v_{\xi\xi} = \frac{\partial x}{\partial \xi} = \frac{1}{(\partial \xi/\partial x)} = -\frac{1}{u_{xx}}$$

on the basis of (4.2).

Our free boundary problem is

$$\begin{aligned} -u_{xx} + u_t &= -k \qquad &&\text{in} \quad \Omega \cap U, \\ u = u_x &= 0 &&\text{on} \quad \Gamma \cap U. \end{aligned} \tag{4.3}$$

In terms of v, this boundary value problem becomes

$$\begin{aligned} (1/v_{\xi\xi}) + v_\tau &= -k \qquad &&\text{in} \quad G, \\ v &= 0 &&\text{on} \quad \Sigma \end{aligned} \tag{4.4}$$

The new problem (4.4) is parabolic because its linearized operator L evaluated at $\zeta(\xi, \tau)$ is

$$L\zeta = \frac{d}{d\varepsilon}\left[\frac{1}{(v + \varepsilon\zeta)_{\xi\xi}} + (v + \varepsilon\zeta)_\tau\right]\Bigg|_{\varepsilon=0}$$

$$= -(\zeta_{\xi\xi}/v_{\xi\xi}{}^2) + \zeta_\tau,$$

a parabolic operator with continuous coefficients. It follows from the theory of nonlinear parabolic equations (Ladyzhenskaya *et al.* [1]) that

$$v \in C^\infty(G \cup \Sigma)$$

so $\Gamma \cap U : x = v_\xi(0, \tau)$, $|\tau - \tau_0|$ small, is a C^∞ parametrization of a portion of Γ. We have shown

Theorem 4.1. *Let u be the solution of Problem* 1.2 *and Γ its free boundary. Then Γ is a C^∞ curve.*

Note that in general Γ cannot be analytic.

Comments and Bibliographical Notes

The classical Stefan problem, for arbitrary dimension, is discussed in Friedman [3]. The treatment of the one phase problem as a variational inequality was proposed in Duvaut [1]. It was further developed and the properties of its solution explored in Friedman and Kinderlehrer [1]. For a general discussion of the free boundary we refer to Kinderlehrer and Nirenberg [2, 3]. An interesting proof of the infinite differentiability of Γ ($N = 1$) is due to Schaeffer [1]. Also in this case, it was shown that Γ is analytic provided the supplied heat is analytic (Friedman [4]).

Recently, Caffarelli and Friedman [2] have shown that the temperature in the N dimensional one phase problem is continuous.

The Stefan problem is a paradigm for many variational and quasi-variational inequalities of parabolic type. These concern, for example, stopping times, optimal control, and impulse control. We refer to Bensoussan and Lions [1] and Friedman [5] as examples.

Bibliography

Agmon, S., Douglis, A., and Nirenberg, L.

[1] Estimates near the bcundary for solutions of elliptic partial differential equations satisfying general boundary conditions. I, *Comm. Pure Appl. Math.* **12** (1959), 623–727.

[2] Estimates near the boundary for solutions of elliptic partial differential equations satisfying general boundary conditions. II, *Comm. Pure Appl. Math.* **22** (1964), 35–92.

Alt, H.

[1] A free boundary problem associated with the flow of groundwater, *Arch. Rational Mech. Anal.* **64** (1977), 111–126.

[2] The fluid flow through porous media: regularity of the free surface, *Manus. Math.* **21** (1977), 255–272.

Baiocchi, C.

[1] Su un problema a frontiera libera connesso a questioni di idraulica, *Ann. Mat. Pura Appl.* **92** (1972), 107–127.

[2] Free boundary problems in the theory of fluid flow through porous media, *Proc. Int. Congr. Math., Vancouver, 1974* Vol. II, pp. 237–343. Canadian Math. Congress.

[3] Problèmes à frontière libre et inéquations variationnelles, *C.R. Acad. Sci. Paris* **283** (1976), 29–32.

[4] Free boundary problems and variational inequalities, *SIAM Rev.*

Baiocchi, C., and Capelo, A.

[1] Disequazioni variazionali e quasi variazionali, Applicazioni a problemi di frontiera libera, I, II, Quaderni dell'U.M.I., Pitagora, Bologna (1978).

Bellman, R.

[1] *Dynamic Programming*. Princeton Univ. Press, Princeton, New Jersey, 1957.

Benci, V.

[1] On a filtration problem through a porous membrane, *Ann. Mat. Pura Appl.* **100**, (1974), 191–209.

Bensoussan, A., and Lions, J. L.

[1] *Temps d'arret optimal et controle impulsionnel*, Vol. I, Dunod, Paris (1977).

Berestycki, H., and Brezis, H.

[1] Sur certains problèmes de frontière libre, *C.R. Acad. Sci. Paris* **283** (1976), 1091–1094.

Bernstein, S.

[1] A limitation of the moduli of the derivatives of equations of parabolic type, *Dokl. Akad. Nauk, U.S.S.R.* **18** (1938), 385–388 (Russian).

Bers, L.

[1] *Mathematical Aspects of Subsonic and Transonic Gas Dynamics.* Chapman, London, 1958.

Bers, L., John, F., and Schechter, M.

[1] *Partial Differential Equations.* Wiley (Interscience), New York, 1962.

Brezis, H.

[1] Problèmes unilatéraux, *J. Math. Pures Appl.* **51** (1972), 1–168.

[2] Multiplicateur de Lagrange en torsion elastroplastique, *Arch. Rational Mech. Anal.* **41** (1971), 254–265.

[3] Monotonicity methods in Hilbert spaces and some applications to non-linear partial differential equations, *in Contributions to Non Linear Functional Analysis* (E. Zarantonello, ed.) pp. 101–156. Academic Press, New York, 1971.

Brezis, H., and Duvaut, G.

[1] Ecoulment avec sillage autour d'un profile symètrique sans incidence, *C.R. Acad. Sci. Paris* **276** (1973), 875–878.

Brezis, H., and Evans, L.

[1] A variational inequality approach to the Bellman-Dirichlet equation for two elliptic operators, *Arch. Rational Mech. Anal.* **71** (1979), 1–13.

Brezis, H., and Kinderlehrer, D.

[1] The smoothness of solutions to nonlinear variational inequalities, *Indiana J. Math.* **23** (1974), 831–844.

Brezis, H., Kinderlehrer, D., and Stampacchia, G.

[1] Sur une nouvelle formulation du problème de l'écoulement à travers une digue, *C.R.A.S. Paris* **287** (1978), 711–714.

Brezis, H., and Stampacchia, G.

[1] Sur la régularité de la solution d'inéquations elliptiques, *Bull. Soc. Math. France* **96** (1968), 153–180.

[2] The hodograph method in fluid-dynamics in the light of variational inequalities, *Arch. Rational Mech. Anal.* **61** (1976), 1–18.

[3] Une nouvelle méthode pour l'étude d'écoulements stationnaires, *C.R. Acad. Sci. Paris* **276** (1973), 129–132.

[4] Remarks on some fourth order variational inequalities, *Ann. Scuola Norm. Sup. Pisa* **4** (1977), 363–371.

Browder, F.

[1] Nonlinear monotone operators and convex sets in Banach spaces, *Bull. Amer. Math. Soc.* **71** (1965), 780–785.

[2] On a theorem of Beurling and Livingston, *Canad. J. Math.* **17** (1965), 367–372.

Caffarelli, L.

[1] The smoothness of the free surface in a filtration problem, *Arch. Rational Mech. Anal.* **63** (1976), 77–86.

[2] The regularity of free boundaries in higher dimensions, *Acta Math.* **139** (1978), 155–184.

[3] Further regularity in the Signorini problem,

Caffarelli, L. A., and Friedman, A.

[1] The obstacle problem for the biharmonic operator, *Ann. Scuola Norm. Sup. Pisa* **6** (1979), 151–184.

[2] The one phase Stefan problem and the porous medium equation: Continuity of the solution in *n* space dimensions, *Proc. Nat. Acad. Sci. U.S.A.* **75** (1978), 2084.

Caffarelli, L. A., and Riviere, N. M.

[1] On the rectifiability of domains with finite perimeter, *Ann. Scuola Norm. Sup. Pisa.* **3** (1976), 177–186.

[2] Smoothness and analyticity of free boundaries in variational inequalities, *Ann. Scuola Norm Sup. Pisa* **3** (1976), 289–310.

[3] Asymptotic behavior of free boundaries at their singular points, *Ann. Math.* **106** (1977), 309–317.

[4] On the lipschitz character of the stress tensor when twisting an elastic plastic bar, *Arch. Rational Mech. Anal.* **69** (1979) 31–36.

Capriz, G., and Cimatti, G.

[1] On some singular perturbation problems in the theory of lubrication, *Appl. Math. Optim.* **4** (1978), 285–297.

Carathéodory, C.

[1] *Conformal Representation.* Cambridge Univ. Press, London and New York, 1932.

Cimatti, G.

[1] On a problem of the theory of lubrication governed by a variational inequality, *Appl. Math. Optim.* **3** (1977), 227–242.

Courant, R., and Hilbert, D.

[1] *Methods of Mathematical Physics*, Vol. II, Partial Differential Equations, especially p. 350. Wiley (Interscience), New York, 1962.

Coppoletta, G.

[1] A remark on existence and uniqueness in the theory of variational inequalities, *Boll. Un. Mat. Ital.* (to appear).

De Giorgi, E.

[1] Sulla differenziabilità e analiticità delle estremali degli integrali multipli regolari, *Mem. Accad. Sci. Torino* **3** (1957), 25–43.

Duvaut, G.

[1] Resolution d'un probleme de Stefan (Fusion d'un bloc de glace a zero degre), *C.R. Acad. Sci. Paris* **276** (1973), 1461–1463.

Duvaut, G., and Lions, J. L.

[1] *Les Inequations en Mécanique et en Physique.* Dunod, Paris, 1972.

Fichera, G.

[1] Problemi elastostatici con vincoli unilaterali: il problema di Signorini con ambigue condizioni al contorno, *Atti Accad. Naz. Lincei Mem. Cl. Sci. Fis. Mat. Natur Sez. Ia*(8) **VII** (1963–1964), 91–140.

Fraenkel, L., and Berger, M.

[1] A global theory of steady vortex rings in an ideal fluid, *Acta Math.* **132** (1974), 13–51.

Frehse, J.

[1] On the regularity of the solution of a second order variational inequality, *Boll. Un. Mat. Ital.* **6** (1972), 312–315.

[2] On the regularity of the solution of the biharmonic variational inequality, *Manuscripta Math.* **9** (1973), 91–103.

[3] Two dimensional varianon problems with thin obstacles, *Math. Z.* **143** (1975), 279–288.

[4] On Signorini's problem and variational problems with thin obstacles, *Ann. Scuola Norm. Sup. Pisa* **4** (1977), 343–362.

Friedman, A.

[1] On the regularity of the solutions of non-linear elliptic and parabolic systems of partial differential equations, *J. Math. Mech.* **7** (1958), 43–60.

[2] *Partial Differential Equations of Parabolic Type.* Prentice-Hall, Englewood Cliffs, New Jersey, 1964.

[3] The Stefan problem in several space variables, *Trans. Amer. Math. Soc.* **133** (1968), 51–87.

[4] Analyticity of the free boundary for the Stefan problem, *Arch. Rational Mech. Anal.* **61** (1976), 97–125.

[5] On the free boundary of a quasi variational inequality arising in a problem of quality control, *Trans. Amer. Math. Soc.* (to appear).

Friedman, A., and Jensen, R.

[1] Convexity of the free boundary in the Stefan problem and in the dam problem, *Arch. Rational Mech. Anal.* **67** (1977), 1–24.

Friedman, A., and Kinderlehrer, D.

[1] A one phase Stefan problem, *Indiana Univ. Math. J.* **24** (1975), 1005–1035.

Friedrichs, K. O.

[1] Über ein Minimumproblem für Potentialstromungen mit freiem Rande, *Math. Ann.* **109** (1934), 60–82.

Garabedian, P.

[1] *Partial Differential Equations*. Wiley, New York, 1964.

Gerhardt, C.

[1] Hypersurfaces of prescribed mean curvature over obstacles, *Math. Z.* **133** (1973), 169–185.

[2] Regularity of solutions of nonlinear variational inequalities, *Arch. Rational Mech. Anal.* **52** (1973), 389–393.

Giaquinta, M., and Pepe, L.

[1] Esistenza e regolarità per il problema dell' area minima con ostacoli in n variabili, *Ann. Scuola Norm. Sup Pisa* **25** (1971), 481–507.

Giusti, E.

[1] Minimal surfaces with obstacles, CIME course on Geometric Measure Theory and Minimal Surfaces, Edizioni Cremonese, Rome, pp. 119–153 (1973).

Hartman, P., and Stampacchia, G.

[1] On some nonlinear elliptic differential functional equations, *Acta Math.* **115** (1966), 153–188.

Hartman, P., and Wintner, A.

[1] On the local behavior of non parabolic partial differential equations, *Amer. J. Math.* **85** (1953), 449–476.

Iordanov, I.V.

[1] Problème non-coercif pour une inéquation variationnelle elliptique à deux contraintes, *Serdica* **1** (1975), 261–268.

Karamardian, S.

[1] The complementarity problem, *Math. Programming* **2** (1972), 107–129.

Kinderlehrer, D.

[1] Variational inequalities with lower dimensional obstacles, *Israel J. Math.* **10** (1971), 339–348.

[2] The coincidence set of solutions of certain variational inequalities, *Arch. Rational Mech. Anal.* **40** (1971), 231–250.

[3] How a minimal surface leaves an obstacle, *Acta Math.* **130** (1973), 221–242.

[4] The free boundary determined by the solution as a differential equation, *Indiana Univ. Math. J.* **25** (1976), 195–208.

[5] Variational inequalities and free boundary problems, *Bull. Amer. Math. Soc.* **84** (1978), 7–26.

Kinderlehrer D., and Nirenberg, L.

[1] Regularity in free boundary value problems, *Ann. Scuola Norm. Sup. Pisa Ser. IV* **4** (1977), 373–391.

[2] The smoothness of the free boundary in the one phase Stefan Problem, *Comm. Pure Appl. Math.* **31** (1978), 257–282.

[3] Analyticity at the boundary of solutions of nonlinear second-order parabolic equations, *Comm. Pure Appl. Math.* **31** (1978), 283–338.

Kinderlehrer, D., Nirenberg, L., and Spruck, J.

[1] Regularité dans les problèmes elliptiques à frontière libre, *C.R. Acad. Sci. Paris* **286** (1978), 1187–1190.

[2] Regularity in elliptic free boundary problems, I, *J. d'Analyse Math.* **34** (1978), 86–119.

[3] Regularity in elliptic free boundary problems, II, *Annali della SNS* (to appear).

Kinderlehrer, D., and Spruck, J.

[1] The shape and smoothness of stable plasma configurations, *Annali della SNS* **5** (1978), 131–148.

Kinderlehrer, D. and Stampacchia, G.

[1] A free boundary problem in potential theory, *Ann. Inst. Fourier* (*Grenoble*) **25** (1975), 323–344.

Ladyzhenskaya, O. A., Solonnikov, B. A., and Ural'tseva, N. N.

[1] Linear and quasilinear equations of parabolic type, Moscow, 1967, *Amer. Math. Soc. Transl.* **23** (1968).

Ladyzhenskaya, O. A., and Ural'tseva, N. N.

[1] *Linear and Quasilinear Elliptic Equations*, Academic Press, New York (1968) (English translation of 1964 Russian edition).

Lewy, H.

[1] A priori limitations for the solutions of the Monge-Ampère equations, II, *Trans. Amer. Math. Soc.* **41** (1937), 365–374.

[2] On the boundary behavior of minimal surfaces, *Proc. Nat. Acad. Sci. U.S.A.* **37** (1951), 103–110. MR 14, 168.

[3] A note on harmonic functions and a hydrodynamical application, *Proc. Amer. Math. Soc.* **3** (1952), 11–113.

[4] On minimal surfaces with partially free boundary, *Comm. Pure Appl. Math.* **4** (1952), 1–13, MR 14, 662.

[5] On the reflection laws of second order differential equations in two independent variables, *Bull. Amer. Math. Soc.* **65** (1959), 37–58.

[6] On a variational problem with inequalities on the boundary, *Indiana J. Math.* **17** (1968), 861–884.

[7] The nature of the domain governed by different regimes, Atti del Convegno Internazionale Metodi valutativi nella fisicamatematica, *Accad. Naz. Lincei* (1975), 181–188.

[8] On the coincidence set in variational inequalities, *J. Diff. Geom.* **6** (1972), 497–501.

[9] Inversion of the obstacle problem and explicit solutions, *Ann. Scuola Norm. Sup. Pisa* (to appear).

Lewy, H., and Stampacchia, G.

[1] On the regularity of the solution of a variational inequality, *Comm. Pure Appl. Math.* **22** (1969), 153–188.

[2] On existence and smoothness of solutions of some noncoercive variational inequalities, *Arch. Rational Mech. Anal.* **41** (1971), 241–253.

Leray, J., and Lions, J. L.

[1] Quelques résultats de Visik sur les problèmes elliptiques non linéaires par les méthodes de Minty-Browder, *Bull. Soc. Math. France* **93** (1965) 97–107.

Lions, J. L.

[1] Some remarks on variational inequalities, *Proc. Tokyo Conf. Functional Anal. Appl.*, Tokyo (1969), 269–281.

Lions, J. L., and Magenes, E.

[1] *Problèmes aux Limites non Homogènes*, Vols. I, II, III. Dunod, Paris, 1968–1970 (*English transl.*: Springer, New York, 1972).

Lions, J. L., and Stampacchia, G.

[1] Variational inequalities, *Comm. Pure Appl. Math.* **20** (1967), 493–519.

Littman, W., Stampacchia, G., and Weinberger, H.
[1] Regular points for elliptic equations with discontinuous coefficients, *Ann. Scuola Norm. Sup. Pisa*, **17** (1963), 43–77.

Magenes, E.
[1] Topics in parabolic equations: some typical free boundary problems, *Proc. NATO Adv. Study Inst., Liege* (1976).

Mancino, O., and Stampacchia, G.
[1] Convex programming and variational inequalities, *J. Opt. Theory Appl.* **9** (1972), 3–23.

Massey, W. S.
[1] *Algebraic Topology, An Introduction.* Springer, New York, 1967.

Mazzone, S.
[1] Existence and regularity of the solution of certain nonlinear variational inequalities with an obstacle, *Arch. Rational Mech. Anal.* **57** (1974), 115–127.
[2] Un problema di disequazioni variazionali per superficie di curvatura media assegnata, *Boll. Un. Mat. Ital.* **7** (1973), 318–329.

Minty, G. J.
[1] Monotone (non linear) operators in Hilbert space, *Duke Math. J.* **29** (1962), 341–346.
[2] On a "monotonicity" method for the solution of nonlinear equations in Banach spaces, *Proc. Nat. Acad. Sci. U.S.A.* **50** (1963), 1038–1041.

Miranda, C.
[1] *Su un Problema di Frontiera Libera* (*Symp. Math. II*), pp. 71–83. Academic Press, New York, 1969.

Miranda, M.
[1] Distribuzioni aventi derivate misure: insieme di perimetro localmente finito, *Ann. Scuola Norm. Pisa* **18** (1964), 27–56.
[2] Frontiere minimali con ostacoli, *Ann. Un. Ferrara* **XVI.2** (1971), 29–37.

Moreau, J. J.
[1] Principles extrémaux pour le problème de la naissance de la cavitation, *J. Mécanique* **5** (1966), 439–470.

Morrey, C. B., Jr.
[1] *Multiple Integrals in the Calculus of Variations.* Springer, New York, 1966.

Murthy, M. K. V., and Stampacchia, G.
[1] A variational inequality with mixed boundary conditions, *Israel J. Math.* **13** (1972), 188–224.

Rockafellar, R. T.
[1] *Convex Analysis.* Princeton Univ. Press, Princeton, New Jersy, 1970.

Schaeffer, D.
[1] A stability theorem for the obstacle problem, *Adv. in Math.* **17** (1975), 34–47.
[2] A new proof of the infinite differentiability of the free boundary in the Stefan problem, *J. Differential Equations* **20** (1976), 266–269.
[3] Some examples of singularities in a free boundary, *Ann. Scuola Norm. Sup. Pisa Cl. Sci.* (*4*) **1** (1977), 133–144.

Schiaffino, A., and Troianiello, G.
[1] Su alcuni problemi di disequazioni variazionali per sistemi variazionali ordinari, *Boll. Un. Mat. Ital.* **3** (1970), 76–103.

Schwarz, L.
[1] *Thèorie des Distributions.* Hermann, Paris, 1966.

Serrin, J.
[1] The problem of Dirichlet for quasilinear elliptic differential equations with many independent variables, *Philos. Trans. Roy. Soc. London Ser. A* **264** (1969), 413–496.

Shamir, E.

[1] Regularisation of mixed second order elliptic problems, *Israel J. Math.* **6** (1968), 150–168.

Stein, E. M.

[1] *Singular Integrals and Differentiability Properties of Functions.* Princeton Univ. Press, Princeton, New Jersey, 1970.

Stampacchia, G.

[1] On some regular multiple integral problems in the calculus of variations, *C.P.A.M.* **16** (1963), 383–421.

[2] Le probleme de Dirichlet pour les equations elliptique du second order à coefficients discontinuous, *Ann. Inst. Fourier* (*Grenoble*) **15** (1965), 189–258.

[3] Formes bilineaires coercitives sur les ensembles convexes, *C.R. Acad. Sci. Paris* **258** (1964), 4413–4416.

[4] Variational inequalities, theory and applications of monotone operators, *Proc. NATO Adv. Study Inst., Oderisi-Gubbio* (1969).

[5] Variational inequalities, *Proc. Internat. Congr. Math., Nice* (1970), 877–883.

[6] On the filtration of a liquid through a porous medium, *Usp. Mat. Nauk* **29** (178) (1974), 89–101.

[7] Su una disequazione variazionale legata al comportamento elastoplastico delle travi appoggiate agli estremi, *Boll. Un. Mat. Ital.* **11** (4) (1975), 444–454.

[8] Le disequazioni variazionali nella dinamica dei fluidi, Atti del convegno Internazionale, "Metodi valutativi nella fisica-matematica," *Accad. Naz. Lincei* (1975) 169–180.

Stampacchia, G., and Vignoli, A.

[1] A remark on variational inequalities for a second order nonlinear differential operator with non-Lipschitz obstacles, *Boll. Un. Mat. Ital.* **5** (1972), 123–131.

Temam, R.

[1] A non-linear eigenvalue problem: the shape at equilibrium of a confined plasma, *Arch. Rational Mech. Anal.* **60** (1975), 51–73.

[2] Remarks on a free boundary problem arising in plasma physics, *Comm. Partial Differential Equations* **2** (1977), 563–585.

Ting, T. W.

[1] Elastic-plastic torsion, *Arch. Rational Mech. Anal.* **24** (1969), 228–244.

Tomarelli, F.

[1] Un probléme de fluidodynamique avec les inèquations variationnelles, *C.R. Acad. Sci. Paris* **286** (1978), 999–1002.

Beirao da Veiga, H.

[1] Sulla Hölderianità della soluzioni di alcune disequazioni variazionali con condizioni unilaterale al bordo, *Ann. Mat. Pura Appl.* **83** (1969), 73–112.

Vergara-Caffarelli, G.

[1] Su un problema al contorno con vincoli per operatori differenziali ordinari, *Boll. Un. Mat. Ital.* **4** (1970), 566–584.

[2] Regolarità di un problema di disequazioni variazionali relativo a due membrane, *Rend. Acad. Linai* **50** (1971), 659–662.

[3] Variational Inequalities for Two Surfaces of Constant Mean Curvature, *Arch. Rational Mech. Anal.* **56** (1974), 334–347.

[4] Superficie con curvatura media assegnata in L_p; Applicazioni ad un problema di disequazioni variazionali, *Boll. Un. Mat. Ital.* **8** (1973), 261–277.

Villaggio, P.

[1] Monodimensional solids with constrained solutions, *Meccanica* **2** (1967), 65–68.

[2] Stability conditions for elastic-plastic Prandtl-Reuss solids, *Meccanica* **2** (1968), 46–47.

Visik, M. I.

[1] Quasi-linear strongly elliptic systems of differential equations of divergence form, *Trudy Moskov. Mat. Obsc.* **12** (1963), 125–184 *English transl.*: *Moscow Math. Soc.* (1963), 140–208.

Zarantonello, E. H.

[1] Solving functional equations by contractive averaging, Tech. Rep. 160, Math. Research Center, Univ. of Wisconsin, Madison, Wisconsin (1960).

Index